Green Technologies

Green Technologies: Bridging Conventional Practices and Industry 4.0

Editors

Pau Loke Show
Suchithra Thangalazhy Gopakumar
Dominic C. Y. Foo

MDPI • Basel • Beijing • Wuhan • Barcelona • Belgrade • Manchester • Tokyo • Cluj • Tianjin

Editors
Pau Loke Show
University of Nottingham Malaysia
Malaysia

Suchithra Thangalazhy Gopakumar
University of Nottingham
Malaysia

Dominic C. Y. Foo
The University of Nottingham Malaysia
Malaysia

Editorial Office
MDPI
St. Alban-Anlage 66
4052 Basel, Switzerland

This is a reprint of articles from the Special Issue published online in the open access journal *Processes* (ISSN 2227-9717) (available at: https://www.mdpi.com/journal/processes/special_issues/green_technology).

For citation purposes, cite each article independently as indicated on the article page online and as indicated below:

LastName, A.A.; LastName, B.B.; LastName, C.C. Article Title. *Journal Name* **Year**, *Article Number*, Page Range.

ISBN 978-3-03936-519-7 (Hbk)
ISBN 978-3-03936-520-3 (PDF)

Contents

About the Editors

Pau Loke Show is the Director of Research in the Department of Chemical and Environmental Engineering, University of Nottingham Malaysia. He also is the Director of the Sustainable Food Processing Research Center and the Co-director of Future Food Malaysia Beacon of Excellence. Currently, he is an Associate Professor in the Faculty of Science and Engineering at University of Nottingham Malaysia. He currently is registered as a Professional Engineer with the Board of Engineers Malaysia and as a Chartered Engineer of the Engineering Council UK. He is also a member of Institution of Chemical Engineers UK and currently serves as an invited member in the IChemE Biochemical Engineering Special Interest Group. Ir. Ts. Dr. Show obtained the Postgraduate Certificate of Higher Education in 2014 and is now a fellow of the Higher Education Academy UK. Since he started his career in September 2012, he has received numerous prestigious domestic and international academic awards, including seven recent Global Top Peer Reviewer Awards from Web of Science and Publons. He is also the winner of ASEAN–India Research and Training Fellowship 2019, the DaSilva Award 2018, JSPS Fellowship 2018 award, Top 100 Asian Scientists 2017, Asia's Rising Scientists Award 2017, and Young Researcher in IChemE Malaysia Award 2016. He has successfully supervised eight PhD students and two MSc students as primary supervisor. Currently, he is the primary supervisor for 11 PhD students and 4 MSc students. He has published more than 200 journal papers.

Suchithra Thangalazhy Gopakumar Gopakumar is currently working as Associate Professor at University of Nottingham Malaysia. Dr. Suchithra's research focuses on the development of liquid biofuels and extraction of chemicals from various biomass feedstocks through thermo-chemical conversions and catalytic upgrading. She has authored 3 book chapters and more than 30 journal papers. Dr. Suchithra has been part of organizing some international conferences and has presented her findings in various international conferences and exhibitions. Her projects have received awards in international exhibitions conducted in Malaysia. She is also the recipient of some research grants at the university and national levels. Suchithra has achieved the status of 'Fellow of the Higher Education Academy', UK. She is also an associate member of Institute of Chemical Engineers (IChemE) and Indian Institute of Chemical Engineers (IIChE).

Dominic C. Y. Foo is a Professor of Process Design and Integration at the University of Nottingham Malaysia and is the Founding Director for the Centre of Excellence for Green Technologies. He is a Fellow of the Institution of Chemical Engineers (IChemE), a Fellow of the Academy of Science Malaysia (ASM), a Chartered Engineer (CEng) with the UK Engineering Council, a Professional Engineer (PEng) with the Board of Engineers Malaysia (BEM), as well as the President for the Asia Pacific Confederation of Chemical Engineering (APCChE). He is a world-renowned scholar in process integration focusing on resource conservation and CO2 reduction. He establishes international collaboration with researchers from various countries in the Asia, Europe, the Americas, and Africa. Professor Foo is an active author, with 8 books and more than 160 journal papers, and he has made more than 220 conference presentations, with more than 30 keynote/plenary speeches. He has served on the International Scientific Committees of many important international conferences (CHISA/PRES, FOCAPD, ESCAPE, PSE, SDEWES, etc.). Professor Foo is the Editor-in-Chief for Process Integration and Optimization for Sustainability (Springer Nature), Subject Editor for Process Safety & Environmental Protection (Elsevier), and is an Editorial Board Member for several other renowned journals. He is the winner of the Innovator of the Year Award 2009 of IChemE, the Young Engineer Award 2010 of IEM, the Outstanding Young Malaysian Award 2012 of Junior Chamber International (JCI), the Outstanding Asian Researcher and Engineer 2013 (Society of Chemical Engineers, Japan), the Vice-Chancellor's Achievement Award 2014 (University of Nottingham), and the Top Research Scientist Malaysia 2016 (ASM). He has conducted close to 100 professional workshops for academics and industrial practitioners worldwide.

Editorial

Special Issue "Green Technologies: Bridging Conventional Practices and Industry 4.0"

Pau Loke Show *, Suchithra Thangalazhy-Gopakumar and Dominic C. Y. Foo

Department of Chemical and Environmental Engineering, Faculty of Science and Engineering, University of Nottingham Malaysia, Broga Road, Semenyih 43500, Malaysia; Suchithra.Thangalazhy@nottingham.edu.my (S.T.-G.); Dominic.Foo@nottingham.edu.my (D.C.Y.F.)

* Correspondence: PauLoke.Show@nottingham.edu.my or showpauloke@gmail.com

Received: 28 April 2020; Accepted: 28 April 2020; Published: 8 May 2020

1. Introduction

Green technologies have been globally accepted as efficient and sustainable techniques for the utilization of natural resources. Currently, Industry 4.0, which is also called a "smart industry", aims for the integration of cyber and physical systems to minimize waste and maximize productivity. Therefore, green technologies can be identified as key components in Industry 4.0. The scope of this Special Issue is to address how conventional green technologies can be a part of smart industries by minimizing waste, maximizing productivity, optimizing the supply chain, or by additive manufacturing (3D printing). This theme focuses on the scope and challenges of integrating current environmental technologies in future industries.

This Special Issue "Green Technologies: Bridging Conventional Practices and Industry 4.0" invites manuscripts from academicians working on green technology-related processes. Authors are invited to submit original research articles covering topics which include, but are not limited to, the following areas: (1) the development of new disease-specific models to guide therapy; (2) air pollution monitoring and control; (3) carbon emission reduction; (4) computational tools for environmental applications; (5) energy and environmental policy; (6) environmental monitoring, assessment and management; (7) Industry 4.0; (8) process system engineering; (9) renewable energy; (10) solid/biomass waste treatment, management, and recycling; and (11) waste minimization, etc. The manuscripts were regularly submitted, selected and reviewed by the regular system and accepted for publication. This Special Issue, "Green Technologies: Bridging Conventional Practices and Industry 4.0", aims to incorporate and introduce the advances in green technologies to the cyber-based industries.

In this Special Issue on "Green Technologies: Bridging Conventional Practices and Industry 4.0", we have accepted and published 17 high-quality and original articles [1–17]. These research papers cover theoretical, numerical, or experimental approaches on green technology that bridge conventional practices and Industry 4.0. The Special Issue operates a rigorous peer-review process with a single-blind assessment and at least two independent reviewers, hence resulting in our final acceptance of these published high-quality papers.

2. Papers Presented in the Special Issue

Borhan et al. [1] researched about the characterization and modelling studies of activated carbon produced from rubber-seed shells using KOH for the CO_2 adsorption. The study experimentally demonstrated that the Freundlich isotherm and pseudo-second kinetic model provided the best fit to the experimental data, suggesting that the rubber-seed shell activated carbon they prepared is an attractive source for CO_2 adsorption applications. Yunus et al. [2] reported that ionic liquids, which are classified as new solvents, have been identified to be potential solvents in the application of CO_2 capture. In this work, six ammonium-based protic ionic liquids, containing ethanolammonium

(EtOHA), tributylammonium (TBA), bis(2-ethylhexyl) ammonium (BEHA) cations, and acetate (AC) and butyrate (BA) anions, were synthesized and characterized.

Pan et al. [3] successfully synthesized an amorphous mesoporous silicon oxycarbide material (SiOC) via a low-cost facile method by using potassium hydroxide activation, high-temperature carbonization, and acid treatment. The precursors were obtained from floating plants (floating moss, water cabbage, and water caltrops). Ali et al. [4] optimized municipal solid waste (MSW) conversion technologies using a process network synthesis tool, the "process graph" (P-graph). The four highest compositions (i.e., food waste, agriculture waste, paper, and plastics) of the MSW generated in Malaysia were optimized using a P-graph. Two types of conversion technologies were considered, namely biological conversion (anaerobic digestion) and thermal conversion (pyrolysis and incinerator). All these conversion technologies were compared with the standard method used: landfilling. One-hundred feasible structures were generated using a P-graph.

There are few excellent examples of research done in enhancing the sustainability of biofuels. Damanik et al. [5] demonstrated the performance and exhaust emissions of a diesel engine fuelled with calophyllum inophyllum—palm biodiesel. Meanwhile, Wan Nurain et al. [6] discussed the sugarcane bagasse-based adsorbent employed for mitigating eutrophication threats and producing biodiesel simultaneously. Further, Bello et al. [7] reported the thermal analysis of Nigerian oil palm biomass with sachet-water plastic wastes for the sustainable production of biofuel. Besides, Xuefei et al. [8] discussed the fabrication of green superhydrophobic and superoleophilic wood flour for an efficient oil separation from water. Wong et al. [9] conducted an in situ fermentation process for improving protein and lipid contents in the larval biomass of the black soldier fly, which can be subsequently converted into nutrients and biofuels. All these collections are important in contributing to the sustainability of biofuel production in Industry 4.0.

Few of the papers published in this Special Issue also investigated the concept of automation and investigations were done on the underlying principles and technologies for implementation in an automated industry. Tran Van et al. [10] studied the hygro-thermo-mechanical responses of balsa wood core to observe the permeability and fire resistance of the composites. Experimental, analytical and numerical methods were applied to understand the moisture impervious barrier significance of the structure. De-la-torre et al. [11] performed a study on a multivariate analysis and machine learning algorithm for the ripeness classification of Cape gooseberry fruits. The work applied sophisticated algorithms to analyze the feature selection and extraction, and combined them to find the best combination for a particular application. The optimization work may be developed to use for measuring the level of ripeness of the Cape gooseberry or any different type of fruit. Moreover, the work by Mohd Aris et al. [12] shows a Gaussian process (GP) methodology for a multi-frequency marine controlled-source electromagnetic profile estimation in an isotropic medium. The Gaussian process proposed can reduce the high computational cost and complexity of the mathematical equations involved, where a 2D forward GP model was developed and the model was validated. Good agreement between the output and estimation was achieved. These works are important as a stepping stone for the creation of an automated industry.

Apart from that, this Special Issue also attracted three quality review papers. The first review article is written by Khoo et al. [13], and this review paper covers the latest developments in bioseparation technology using a liquid biphasic system (LBS). The review article begins with an in-depth discussion on the fundamental principle of LBS and this is followed by the discussion on the further developments of the various phase-forming components in LBS. Additionally, the implementation of various advance technologies to the LBS that is beneficial towards the efficiency of LBS for the extraction, separation, and purification of biomolecules was discussed. The key parameters affecting the LBS were presented and evaluated. Moreover, future prospects and challenges were highlighted to be a useful guide for the future development of LBS. The efforts presented in this review will provide an insight for future research in liquid–liquid separation techniques. In the Special Issue, there are works by Tham et al. [14,15], where the article critically discussed the recovery of protein from dairy milk waste products using an alcohol–salt liquid biphasic flotation, which is one of the latest technologies in LBS that can be potentially applied in Industry 4.0.

On the other hand, the second review paper was written by Chow et al. [16] and is about the potential co-substrates and operating factors for an improved methane yield from the perspective of anaerobic co-digestion of wastewater sludge. This review summarizes the results from numerous laboratory, pilot, and full-scale anaerobic co-digestion (ACD) studies of wastewater sludge with the co-substrates of organic fractions of municipal solid waste, food waste, crude glycerol, agricultural waste, and fat, oil and grease. The critical factors that influence the ACD operation are also discussed. The ultimate aim of this review is to identify the best potential co-substrate for wastewater sludge anaerobic co-digestion and to provide a recommendation for future reference. By adding co-substrates, a gain ranging from 13% to 176% in the methane yield was accomplished compared with mono-digestion.

In the third review paper contributed by Yong et al. [17], a comprehensive review of the appraisal of the environmental, financial, and public issues related to the energy recovery from municipal solid waste in the view of sustainable waste-to-energy (WTE) development in Malaysia is offered. This review article mainly discusses the various WTE technologies in Malaysia by considering the energy potentials from all the existing incineration plants and landfill sites as an effective MSW management in Malaysia. Furthermore, to promote local innovation and technology development and to ensure the successful long-term sustainable economic viability, social inclusiveness, and environmental sustainability in Malaysia, the four faculties of sustainable development, namely technical, economic, environmental, and social issues affiliated with MSW-to-energy technologies, were compared and evaluated.

3. Conclusions

It is hope that the novel green technologies presented in this issue are useful in assisting the global community in working towards fulfilling the Sustainable Development Goals of United Nation. The guest editors thank the authors for their contribution to the new knowledge and the reviewers for their valuable time and efforts in the review process. Besides, we would like to thank the editorial office and Dr Unai Vicario for their help and support in completing this Special Issue, especially during the pandemic of COVID-19.

Conflicts of Interest: The authors declare no conflict of interest.

References

1. Borhan, A.; Yusup, S.; Lim, J.-W.; Show, P.L. Characterization and Modelling Studies of Activated Carbon Produced from Rubber-Seed Shell Using KOH for CO_2 Adsorption. *Processes* **2019**, *7*, 855. [CrossRef]
2. Yunus, N.M.; Halim, N.H.; Wilfred, C.D.; Murugesan, T.; Lim, J.W.; Show, P.L. Thermophysical Properties and CO_2 Absorption of Ammonium-Based Protic Ionic Liquids Containing Acetate and Butyrate Anions. *Processes* **2019**, *7*, 820. [CrossRef]
3. Pan, G.; Chong, S.; Chan, Y.J.; Tiong, T.J.; Lim, J.-W.; Huang, C.-M.; Shukla, P.; Yang, T.C. Physical and Thermal Studies of Carbon-Enriched Silicon Oxycarbide Synthesized from Floating Plants. *Processes* **2019**, *7*, 794. [CrossRef]
4. Ali, R.A.; Ibrahim, N.N.L.N.; Lam, H.L. Conversion Technologies: Evaluation of Economic Performance and Environmental Impact Analysis for Municipal Solid Waste in Malaysia. *Processes* **2019**, *7*, 752. [CrossRef]
5. Damanik, N.; Ong, H.C.; Mofijur, M.; Chong, W.T.; Silitonga, A.S.; Shamsuddin, A.; Sebayang, A.H.; Mahlia, T.M.I.; Wang, C.-T.; Jang, J.-H. The Performance and Exhaust Emissions of a Diesel Engine Fuelled with Calophyllum inophyllum—Palm Biodiesel. *Processes* **2019**, *7*, 597. [CrossRef]
6. Wan Basri, W.N.F.; Daud, H.; Lam, M.K.; Cheng, C.K.; Oh, W.D.; Tan, W.N.; Shaharun, M.S.; Yeong, Y.F.; Paman, U.; Kusakabe, K.; et al. A Sugarcane-Bagasse-Based Adsorbent Employed for Mitigating Eutrophication Threats and Producing Biodiesel Simultaneously. *Processes* **2019**, *7*, 572. [CrossRef]
7. Salman, B.; Ong, M.Y.; Nomanbhay, S.; Salema, A.; Sankaran, R.; Show, P.L. Thermal Analysis of Nigerian Oil Palm Biomass with Sachet-Water Plastic Wastes for Sustainable Production of Biofuel. *Processes* **2019**, *7*, 475. [CrossRef]

8. Tan, X.; Zang, D.; Qi, H.; Liu, F.; Cao, G.; Ho, S.-H. Fabrication of Green Superhydrophobic/Superoleophilic Wood Flour for Efficient Oil Separation from Water. *Processes* **2019**, *7*, 414. [CrossRef]
9. Kalnik, M.W.; Kouchakdjian, M.; Li, B.F.; Swann, P.F.; Patel, D.J. Base pair mismatches and carcinogen-modified bases in DNA: An NMR study of G.T and G.O4meT pairing in dodecanucleotide duplexes. *Biochemistry* **1988**, *27*, 337. [CrossRef] [PubMed]
10. Tranvan, L.; Legrand, V.; Casari, P.; Sankaran, R.; Show, P.L.; Berenjian, A.; Lay, C.-H. Hygro-Thermo-Mechanical Responses of Balsa Wood Core Sandwich Composite Beam Exposed to Fire. *Processes* **2020**, *8*, 103. [CrossRef]
11. De-La-Torre, M.; Zatarain, O.; Avila-George, H.; Muñoz, M.; Cruz, J.O.; Lozada, R.; Mejía, J.; Castro, W. Multivariate Analysis and Machine Learning for Ripeness Classification of Cape Gooseberry Fruits. *Processes* **2019**, *7*, 928. [CrossRef]
12. Aris, M.N.M.; Daud, H.; Dass, S.C.; Noh, K.A.M. Gaussian Process Methodology for Multi-Frequency Marine Controlled-Source Electromagnetic Profile Estimation in Isotropic Medium. *Processes* **2019**, *7*, 661. [CrossRef]
13. Khoo, K.S.; Leong, H.; Chew, K.W.; Lim, J.-W.; Ling, T.C.; Show, P.L.; Yen, H.-W. Liquid Biphasic System: A Recent Bioseparation Technology. *Processes* **2020**, *8*, 149. [CrossRef]
14. Tham, P.E.; Ng, Y.J.; Sankaran, R.; Khoo, K.S.; Chew, K.W.; Yap, Y.J.; Malahubban, M.; Aziz Zakry, F.A.; Show, P.L. Recovery of Protein from Dairy Milk Waste Product Using Alcohol-Salt Liquid Biphasic Flotation. *Processes* **2019**, *7*, 875. [CrossRef]
15. Tham, P.E.; Ng, Y.J.; Sankaran, R.; Khoo, K.S.; Chew, K.W.; Yap, Y.J.; Malahubban, M.; Aziz Zakry, F.A.; Show, P.L. Correction: Tham, P.E., et al. Recovery of Protein from Dairy Milk Waste Product Using Alcohol–Salt Liquid Biphasic Flotation. *Processes* **2019**, *7*, 875. *Processes* **2020**, *8*, 381. [CrossRef]
16. Chow, W.; Chong, S.; Lim, J.-W.; Chan, Y.J.; Chong, M.; Tiong, T.J.; Chin, J.; Pan, G.-T. Anaerobic Co-Digestion of Wastewater Sludge: A Review of Potential Co-Substrates and Operating Factors for Improved Methane Yield. *Processes* **2020**, *8*, 39. [CrossRef]
17. Yong, Z.J.; Bashir, M.J.; Ng, C.A.; Sethupathi, S.; Lim, J.W.; Show, P.L. Sustainable Waste-to-Energy Development in Malaysia: Appraisal of Environmental, Financial, and Public Issues Related with Energy Recovery from Municipal Solid Waste. *Processes* **2019**, *7*, 676. [CrossRef]

Article

Robust Design of PC/ABS Filled with Nano Carbon Black for Electromagnetic Shielding Effectiveness and Surface Resistivity

Wipoo Sriseubsai [1,*], Arsarin Tippayakraisorn [1] and Jun Wei Lim [2]

[1] Department of Industrial Engineering, Faculty of Engineering, King Mongkut's Institute of Technology Ladkrabang, Bangkok 10520, Thailand; snoopynest_q@hotmail.com

[2] Department of Fundamental and Applied Sciences, HICoE-Centre for Biofuel and Biochemical Research, Institute of Self-Sustainable Building, Universiti Teknologi PETRONAS, Perak 32610, Malaysia; junwei.lim@utp.edu.my

* Correspondence: wipoo.sr@kmitl.ac.th; Tel.: +66-2-329-8339

Received: 31 March 2020; Accepted: 11 May 2020; Published: 21 May 2020

Abstract: This study focuses on the electromagnetic interference shielding effectiveness (EMI SE), dissipation of electrostatic discharge (ESD), and surface resistivity of polymer blends between polycarbonate (PC) and acrylonitrile–butadiene–styrene (ABS) filled with carbon black powder (CBp) and carbon black masterbatch (CBm). The mixtures of PC/ABS/CB composites were prepared by the injection molding for the 4-mm thickness of the specimen. The D-optimal mixture design was applied in this experiment. The EMI SE was measured at the frequency of 800 and 900 MHz with a network analyzer, MIL-STD-285. The result showed that the EMI SE was increased when the amount of filler increased. The surface resistivity of the composites was determined according to the ASTM D257. It was found that the surface resistivity of the plastic with no additives was 10^{12} Ω/ square. When the amount of fillers was added, the surface resistivity of plastic composites decreased to the range of 10^6–10^{11} Ω/square, which was suitable for the application without the electrostatic discharge. The optimization of multi-response showed using high amounts of PC and CB was the best mixture of this research.

Keywords: PC/ABS; carbon black; electromagnetic shielding effectiveness; dissipation of electrostatic discharge; surface resistivity

1. Introduction

Nowadays, plastics, especially thermoplastics, are formed and used for many applications such as parts of automotive, electronic devices, and packaging. Some electronic devices generate and/or transmit electromagnetic waves that affect other devices, e.g., noise, an error operation, or the malfunction of electronic components [1]. An example is the capacitor in amplifiers that can generate electromagnetic waves that affect the quality of sound because of electromagnetic interference. Moreover, the electrostatic discharge transmitted from humans or tools may destroy some electronic parts. In order to prevent those problems, there were many researchers that have studied and developed electromagnetic interference shielding and dissipative material.

Generally, the material which has high performance for electromagnetic interference shielding effectiveness (EMI SE) is metal, due to high conductive properties. However, it has limitations such as weight, cost, processability, and corrosion [2]; then, plastic becomes the material of choice. There are many researchers who have developed and improved the EMI SE and dissipative plastic composites instead of metal, although normally, the plastic is electrically insulated and does not contribute to electromagnetic interference shielding. Plastic that is the matrix of the composite can connect the conductive filler. Plastic composites having conductive filler is one method to make an EMI shielding material. The filler can be aluminum flakes, steel fiber, or carbon fiber [3]. There are high demands of electrically conductive polymer, but it is not the same as plastic composites because of the poor processability. The conductive polymer does not require conductive filler in order to provide the shielding, so plastic composites with conductive filler are concerned and studied [3], with the increasing demand of customers for the reliability of electronic equipment [4–11]. Nanofillers that have been investigated by a number of researchers for EMI shielding were reviewed by Wanasinghe D. et al. [12]. It showed that nanocarbon black mixed with plastic made good shielding effectiveness, and the composite could have potential application in industry. However, the cost of the entire composite was high due to the nanoparticle production and additional material preparation process. Yangyong Wang and Xinli Jing studied EMI shielding by using polypyrrole (PPy) and polyaniline (PANI) [13], and the results showed the high performance of the shielding. Silver-palladium (AdPd) was coated to polyethylene terephthalate (PET) to be EMI shielding, and it was found that the shielding effectiveness depended on the conductive properties [14]. Quinton J. studied EMI and radio wave shielding with three additives, i.e., carbon, graphite, and carbon fiber, mixed with 2 types of polymer matrix, PA6.6 and polycarbonate (PC) [15]. The results showed that carbon black was more effective than other additives. Moreover, when using multiple additives, the shielding effectiveness was higher than using only one additive and related to the study of Pramanik et al. [16].

In addition, electrostatic discharge (ESD) is another problem when the insulation polymer has conductive property; it can cause the electrical equipment to be damaged. The resistance of the polymer is between conduction and insulation material, which is called static dissipative material. It has the surface resistivity between 10^4 and 10^{11} ohm/square, and it is used to make a product and prevent the electrostatic discharge [17].

PC is a high impact- and heat resistance, fair chemical resistance, and is transparent. ABS is a low-cost as well as flexible material. Both of them are widely used in many applications. Moreover, PC and ABS have been blended to get the advantages of both material properties for applications such as automotive, electronics and telecommunication, and medical devices. This research investigates PC/ABS mixed with carbon black powder (CBp) and carbon black masterbatch (CBm) as electromagnetic interference shielding, the dissipation of electrostatic discharge (ESD) material, and surface resistivity. Carbon black powder is used as a filler for EMI, and it has been studied by many researchers for many applications, such as mixing with rubber to increase friction resistance and strain. Carbon black masterbatch is ready-mixed carbon black plastic. It can be added to compatible plastic during the forming of the product. It is easy to use compared with carbon black powder. The powder has to be compounded with a plastic matrix before forming, but the masterbatch can be added directly to the production process. However, the mixing ratio of the carbon black when using the masterbatch is more difficult to adjust than when using the powder grade. While a number of researchers have studied the effect of filler to EMI, this research studies the mixing ratio of each material, which is discussed and determined by the mixture design and statistical method to analyze and optimize the mixture of those materials.

2. Materials and Methods

Basically, plastic will have electromagnetic interference shielding effectiveness (EMI SE) property when it can act as the wave impedance and effect to the discontinuous electromagnetic field. When the electromagnetic waves attack the material, there are three mechanisms that polymeric material should have as shielding, such as reflection and absorption, so that little of the electromagnetic waves pass through that material [1] (as shown in Figure 1). This is defined as shielding effectiveness and can be determined by the following equation.

$$SE = 20\ log\,\frac{E_1}{E_2} = 20\ log\,\frac{H_1}{H_2} \tag{1}$$

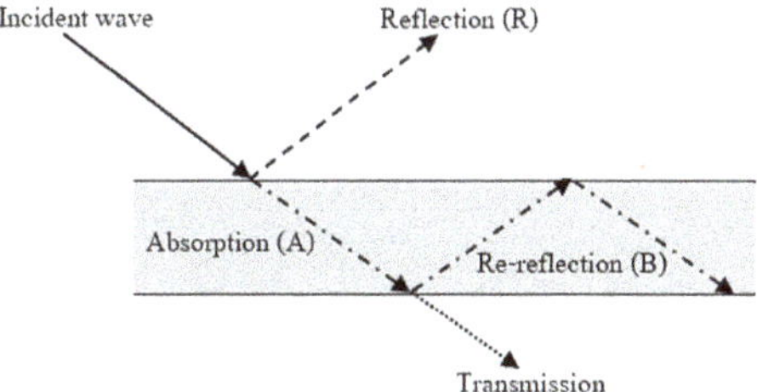

Figure 1. The mechanism of electromagnetic interference shielding effectiveness (EMI SE).

In Equation (1), SE is the shielding effectiveness, dB; E_1, E_2 are the amplitudes of the incident wave and transmitted wave (V/m), respectively; H_1, H_2 are incident and transmitted magnetic field strengths (H/m), respectively.

The development of composited plastic by conductive filler is one of the methods to get the electromagnetic interference shielding property. The mixtures of PC, ABS, and carbon black were prepared with the design of the experiment called a mixture design with the D-optimal method. This method is recommended when there are constraints in the proportions of the mixture components [18]. This research was limited to the mixture ratio by the viscosity of the mixture. When mixing with a high amount of carbon black, the viscosity of the composite material is increased. This would cause damage to the injection molding machine when the viscosity of the material is too high. Then, the mixture ratio of carbon black was limited by the mixture melt flow rate of 5 g/10 min, which was performed following the ASTM D1238. Then, the mixing of each composition by the mixture design with D-optimal was designed and is shown in Figure 2 and Table 1. The PC and ABS used in this research were commercial-grade 110 and PA 707, which were manufactured by CHIMEI. Two types of carbon black were used as the additive, i.e., 22-nm powder grade N220 manufactured by Thai Tokai Carbon Product and 26-nm commercial masterbatch PLASBLAK® UN2014 from COBOT.

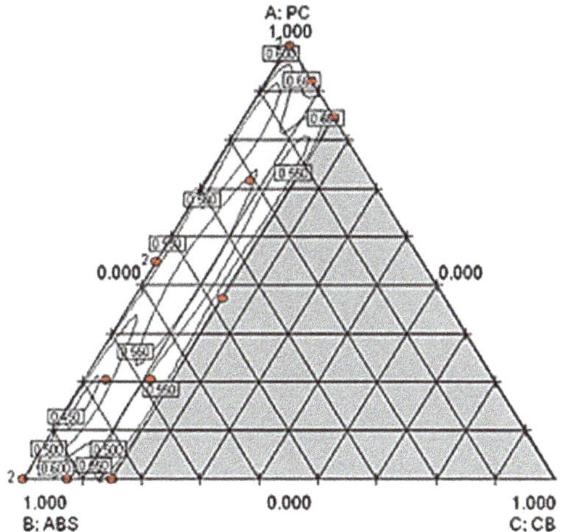

Figure 2. Mixture design.

Table 1. Percentage of the composition of polycarbonate (PC)/ acrylonitrile–butadiene–styrene (ABS) and carbon black (CB).

run	PC	ABS	CB (CBm, CBp)
1	0.00	0.83	0.17
2	0.50	0.50	0.00
3	0.00	0.83	0.17
4	0.23	0.65	0.13
5	0.23	0.73	0.04
6	0.83	0.00	0.17
7	0.42	0.42	0.17
8	0.83	0.00	0.17
9	0.69	0.23	0.08
10	1.00	0.00	0.00
11	0.00	0.92	0.08
12	0.00	1.00	0.00
13	1.00	0.00	0.00
14	0.92	0.00	0.08
15	0.00	1.00	0.00
16	0.50	0.50	0.00

All 16 combinations were mixed, and the plaque specimens performed with the dimension of 180 × 100 mm and 4 mm thickness by an injection molding machine, Toshiba 80 Tons. The specimen was used to study electromagnetic interference shielding effectiveness by using the network analyzer MIL-STD-285, with the electromagnetic frequency of 800 and 900 MHz; the experimental setup is shown in Figure 3. The shielding effectiveness was determined by the following equation:

$$\text{Shielding effectiveness (SE)} = \text{P1} - \text{P2} \tag{2}$$

where P is the power level at Points 1 and 2, respectively.

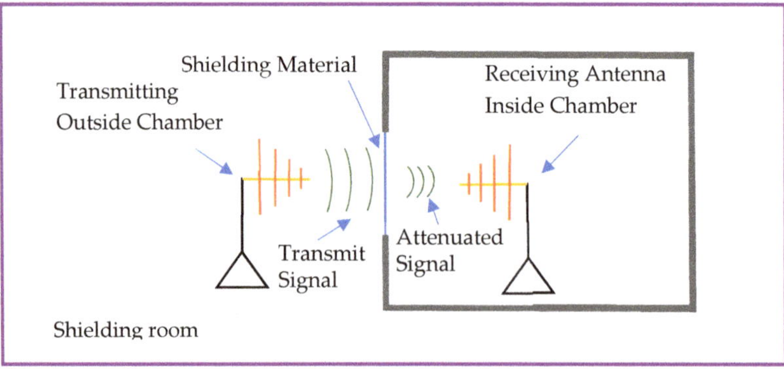

Figure 3. Source and receiver of the network analyzer.

The dielectric constant was performed with the specimen dimension of 70 × 100 mm and 4 mm thickness by using the Agilent 4263B with 100 kHz and 1000 mV. The parallel capacitance, C_p, was measured, and the dielectric constant was determined by the following equation:

$$\varepsilon_r = \frac{tC_p}{A\varepsilon_0}\varepsilon = \varepsilon_r\varepsilon_0 \quad (3)$$

where

ε is the dielectric constant (F)
ε_0 is 8.854×10^{-12} (F/m)
ε_r is the relative dielectric constant
C_p is the capacitance (F)
A is the cross-section area (m)
t is the thickness (m)

The surface resistivity was performed following the ASTM D257, as shown in Figure 4. The specimens were prepared as the plaque of 100 × 100 × 4 mm. The surface resistance was measured, and the surface resistivity was determined by

$$\sigma = \frac{RP}{g} \quad (4)$$

where

σ is the surface resistivity (Ω/square)
R is the surface resistance (Ω)
P is the distance between electrodes (cm)
g is the electrode circumference (cm)

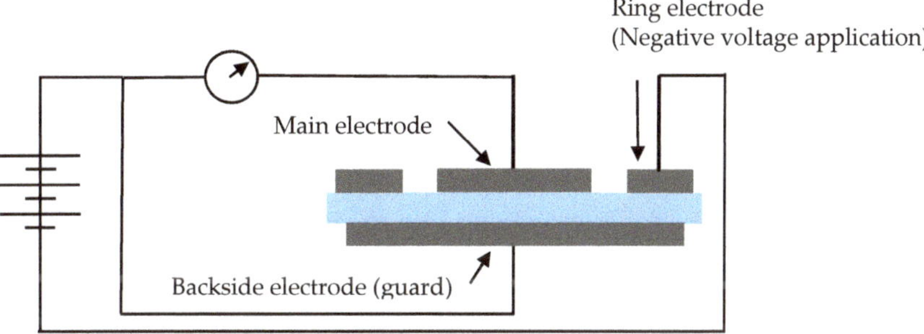

Figure 4. Surface resistivity measurement following the ASTM D257.

3. Results and Discussion

According to the mixture design of the experiment, the electromagnetic interference shielding effectiveness of the mixture between PC/ABS and carbon black masterbatch and carbon black power for each testing frequency are shown in Figures 5 and 6, respectively. The results showed that when using a higher carbon black mixing ratio, the SE was increased by both testing frequencies because the additive is the conductive material, allowing the plastic composite to reflect and absorb the electromagnetic wave. The SE of the composite also showed a maximum value of about 9 dB at 800 MHz, and about 5 dB at 900 MHz had been obtained for the mixing containing 17 wt % carbon black. Moreover, the results showed that both plastic composites that used different carbon blacks had a slightly different effect on the SE because the size of the carbon black used was a small difference in size.

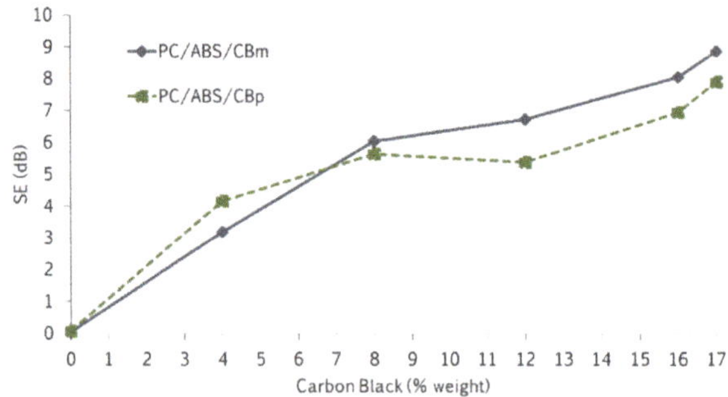

Figure 5. Shielding effectiveness (SE) at 800 MHz with the carbon black.

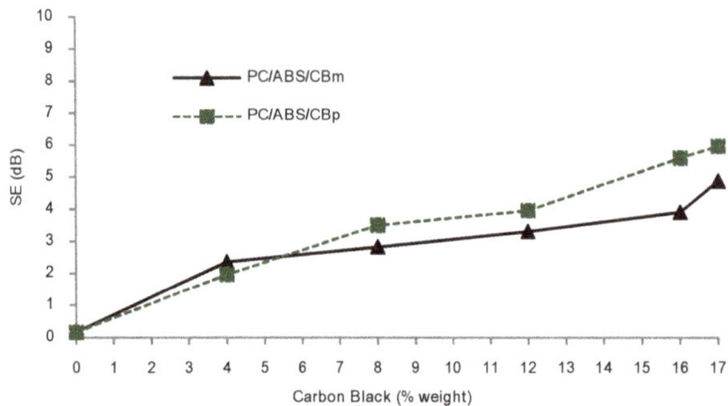

Figure 6. SE at 900 MHz with the carbon black.

The morphology studied of PC/ABS/CBp (0.42/0.42/0.16) and PC/ABS/CBm (0.69/0.23/0.08) conducted through SEM images is given in Figures 7 and 8, respectively, showing the proper distribution of carbon black within the plastic composite.

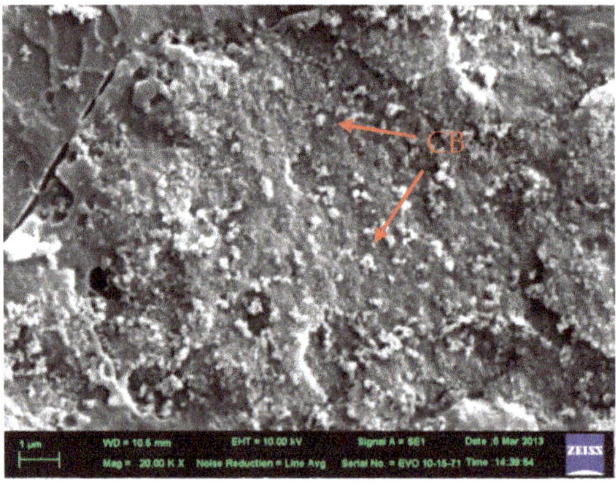

Figure 7. SEM image of the 16 wt % carbon black powder (CBp) in the PC/ABS.

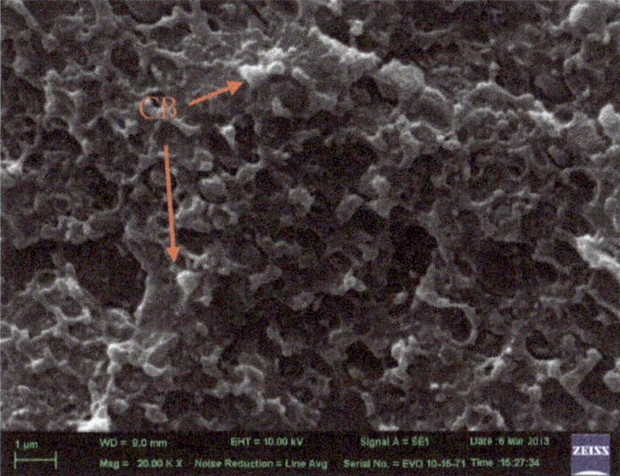

Figure 8. SEM image of the 8 wt % carbon black masterbatch (CBm) in the PC/ABS.

The dielectric constant is the ability of a substance to store electric charge or electrostatic field energy [19]. When the dielectric is high, the material has low electrical insulation. The dielectric constant of mixing PC and ABS without carbon black in this research was between 3.04–3.34. After mixing PC and ABS with carbon black, the plastic composites were measured the dielectric constant by using the Agilent 4263B with 100 kHz and 1000 mV. The results showed the dielectric constant was increased when the amount of carbon black in the mixture increased, as shown in Figure 9.

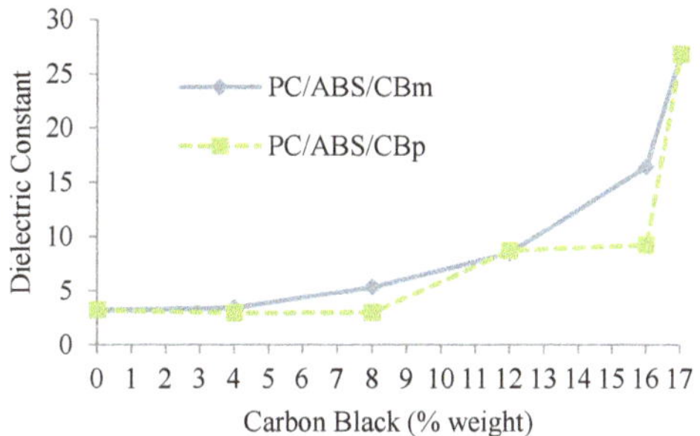

Figure 9. The relationship between dielectric constant and percentage of carbon black.

The maximum of the dielectric was about 25 when the plastic composite contained 17 wt % of carbon black, which was the upper limit of mixing carbon black for this research due to the high viscosity of the composite polymer. In contrast, the surface resistivity of the composite was the resistance to leakage current along the surface of an insulating material, which was decreased when the amount of carbon black filler increased. The surface resistivities were measured in both horizontal and vertical directions. The average surface resistivities of the composite are shown in Figure 10. The results show that when the composite contained 17 wt % of carbon black, the composite had the surface resistivity between 10^7–10^8 Ω/square, while the suitable surface resistivity for reducing the ESD of the plastic composite is between 10^4–10^{11} Ω/square [17]. This is confirmation that carbon black, the conductive filler, is effective on the surface resistivity of the composite. The composite becomes a dissipative material when at least 5 wt % of carbon black is mixed.

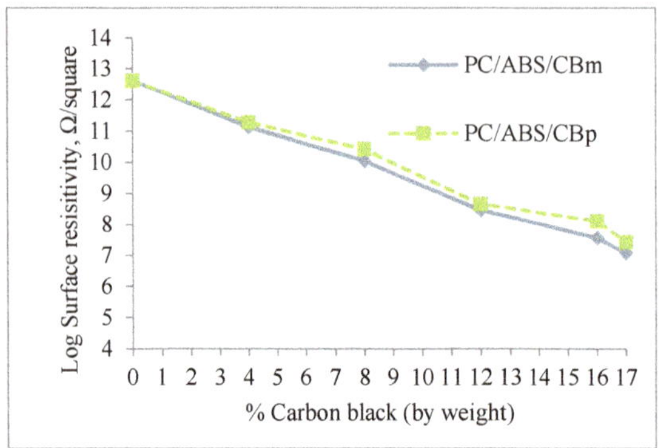

Figure 10. The relationship between surface resistivity and percentage of carbon black.

The EMI SE recorded data from the above experiments were used and analyzed by a statistical method. This method is a response surface methodology to determine the suitable regression model for the prediction of the EMI SE of the mixture. The statistical results, such as standard deviation, R-square, adjusted R-square, and PRESS, were analyzed for linear, quadratic, special cubic, and cubic models. When compared to those results, the adjusted R-square and R-square of the cubic model was higher than other models. Moreover, the standard deviation and PRESS of the cubic model were the lowest values when compared with other models. Then, the cubic model was selected for the prediction of the EMI SE of the mixture. The suitable regression model hypothesis was tested with ANOVA as well. The p-value and p-value of the lack of fit were statistically significant with $\alpha = 0.05$. The model of those experiments is shown as the following:

At 800 MHz

Masterbatch:

$$SE = -0.061A + 0.10B + 502.72C + 0.14AB - 819.99AC - 816.79BC + 811.86ABC + 2.16AB(A - B) + 426.13AC(A - C) + 410.22BC(B - C) \quad (5)$$

Powder:

$$SE = 0.088A + 0.081B + 4451.18C - 0.20AB - 7296.87AC - 7286.17BC + 6009.48ABC + 2.72AB(A - B) + 3019.66AC(A - C) + 2999.24BC(B - C) \quad (6)$$

At 900 MHz

Masterbatch:

$$SE = 0.083A + 0.29B + 1483.98C - 0.27AB - 2402.47AC - 2414.15BC + 1961.35ABC - 4.65AB(A - B) + 988.02AC(A - C) + 996.78BC(B - C) \quad (7)$$

Powder:

$$SE = 0.035A + 0.17B + 1349.77C + 0.75AB - 2188.58AC - 2148.86BC + 1769.32ABC + 2.12AB(A - B) + 918.63AC(A - C) + 855.09BC(B - C) \quad (8)$$

In Equations (5)–(8), A is PC, B is ABS, and C is carbon black, respectively. Those equations showed the independence and interaction of the factors. When considering the independent term, they show that carbon black (C) is more effective to the SE than other factors. That was the reason why an increase in carbon black increased SE.

There were three data sets of the test: EMI SE, dielectric constant and surface resistivity. The dielectric constant and surface resistivity were related together and depended on each other.

The optimization of the multi responses, EMI shielding effectiveness, and surface resistivity of each testing frequency was determined by using Design Expert software, while the level of PC/ABS/CB was the factor. The minimized parameters of the composite were determined by using the overlaid contour plot method and are shown in Figures 11–14. The results of SE and surface resistivity of the optimized PC/ABS/CB are also shown in Tables 2 and 3.

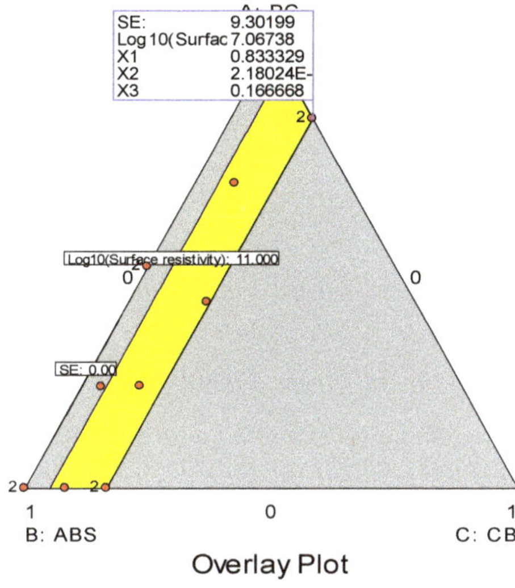

Figure 11. Overlay mapping @ 800 MHz with carbon black masterbatch.

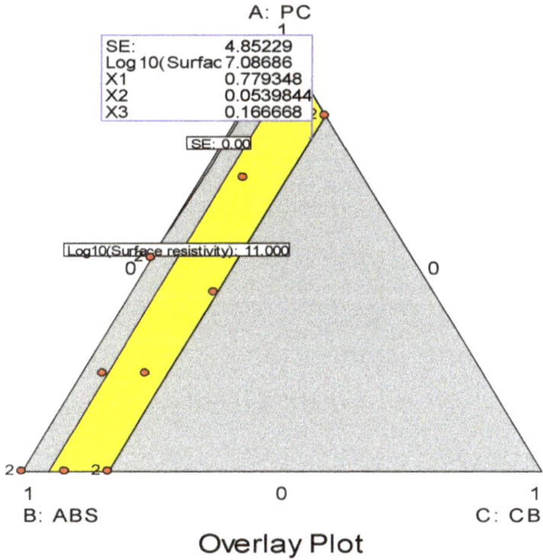

Figure 12. Overlay mapping @ 900 MHz with carbon black masterbatch.

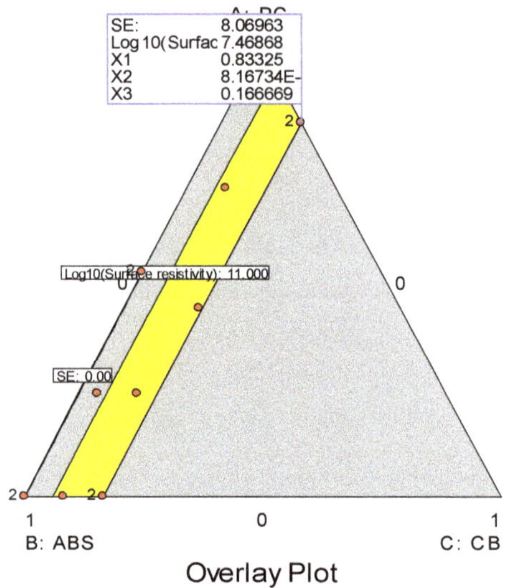

Figure 13. Overlay mapping @ 800 MHz with carbon black particles.

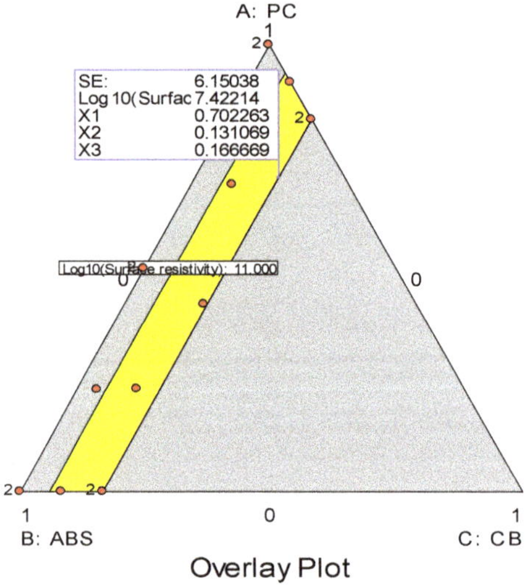

Figure 14. Overlay mapping @ 900 MHz with carbon black particles.

Table 2. Optimized mixing ratio among PC, ABS, and CBm.

Frequency	PC	ABS	CBm	SE	Log_{10}(Surface Resistivity)
@ 800	0.83	0	0.17	9.31	7.09
@ 900	0.78	0.05	0.17	4.86	7.08

Table 3. Optimized mixing ratio among PC, ABS, and CBp.

Frequency	PC	ABS	CBp	SE	Log_{10}(Surface Resistivity)
@ 800	0.83	0	0.17	8.06	7.46
@ 900	0.7	0.13	0.17	6.15	7.42

The results showed that using a high amount of PC and CB optimized the mixture, which gave the high EMI shielding effectiveness for each testing frequency but gave the low surface resistivity that was about 10^7 Ω/square. It was also between the suitable range for reducing the ESD, 10^4–10^{11} Ω/square [17]. At 800 MHz, the best composition of PC/ABS/CB was 0.83/0/0.17 when using carbon black masterbatch or powder. At 900 MHz, the best composition of PC/ABS/CBm was 0.78/0.05/0.17, but when using the powder carbon black, the best composition was 0.7/0.13/0.17. While all the optimized compositions used a high percentage of PC, the EMI SE was high because the PC had a high polarity than ABS. The PC has polar side groups and regularity in the chain, while ABS has the polarity from the nitrile group. The polarity of the material may influence the shielding effectiveness of the composite as well.

4. Conclusions

The mixture of PC/ABS/CB that was studied in this research showed that CB influenced the EMI SE. The increasing CB in the mixture affected the increasing electromagnetic interference shielding effectiveness and dielectric constant, but the surface resistivity was decreased. The design of experiments with the response surface method gave the suitable cubic regression model, which could predict those properties. The optimization of the mixture showed that a high amount of PC and CB gave better EMI SE. However, the 17 wt % of CB was the maximum level of this research due to the limitation of high viscosity. The electromagnetic field was reflected or absorbed by the composite due to the shielding property. The high polarity polymer was more significant than the low one. The size of carbon black from masterbatch and powder was not significant in this research. Both filler materials can be used to make the shielding polymer. However, the carbon black masterbatch is commercial-grade and easier to use than the powder. The powder grade is suitable when adjusting the mixture is often required. When PC/ABS is required for shielding properties such as car audio components, it is recommended to add a high amount of CB, and the ratio of PC should higher than ABS to get high EMI SE. However, this research suggests that the mechanical properties of the composite should be considered as an additional response because PC and ABS are blended to get the advantage of both material properties.

Author Contributions: Conceptualization, W.S.; methodology, W.S.; software, A.T.; validation, W.S. and A.T.; formal analysis, W.S.; investigation, W.S. and A.T.; resources, A.T.; data curation, A.T.; writing—original draft preparation, W.S. and A.T.; writing—review and editing, W.S. and J.W.L.; visualization, A.T.; supervision, W.S.; project administration, W.S. All authors have read and agreed to the published version of the manuscript.

Funding: This research was funded by King Mongkut's Institute of Technology, Ladkrabang, grant number CRT29-2561.

Acknowledgments: The authors wish to thank King Mongkut's Institute of Technology, Ladkrabang (Grant No. CRT29-2561), and the Faculty of Engineering, King Mongkut's Institute of Technology, Ladkrabang, where the experiments were performed.

Conflicts of Interest: The authors declare no conflict of interest.

References

1. Shuying, Y.; Lozano, K.; Lomeli, A.; Foltz, H.D.; Jones, R. Electromagnetic interference shielding effectiveness of carbon. *Compos. Part A* **2005**, *36*, 691–697.
2. Geetha, S.; Satheesh Kumar, K.K.; Rao, C.R.; Vijayan, M.; Trivedi, D.C. EMI Shilding: Method and Materials—A Review. *J. Appl. Polym. Sci.* **2009**, *112*, 2073–2086. [CrossRef]
3. Chung, D. Electromagnetic interference shielding effectiveness of carbon. *Carbon* **2001**, *39*, 279–285. [CrossRef]
4. Bjorklof, D. EMC Fundamentals Part Six: EMI filters and transient. *Compliance Eng.* **1998**, *15*, 10.
5. Brewer, R.; Fenical, G. Shielding: The hole problem. *Eval. Eng.* **1998**, *37*, S4–S10.
6. O'Shea, P. How to meet the shielding needs of a 500-MHz PC. *Eval. Eng.* **1998**, *37*, 40–46.
7. Ramasamy, S.R. Review of EMI shielding and suppression materials. In Proceedings of the International Conference Electromagnetic Interference and Compatibility, Piscataway, NJ, USA, 3–5 December 1997; pp. 459–466.
8. Geddes, B. Putting a Lid on EMI/RFI. *Control (Chicago Ill)* **1996**, *9*, 4.

9. Hempelmann, S. Surface engineering for EMI compliance. Process and practical examples. *Galvanotechnik* **1997**, *88*, 418–424.
10. Kimmel, W.D.; Gerke, D.D. Controlling EMI with cable shields. *Med. Device Diagn. Ind.* **1995**, *17*, 112–115.
11. Markstein, H.W. Effective shielding defeats EMI. *Electron. Packag. Prod.* **1995**, *35*, 4.
12. Wanasinghe, D.; Aslani, F.; Ma, G.; Habibi, D. Review of Polymer Composites with Diverse Nanofillers for Electromagnetic Interference Shielding. *Nanomaterials* **2020**, *10*, 541. [CrossRef] [PubMed]
13. Yangyong Wang and Xinli Jing, Intrinsically Conducting Polymers for Electromagnetic Interference Shielding. *Polym. Adv. Technol.* **2005**, *16*, 344–351. [CrossRef]
14. Lee, C.Y.; Lee, D.E.; Jeong, C.K.; Hong, Y.K.; Shim, J.H.; Joo, J.; Kim, M.S.; Lee, J.Y.; Jeong, S.H.; Byun, S.W.; et al. Electromagnetic Interference Shielding by Using Conductive Polypyrrole and Metal Compound Coated on Fabrics. *Polym. Adv. Technol.* **2002**, *13*, 577–583. [CrossRef]
15. Krueger, Q.J. Electromagnetic Interference and Radio Frequency Interference Shielding of Carbon-Filled Conductive Resins. Master's Thesis, Michigan Technological University, Houghton, MI, USA, 2002.
16. Pramanik, P.K.; Khastgir, D.; Saha, T.N. Electromagnatic Interference Shielding by Conductive Nitrile Rubber Composite Containing Carbon Filler. *J. Elastomer Plast.* **1991**, *23*, 345–361. [CrossRef]
17. Fundamentals of Electrostatic Discharge. Available online: www.https://incompliancemag.com/ (accessed on 1 March 2020).
18. Cornell, J.A. *Experiments with Mixtures: Design, Models, and the Analysis of Mixture Data*, 3rd ed.; John Wiley: Hoboken, NJ, USA, 2002; p. 400.
19. Plastics Design Library. *Handbook of Plastics Joining A Practical Guide*; Plastics Design Library: New York, NY, USA, 1997; p. 79.

 processes

Article

In-Situ Yeast Fermentation Medium in Fortifying Protein and Lipid Accumulations in the Harvested Larval Biomass of Black Soldier Fly

Chung Yiin Wong [1], Yeek Chia Ho [2], Jun Wei Lim [1,*], Pau Loke Show [3], Siewhui Chong [3], Yi Jing Chan [3], Chii Dong Ho [4,*], Mardawani Mohamad [5], Ta Yeong Wu [6,7], Man Kee Lam [8] and Guan Ting Pan [9]

1 Department of Fundamental and Applied Sciences, HICoE-Centre for Biofuel and Biochemical Research, Institute of Self-Sustainable Building, Universiti Teknologi PETRONAS, Seri Iskandar 32610, Malaysia; johnsonwcy@gmail.com

2 Department of Civil and Environmental Engineering, Centre of Urban Resource Sustainability, Institute of Self-Sustainable Building, Universiti Teknologi PETRONAS, Seri Iskandar 32610, Malaysia; yeekchia.ho@utp.edu.my

3 Department of Chemical and Environmental Engineering, University of Nottingham Malaysia, Broga Road, Semenyih 43500, Malaysia; PauLoke.Show@nottingham.edu.my (P.L.S.); faye.chong@nottingham.edu.my (S.C.); Yi-jing.chan@nottingham.edu.my (Y.J.C.)

4 Department of Chemical and Materials Engineering, Tamkang University, Tamsui, New Taipei City 251, Taiwan

5 Faculty of Bioengineering and Technology, Universiti Malaysia Kelantan, Jeli Campus, Jeli 17600, Malaysia; mardawani.m@umk.edu.my

6 Chemical Engineering Discipline, School of Engineering, Monash University, Jalan Lagoon Selatan, Bandar Sunway 47500, Malaysia; wu.ta.yeong@monash.edu

7 Monash-Industry Palm Oil Education and Research Platform (MIPO), School of Engineering, Monash University, Jalan Lagoon Selatan, Bandar Sunway 47500, Malaysia

8 Department of Chemical Engineering, HICoE-Centre for Biofuel and Biochemical Research, Institute of Self-Sustainable Building, Universiti Teknologi PETRONAS, Seri Iskandar 32610, Malaysia; lam.mankee@utp.edu.my

9 Department of Chemical Engineering and Biotechnology, National Taipei University of Technology, Taipei 106, Taiwan; gtpan@mail.ntut.edu.tw

* Correspondence: junwei.lim@utp.edu.my (J.W.L.); cdho@mail.tku.edu.tw (C.D.H.); Tel.: +60-5368-7664 (J.W.L.); +886-2-2621-5656 (C.D.H.)

Received: 2 January 2020; Accepted: 7 February 2020; Published: 14 March 2020

Abstract: Recently, worldwide researchers have been focusing on exploiting of black soldier fly larval (BSFL) biomass to serve as the feed mediums for farmed animals, including aquaculture farming, in order to assuage the rising demands for protein sources. In this study, yeast was introduced into coconut endosperm waste (CEW) whilst serving as the feeding medium to rear BSFL in simultaneously performed in situ fermentation. It was found that at a 2.5 wt% yeast concentration, the total biomass gained, growth rate and rearing time were improved to 1.145 g, 0.085 g/day and 13.5 days, respectively. In terms of solid waste reduction, the inoculation of yeast over 0.5 wt% in CEW was able to achieve more than 50% overall degradation, with the waste reduction indexes (WRIs) ranging from 0.038 to 0.040 g/day. Disregarding the concentration of yeast introduced, the protein productivity from 20 BSFL was enhanced from only 0.018 g/day (the control) to 0.025 g/day with the presence of yeast at arbitrary concentrations. On the other hand, the larval protein yield was fortified from the control (28%) to a highest value of 35% with the presence of a mere 0.02 wt% yeast concentration. To summarize, the inclusion of a minimal amount of yeast into CEW for in situ fermentation ultimately enhanced the growth of BSFL, as well as its protein yield and productivity.

Keywords: black soldier fly; yeast; fermentation; protein; larvae; organic waste; coconut endosperm waste

1. Introduction

The black soldier fly (BSF) thrived in North America before it migrated to tropical other countries during WWII. It mimics the appearance of a wasp, confusing the public with its appearance. BSF larvae (BSFL) are intrinsically polyphagous as well as saprophagous, since the larvae only consume organic matter during this stage and can ingest different kinds of decaying organic matters such as animal manure, animal carcasses or sometimes even decaying wood matters. Unlike houseflies, the BSF does not carry any transmitted diseases, as the adult fly does not feed and only relies on body fat or the energy accumulated during the larval stage for metabolism. Upon maturing sexually, the female BSF will oviposit eggs at the cracks near to food sources to ensure the newly ecloded BSF larvae (neonates) have enough food to complete their life cycle [1]. Generally, after the copulation process, the female black soldier fly will oviposit the eggs after two to three days. The whole life cycle of a black soldier fly from egg to adult will take up to around 40 to 44 days [2].

Owing to its high protein content, the direct introduction of BSFL biomass into animal feed has been explored as an alternative fishmeal, which is growing in cost. From previous research studies, the inclusion of BSFL biomass at 17%, 33%, 49%, 64% and 75% into aquaculture feed was found to decrease feed consumption due to its low digestibility. In this case, the highest protein retention in fed fish was obtained when 33% of BSFL biomass was used, thereafter decreasing as BSFL biomass was incorporated. From the study, the inclusion of BSFL biomass into aquaculture feed was feasible at low percentages, and it has been suggested that the presence of chitin in BSFL biomass contributes certain benefits to the growth performance of the turbot from the feed intake, including the availability and digestibility of nutrients [3]. The BSFL protein was also introduced to rainbow trout as a replacement meal with the partial inclusion at 25% and 50%, and the outcome showed that the BSFL biomass degraded the lipid health indexes of the rainbow trout while negatively impacting the contents of polyunsaturated fatty acids with increases of BSFL biomass. In order to prevent the negative impacts of BSFL inclusion on trout, it was suggested that a 40% inclusion level of BSFL biomass could be used without impacting the survival, growth performance, condition factor and so on [4]. Apart from the aquaculture field, BSFL biomass can also be introduced as animal feed for broilers in either a partial or highly defatted form. From the past study, an inclusion of partially defatted BSFL biomass into broilers' feed showed higher digestibility by the chicken. [5]. According to Schiavone et al. [6], an inclusion of defatted BSFL in broiler chicken diets at 10% showed improvements in carcass and meat quality parameters as well as the heavy metal contents, and there were no negative consequences. Moreover, when the BSFL biomass was incorporated into quail feed to replace fishmeal, the outcome showed a similar result as with the fishmeal. When 25% to 50% BSFL biomass at 25% and 50% was included, no impact on the palatability of ration or quail appetites was detected. In short, the 50% replacement of fishmeal with BSFL biomass was generally recommended, as no negative impact was demonstrated on the growth performance of most of the farmed animals [7].

The study by Loponte et al. [8] showed that the corn-soybean meal diet used for *Barbary partridge* rearing could be replaced with *Tenebrio molitor* and *Hermetia illucens* biomass at 25% and 50%. Even though the control group had heavier weight of partridges fed and longer intestinal and caecal lengths, the live weights of the birds that were fed *T. molitor* and *H. illucens* meals were significantly higher than the control due to improved nutrient digestibility. Apart from these, several studies were carried out to determine the impacts of insect meal on the egg characteristics of laying hens. With the inclusion of *H. illucens* into laying hens' diets, lay percentage and egg mass were found to be affected only at 25% replacement, owning to higher methionine and lysine. A replacement by insect meal more than 50% negatively impacted dry matter, organic matter and crude protein digestibility due to the presence of chitin; hence, a 25% insect meal replacement was recommended for the diets of laying hens [9]. A 100% soybean meal replacement by *H. illucens* was found feasible in Lohmann Brown Classic laying hens during 21 weeks of rearing. Eggs laid by the hens fed with the insect diet were found to possess higher quality of yolks than the control group, which was fed soybean meal. Also, the red index of the eggs laid was found to be higher in the insect treatment group (5.63) compared with the control (1.36). Moreover, the insect treatment group laid eggs with higher γ-tocopherol (4.0 against 2.4 mg/kg), lutein (8.6 against 4.9 mg/kg), β-carotene (0.33 against 0.19 mg/kg) and total carotenoids (15 against 10.5 mg/kg) than the control. Nonetheless, the insect treatment group eggs contained 11% less cholesterol than the control group, and no differences were found in fatty acid composition [10].

Recently, worldwide researchers have focused on exploiting BSFL biomass to serve as a feed medium for farmed animals, including aquaculture farming, in order to sustain the rising demands for a protein source. In this regard, various low-cost organic wastes had been employed to farm BSFL without truly optimizing its larval protein content. It has been hypothesized that increasing the protein content of BSFL would directly permit a higher inclusion of larval biomass in animal feeds whilst reducing the costs attributed mostly as a result of the unsustainable use of fishmeals. BSFL is currently proposed as the best protein source for animal farming and aquafarming, since the cost of animal feed and fishmeal continue increasing year after year due to marine overexploitation and a limited availability of lands. Animal feeds consist mainly of fishmeal and soybean, which serve as the protein alimentation, in addition to fish oils, seed cakes and other grains [11]. Thus, the main objective of this study was to enhance the protein content of BSFL by introducing yeast to execute fermentation on low-cost organic waste for larval feeding (i.e., coconut endosperm waste). The presence of yeast to ferment coconut endosperm waste would improve the nutritional content of larval feeding medium and eventually the larval protein content upon feeding. The degree of fermented coconut endosperm waste valorization by BSFL has also been reported to unveil organic waste treatment potentiality.

2. Materials and Methods

2.1. Acquisition of Coconut Endosperm Waste

The grated fresh coconut endosperm waste (CEW) was initially acquired from a local stall selling coconut milk and kept within 2 to 4 °C in a refrigerator. The moisture content of the CEW was determined through a gravimetric method and adjusted to 70% by homogenizing with sterile distilled water as calculated using Equation (1) prior to being used in the experiment.

$$V_{H_2O} = \frac{(\%_{H_2O})(M_S)}{1 - (\%_{H_2O})} - M_{H_2O} \quad (1)$$

where V_{H_2O} represents the total volume of sterile distilled water to be added (in g considering the density of water 1 g/mL), $\%_{H_2O}$ represents the percentage of desired moisture (which was 70% (0.7 was inserted into the equation) in this study), M_S represents the total dry weight of the CEW (in g) and M_{H_2O} represents the initial moisture content of the CEW (in g).

2.2. Attainment of Black Soldier Fly Larvae (BSFL)

We weighed 200 g of fresh CEW and transferred it into a plastic container with a size of 35 × 25 cm (height × diameter). We left the ventilated container in a sun-shaded area, serving as a bait to lure female BSFs. Several pieces of paper box cardboard with a size of 8 cm × 3 cm (length × width) were attached to the inner wall of the plastic container about 3 to 5 cm above the CEW medium, acting as a platform for the female BSF to oviposit her eggs. This cardboard was checked daily for BSF eggs. The attained eggs were then transferred into sterile Petri dishes and incubated until the larvae emerged. The new BSFL (neonate) were reared on CEW until 6 days old prior to being used in the experiments [12].

2.3. Rearing of BSFL Using CEW Inoculated with Yeast

Figure 1 presents the schematic flow of the reported works. Different quantities of dry yeast powder (commercial brand: Bunga Raya) with 0.02, 0.1, 0.5, 1.0 and 2.5 wt% were separately homogenized with CEW to serve as an initial inoculum for fermentation to take place. A 10 g, dry weight basis of each CEW that had been inoculated with yeast medium was then immediately administered to 20 six-day-old BSFLs. The larval rearing using each CEW medium inoculated with different percentages of yeast was stopped once the BSFL reached its fifth instar, as determined by head size and body color [1,13]. Each batch of harvested BSFL was deactivated at 105 °C for 5 min then dried at 60 °C until reaching a constant weight. This was followed by grinding the BSFL into powder and storing it at −20 °C prior to the chemical analyses [14]. All CEW residues were also separately collected and dried at 105 °C until reaching a constant weight. All setups were (at least) duplicated to verify the statistical reproducibility.

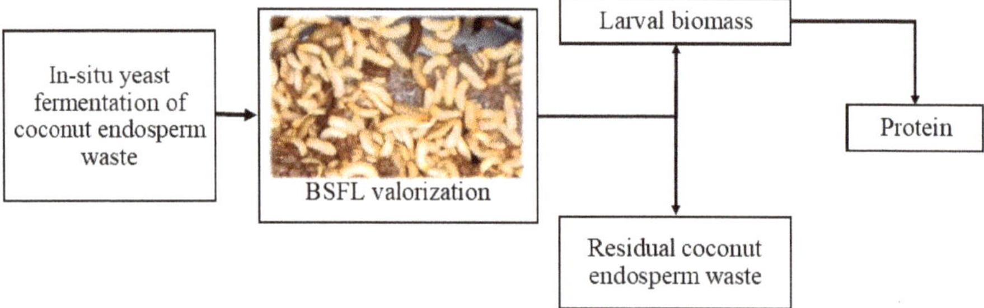

Figure 1. Schematic flow of the experimental procedures.

2.4. Growth Performance of the BSFL

Upon the completion of experiments, growth of the BSFL was evaluated using Equation (2) for the total biomass gained and Equation (3) for the BSFL growth rate [15], as shown below:

$$\text{Total biomass gained (g)} = \text{Final BSFL dried mass (g)} - \text{Initial BSFL dried mass (g)} \tag{2}$$

$$\text{BSFL growth rate (g/day)} = \text{Total biomass gained (g)/Rearing time (day)} \tag{3}$$

2.5. Treatment of CEW Via Valorization by BSFL

In order to determine the degree of CEW reduction, two parameters were measured including Equation (4) for overall degradation (OD) and Equation (5) for the waste reduction index (WRI) [16], as shown below:

$$\text{Overall degradation} = \text{Total feed consumed (g)/Total feed offered (g)} \tag{4}$$

$$\text{WRI (g/day)} = \text{Total feed consumed (g)/Rearing time (day)} \tag{5}$$

2.6. Nitrogen, Chitin and Protein Analyses

Nitrogen contents of dried BSFL biomass were determined through the Dumas combustion method (Perkin Elmer, CHNS/O 2400). The sample was weighed in the range of 1 to 1.5 mg then transferred into a tin capsule, wrapped and combusted at 925 °C. The nitrogen compounds were then converted into NO_x, further reduced to nitrogen gas at 640 °C and detected by a thermal conductivity detector (TCD) [17]. In this study, the larval protein contents were estimated with a multiplication factor of 6.25 [18]. However, the presence of chitin in BSFL biomass will influence the larval protein content and, hence, nitrogen from chitin has to be deducted from the total larval nitrogen content prior to protein conversion in order to avoid over-estimation [19]. Chitin is a polysaccharide that can be found in yeast, fungi, crustaceans and insects [20], as well as being present in the exoskeleton of BSFL, where it accounts for 6.89% of the nitrogen content [16]. The formic acid method was applied for chitin determination in this study [19,21], with modification to suit a small sample size. We mixed 10 mL of 90% formic acid with 1 g of BSFL dried fat-free biomass (the initial mass prior to being defatted had been recorded) at room temperature for 24 h. Then, the mixture was centrifuged, and the supernatant was decanted. The residue was washed with 10 mL of 100% acetone, followed by 10 mL of 70% acetone before being recentrifuged to separate the acetone. The residue was refluxed with 5% of 10 mL sodium hydroxide for 90 min before being filtered and washed with distilled water on ashless filter paper (Whatman No. 1 with a 55 mm diameter). Next, the residue was dried in the oven to a constant weight at 105 °C, then later it was ashed at 550 °C for 24 h. The final weight of the sample was recorded and assumed to be intact chitin.

$$\text{Chitin content (\%)} = \text{Mass of residues after ashing (g)/Initial mass of BSFL (g)} \times 100\% \quad (6)$$

$$TN_{Chitin}\ \text{(\%)} = [\text{Chitin content (\%)} \times \text{Nitrogen content in chitin (\%) (which is 6.89\%)}]/100\% \quad (7)$$

$$\text{Corrected protein yield for BSFL (\%)} = [TN_{BSFL}\ \text{(\%)} - TN_{Chitin}\ \text{(\%)}] \times 6.25 \quad (8)$$

$$\text{Protein productivity (g/day)} = \text{Protein content (g)/Rearing time (day)} \quad (9)$$

where TN_{BSFL} is the total nitrogen from the BSFL biomass and TN_{Chitin} is the total nitrogen from the chitin.

3. Results and Discussion

3.1. Growth Performances of BSFL

Initially, 10 g of yeast-inoculated feed was introduced to 20 BSFL at different concentrations. The total biomasses gained for the BSFL were recorded once every setup had reached the fifth instar, as shown in Table 1. Under the control condition, the total biomass gained by the BSFL was attained at only 0.998 g from a total of 20 BSFL. This value increased with the increment of yeast concentrations rising from 0.02 to 2.5 wt%, and it attained its highest point at 1.145 g. As compared with a previous study by Zheng et al. [22], the performance of in situ yeast fermentation at the highest concentration in this study was comparable to the best RID-X dosage (w/w), which was equivalent to 1.228 g per 20 BSFL with a difference of merely 0.08 g per 20 BSFL. RID-X was the active bacterial product introduced into the larval feeding medium in the study by Zheng et al. [22]. On the other hand, besides changing the nutritional properties of larval feed by introducing microorganisms, the growth of the BSFL could also be altered by feeding with a protein-rich medium, as suggested by Rehman et al. [23]. At a 1:4 ratio of dairy manure to protein-rich soybean curd residue, the total dry larval mass that could be attained was 28.1 g, which is equivalent to 0.56 g from 20 BSFL. This showed that the performance of BSFL growth through the co-digestion treatment was still lower compared to the microorganism inoculation treatment (i.e., yeast in this study). Thus, the inoculation of microorganisms into larval feed is strongly recommended for better BSFL growth.

Table 1. Growth performances of BSFL fed with CEW having been inoculated with different yeast concentrations.

Yeast Concentration (wt %)	Total Biomass Gained (g)	Growth Rate (g/day)	Rearing Time (day)
0 (Control)	0.998 ± 0.125	0.065 ± 0.011	15.5 ± 0.7
0.02	1.013 ± 0.115	0.070 ± 0.011	14.5 ± 0.7
0.10	1.082 ± 0.019	0.077 ± 0.001	14.0 ± 0
0.50	1.088 ± 0.014	0.081 ± 0.005	13.5 ± 0.7
1.00	1.064 ± 0.030	0.079 ± 0.006	13.5 ± 0.7
2.50	1.145 ± 0.099	0.085 ± 0.012	13.5 ± 0.7

Moreover, the growth rate of the BSFL also increased in parallel to the increasing concentrations of yeast from an initial 0.065 g/day to a maximum of 0.085 g/day. This phenomenon can be explained by the shortening of the rearing time of the BSFL. The in situ yeast fermentation of feeding medium had a reduced rearing time from 15.5 days to 13.5 days. This occurrence could have been due to the introduction of yeast that favored the digestibility of carbohydrate compounds in CEW [24] and thus improved the assimilation of nutrients into the BSFL body mass in the form of lipids. Also, Yoon et al. [25] reported that the yeast was capable of breaking down carbohydrates through fermentation, especially common monosaccharides such as D-glucose, D-fructose, D-mannose and D-galactose. On the other hand, it has been proven that the BSFL was also able to convert additional glucose into lipids upon excess availability [26]. Indeed, the measured lipid content increased from about 40% for the control to 50% for a 1.0 wt% yeast concentration. The lipids could later serve as a potential source for biodiesel production, which is something that could be explored further.

3.2. CEW Valorization by BSFL

Due to its polyphagous nature, BSFL is able to reduce solid organic wastes during the rearing process. In this study, the overall degradation of CEW was 0.48 under the control, and this value was maintained for low yeast concentrations of 0.02 and 0.1 wt%. With the addition of yeast at more than 0.5 wt%, the overall degradation of CEW increased to a range of 0.51 to 0.53. Thus, it could be concluded that the 20 BSFL were able to degrade about half of the CEW upon completion of the rearing process, disregarding the concentrations of yeast inoculated. With the introduction of yeast at different concentrations in the feeding medium, it was shown that the WRI increased from 0.31 g/day under the control, to 0.33 g/day with a 0.02 wt% of yeast and 0.38 g/day with a 0.5 wt% yeast concentration. At last, the WRI reached its highest point of 0.40 g/day with a 2.5 wt% yeast concentration. The WRI increment was about 15% faster in 0.5 wt% compared to the 0.02 wt%. This could plausibly be because the addition of 0.5 wt% yeast reached the concentration threshold for maximizing the in situ fermentation to spur the ingestion of CEW by BSFL [27]. Also, it can be observed from Table 1 that the rearing duration for BSFL decreased from 15.5 days and reached a plateau at 13.5 days when the 0.5 wt% yeast concentration (and beyond) were employed for in situ fermentation. Above the 0.5 wt% yeast concentration, the effect on WRI was not significant, if not deteriorating, as reported by Palma et al. [28]. In their study of managing high fiber food waste using BSFL, incremental larval growth led to a decrease in almond hull consumption and vice-versa. The authors presumed that the occurrence was the result of a competition for resources between the BSFL and microbial communities, or because of enhanced synergy between the larvae and their associated microbiota.

3.3. Protein Contents in BSFL

The chitin content from the BSFL was determined to be around 8%, and the nitrogen from the chitin was deducted from the total nitrogen of the BSFL to prevent the over-estimation of BSFL protein content. Figure 2 shows that the corrected protein of the BSFL was only attained around 28% under the control system, and that this value increased to its peak at about 35% when the lowest yeast concentration was used for fermentation. The corrected protein value dropped to around 30% and remained at that level with yeast concentrations from 0.5 to 2.5 wt%. Looking into the protein productivity from 20 BSFL, the value was attained at around 0.02 g/day under the control system and increased to around 0.025 g/day with the introduction of yeast at 0.02 wt%. The value fluctuated within the range of 0.023 to 0.025 g/day with higher yeast concentrations from 0.5 to 2.5 wt%.

As reported by Diener et al. [19], a daily feeding rate of 100 mg of chicken feed per larva was proposed to produce better larval quality and higher waste reduction in the shortest period of time. At this rate, the corrected protein content of BSFL was 34.4%, which is comparable with the current study in which an average of 34.0 ± 3.4% was attained. This result shows that it is possible to attain an output with a similar larval protein content through the initial "one-off feeding method" by using microorganisms to execute fermentation. The introduction of microorganisms into larval feeding media has been widely practiced as a means to improve the growth of BSFL. According to Gao et al. [29], the addition of *Aspergillus oryzae* into maize straw for fermentation ultimately improved the growth of BSFL and was able to obtain approximately 42% of larval crude protein. At the same time, the BSFL reared on fermented maize straw were found to contain higher amounts of monounsaturated fatty acids and polyunsaturated fatty acids, and were lower in saturated fatty acids as opposed to the control medium without exo-microorganisms. Concisely, it could be confidently deduced that the introduction of microorganisms into BSFL media through larval farming systems could promisingly enhance larval growth and, eventually, achieve more harvested larval biomasses.

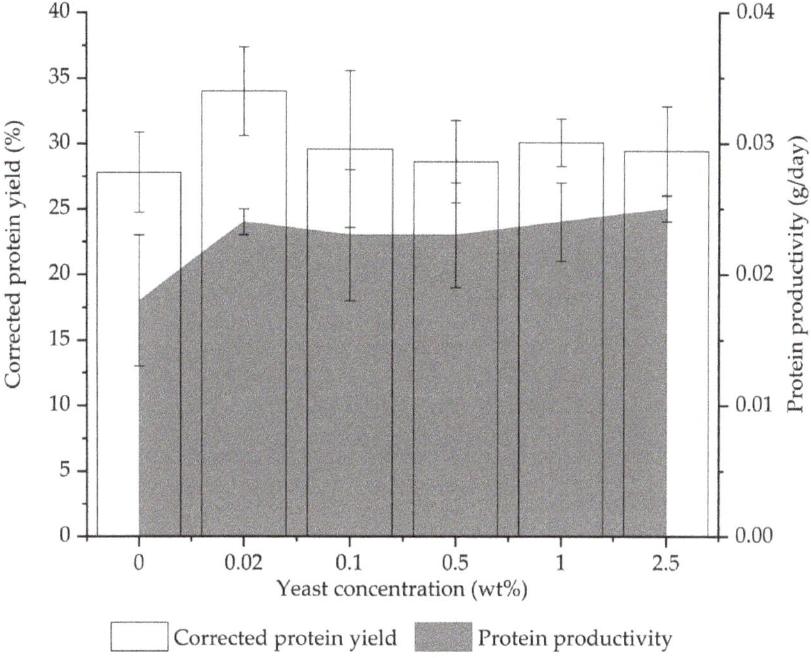

Figure 2. Impact of different yeast concentrations inoculating CEW on corrected protein yields and protein productivities from BSFL.

The introduction of BSFL biomass into animal feed could plausibly replace the exploitation of unsustainable soybean and fishmeal. Indeed, BSFL could serve as the sole protein source, since larval biomass is generally fortified with high protein as well as fat [30]. The inclusion of BSFL into animal feed for laying hens had been found to significantly increase the production of both day and house eggs. At the same time, it has also positively impacted the characteristics of eggs and the growth of laying hens [31]. In the case of aquaculture cultivation, a partial inclusion of BSFL into feed at 25%, serving as fishmeal protein has been shown to increase the growth performances of yellow catfish by 21.7%, while also improving their immune indexes [32]. Moreover, it was reported that the replacement of fishmeal by BSFL between 28.4% and 50% into the diets of juvenile barramundi could promote fish growth, fish whole body proximate and amino acid composition [33]. A 100% replacement of fishmeal by BSFL was also possible in Jian carp cultivation, as it had been reported that there was no unfavorable impact on the growth of Jian carp. BSFL meal could be an economic and sustainable replacement for current fish diets that could circumvent both feed shortages and the increasing price of fishmeal [34]. Thus, it is recommended that BSFL biomass meal be utilized as a substitution for protein alimentation in animal feed and fishmeal in the long-term, whilst also advocating for the green and sustainable farming of land and aquatic animals, respectively.

4. Conclusions

The inoculation of yeast at different concentrations into CEW to serve as the feeding medium for BSFL rearing enhanced larval growth. For a setup initially containing 20 neonates of BSFL, a final weight of 1.145 g, a growth rate at 0.085 g/day and a rearing period of 13.5 days were achieved when BSFL were fed with fermented CEW inoculated with 2.5 wt% yeast. With an increase in yeast concentrations, the overall degradation of CEW was found to improve from 0.48 to 0.53, with the waste reduction indexes fluctuating between 0.38 and 0.40 g/day. Likewise, the protein yield from BSFL was boosted from the control (28%) to its highest value of 35% in the presence of merely 0.02 wt% yeast concentration. On the other hand, protein productivity was increased from 0.018 g/day for the control to around 0.025 g/day across 0.02 to 2.5 wt% yeast concentrations. To conclude, the growth of BSFL was promoted with the inclusion of yeast as the fermentation precursor, and the harvested larval biomass can potentially be used as a replacement of protein sources in animal feeds and fishmeals.

Author Contributions: Conceptualization, C.Y.W. and J.W.L.; methodology, C.Y.W. and Y.C.H.; formal analysis, P.L.S. and S.C.; investigation, Y.J.C. and M.M.; resources, C.D.H.; writing—original draft preparation, C.Y.W.; writing—review and editing, T.Y.W. and G.T.P.; supervision, J.W.L. and M.K.L.; project administration, C.Y.W.; funding acquisition, J.W.L. All authors have read and agreed to the published version of the manuscript.

Funding: This research was funded by The Murata Science Foundation with the cost center of 015ME0-104 and the Ministry of Education Malaysia under HICoE with the cost center of 015MA0-052.

Acknowledgments: The administrative and technical supports provided by the members from the HICoE-Centre for Biofuel and Biochemical Research, Universiti Teknologi PETRONAS are greatly acknowledged.

Conflicts of Interest: The authors declare no conflict of interest.

References

1. Diclaro, J.; Kaufman, P.E. Black soldier fly hermetia illucens linnaeus (insecta: Diptera: Stratiomyidae). *EENY* **2009**, *461*, 1–3.
2. Win, S.S.; Ebner, J.H.; Brownell, S.A.; Pagano, S.S.; Cruz-Diloné, P.; Trabold, T.A. Anaerobic digestion of black solider fly larvae (bsfl) biomass as part of an integrated biorefinery. *Renew. Energy* **2018**, *127*, 705–712. [CrossRef]
3. Kroeckel, S.; Harjes, A.-G.; Roth, I.; Katz, H.; Wuertz, S.; Susenbeth, A.; Schulz, C. When a turbot catches a fly: Evaluation of a pre-pupae meal of the black soldier fly (hermetia illucens) as fish meal substitute—growth performance and chitin degradation in juvenile turbot (psetta maxima). *Aquaculture* **2012**, *364*, 345–352. [CrossRef]
4. Renna, M.; Schiavone, A.; Gai, F.; Dabbou, S.; Lussiana, C.; Malfatto, V.; Prearo, M.; Capucchio, M.T.; Biasato, I.; Biasibetti, E.; et al. Evaluation of the suitability of a partially defatted black soldier fly (hermetia illucens l.) larvae meal as ingredient for rainbow trout (oncorhynchus mykiss walbaum) diets. *J. Anim. Sci. Biotechnol.* **2017**, *8*, 57. [CrossRef] [PubMed]
5. Schiavone, A.; De Marco, M.; Martínez, S.; Dabbou, S.; Renna, M.; Madrid, J.; Hernandez, F.; Rotolo, L.; Costa, P.; Gai, F.; et al. Nutritional value of a partially defatted and a highly defatted black soldier fly larvae (hermetia illucens l.) meal for broiler chickens: Apparent nutrient digestibility, apparent metabolizable energy and apparent ileal amino acid digestibility. *J. Anim. Sci. Biotechnol.* **2017**, *8*, 51. [CrossRef] [PubMed]
6. Schiavone, A.; Dabbou, S.; Petracci, M.; Zampiga, M.; Sirri, F.; Biasato, I.; Gai, F.; Gasco, L. Black soldier fly defatted meal as a dietary protein source for broiler chickens: Effects on carcass traits, breast meat quality and safety. *Animal* **2019**, *13*, 2397–2405. [CrossRef] [PubMed]
7. Widjastuti, T.; Wiradimadja, R.; Rusmana, D. The effect of substitution of fish meal by black soldier fly (hermetia illucens) maggot meal in the diet on production performance of quail (coturnix coturnix japonica). *Anim. Sci.* **2014**, *57*, 125–129.
8. Loponte, R.; Nizza, S.; Bovera, F.; De Riu, N.; Fliegerova, K.; Lombardi, P.; Vassalotti, G.; Mastellone, V.; Nizza, A.; Moniello, G. Growth performance, blood profiles and carcass traits of barbary partridge (alectoris barbara) fed two different insect larvae meals (tenebrio molitor and hermetia illucens). *Res. Vet. Sci.* **2017**, *115*, 183–188. [CrossRef]

9. Bovera, F.; Loponte, R.; Pero, M.E.; Cutrignelli, M.I.; Calabrò, S.; Musco, N.; Vassalotti, G.; Panettieri, V.; Lombardi, P.; Piccolo, G. Laying performance, blood profiles, nutrient digestibility and inner organs traits of hens fed an insect meal from hermetia illucens larvae. *Res. Vet. Sci.* **2018**, *120*, 86–93. [CrossRef]
10. Secci, G.; Bovera, F.; Nizza, S.; Baronti, N.; Gasco, L.; Conte, G.; Serra, A.; Bonelli, A.; Parisi, G. Quality of eggs from lohmann brown classic laying hens fed black soldier fly meal as substitute for soya bean. *Animal* **2018**, *12*, 2191–2197. [CrossRef]
11. Onsongo, V.; Osuga, I.; Gachuiri, C.; Wachira, A.; Miano, D.; Tanga, C.; Ekesi, S.; Nakimbugwe, D.; Fiaboe, K. Insects for income generation through animal feed: Effect of dietary replacement of soybean and fish meal with black soldier fly meal on broiler growth and economic performance. *J. Econ. Entomol.* **2018**, *111*, 1966–1973. [CrossRef] [PubMed]
12. Mohd-Noor, S.-N.; Wong, C.-Y.; Lim, J.-W.; Uemura, Y.; Lam, M.-K.; Ramli, A.; Bashir, M.J.; Tham, L. Optimization of self-fermented period of waste coconut endosperm destined to feed black soldier fly larvae in enhancing the lipid and protein yields. *Renew. Energy* **2017**, *111*, 646–654. [CrossRef]
13. Kim, W.-T.; Bae, S.-W.; Park, H.-C.; Park, K.-H.; Lee, S.-B.; Choi, Y.-C.; Han, S.-M.; Koh, Y.-H. The larval age and mouth morphology of the black soldier fly, hermetia illucens (diptera: Stratiomyidae). *Int. J. Indust. Entomol.* **2010**, *21*, 185–187.
14. Li, Q.; Zheng, L.; Qiu, N.; Cai, H.; Tomberlin, J.K.; Yu, Z. Bioconversion of daiy manure by black soldier fly (*diptera: Stratiomyidae*) for biodiesel and sugar production. *Waste Manag.* **2011**, *31*, 1316–1320. [CrossRef] [PubMed]
15. Lalander, C.; Diener, S.; Zurbrügg, C.; Vinnerås, B. Effects of feedstock on larval development and process efficiency in waste treatment with black soldier fly (hermetia illucens). *J. Clean. Prod.* **2019**, *208*, 211–219. [CrossRef]
16. Parra, J.R.; Panizzi, A.R.; Haddad, M.L. Nutritional indices for measuring insect food intake and utilization. In *Insect Bioecology and Nutrition for Integrated Pest Management*; Panizzi, A.R., Parra, J.R., Eds.; CRC Press: London, UK; New York, NY, USA, 2012; p. 13.
17. Culmo, R.F.; Shelton, C. *The Elemental Analysis of Various Classes of Chemical Compounds Using Chn Application Note*; Perkin Elmer: Waltham, MA, USA, 2013.
18. Jones, D.B. *Factors for Converting Percentages of Nitrogen in Foods and Feeds into Percentages of Proteins*; US Department of Agriculture: Washington, DC, USA, 1941.
19. Diener, S.; Zurbrügg, C.; Tockner, K. Conversion of organic material by black soldier fly larvae: Establishing optimal feeding rates. *Waste Manag. Res.* **2009**, *27*, 603–610. [CrossRef] [PubMed]
20. Rinaudo, M. Chitin and chitosan: Properties and applications. *Prog. Polym. Sci.* **2006**, *31*, 603–632. [CrossRef]
21. Lovell, R.T.; Lafleur, J.R.; Hoskins, F.H. Nutritional value of freshwater crayfish waste meal. *J. Agric. Food Chem.* **1968**, *16*, 204–207. [CrossRef]
22. Zheng, L.; Hou, Y.; Li, W.; Yang, S.; Li, Q.; Yu, Z. Biodiesel production from rice straw and restaurant waste employing black soldier fly assisted by microbes. *Energy* **2012**, *47*, 225–229. [CrossRef]
23. Rehman, K.U.; Rehman, A.; Cai, M.; Zheng, L.; Xiao, X.; Somroo, A.A.; Wang, H.; Li, W.; Yu, Z.; Zhang, J. Conversion of mixtures of dairy manure and soybean curd residue by black soldier fly larvae (*hermetia illucens* l.). *J. Clean. Prod.* **2017**, *154*, 366–373. [CrossRef]
24. Wiedmeier, R.D.; Arambel, M.J.; Walters, J.L. Effect of yeast culture and aspergillus oryzae fermentation extract on ruminal characteristics and nutrient digestibility1. *J. Dairy Sci.* **1987**, *70*, 2063–2068. [CrossRef]
25. Yoon, S.-H.; Mukerjea, R.; Robyt, J.F. Specificity of yeast (saccharomyces cerevisiae) in removing carbohydrates by fermentation. *Carbohydr. Res.* **2003**, *338*, 1127–1132. [CrossRef]
26. Li, W.; Li, M.; Zheng, L.; Liu, Y.; Zhang, Y.; Yu, Z.; Ma, Z.; Li, Q. Simultaneous utilization of glucose and xylose for lipid accumulation in black soldier fly. *Biotechnol. Biofuels* **2015**, *8*, 1–6. [CrossRef] [PubMed]
27. Arcos-García, J.L.; Castrejón, F.A.; Mendoza, G.D.; Pérez-Gavilán, E.P. Effect of two commercial yeast cultures with saccharomyces cerevisiae on ruminal fermentation and digestion in sheep fed sugar cane tops. *Livest. Prod. Sci.* **2000**, *63*, 153–157. [CrossRef]
28. Palma, L.; Fernandez-Bayo, J.; Niemeier, D.; Pitesky, M.; VanderGheynst, J.S. Managing high fiber food waste for the cultivation of black soldier fly larvae. *npj Sci. Food* **2019**, *3*, 1–7. [CrossRef] [PubMed]
29. Gao, Z.; Wang, W.; Lu, X.; Zhu, F.; Liu, W.; Wang, X.; Lei, C. Bioconversion performance and life table of black soldier fly (hermetia illucens) on fermented maize straw. *J. Clean. Prod.* **2019**, *230*, 974–980. [CrossRef]

30. Wang, Y.-S.; Shelomi, M. Review of black soldier fly (hermetia illucens) as animal feed and human food. *Foods* **2017**, *6*, 91. [CrossRef]
31. Al-Qazzaz, M.F.A.; Ismail, D.; Akit, H.; Idris, L.H. Effect of using insect larvae meal as a complete protein source on quality and productivity characteristics of laying hens. *Rev. Bras. Zootec.* **2016**, *45*, 518–523. [CrossRef]
32. Xiao, X.; Jin, P.; Zheng, L.; Cai, M.; Yu, Z.; Yu, J.; Zhang, J. Effects of black soldier fly (hermetia illucens) larvae meal protein as a fishmeal replacement on the growth and immune index of yellow catfish (pelteobagrus fulvidraco). *Aquac. Res.* **2018**, *49*, 1569–1577. [CrossRef]
33. Katya, K.; Borsra, M.; Ganesan, D.; Kuppusamy, G.; Herriman, M.; Salter, A.; Ali, S.A. Efficacy of insect larval meal to replace fish meal in juvenile barramundi, lates calcarifer reared in freshwater. *Int. Aquat. Res.* **2017**, *9*, 303–312. [CrossRef]
34. Zhou, J.; Liu, S.; Ji, H.; Yu, H. Effect of replacing dietary fish meal with black soldier fly larvae meal on growth and fatty acid composition of jian carp (cyprinus carpio var. Jian). *Aquac. Nutr.* **2018**, *24*, 424–433. [CrossRef]

Article

Hygro-Thermo-Mechanical Responses of Balsa Wood Core Sandwich Composite Beam Exposed to Fire

Luan TranVan [1], Vincent Legrand [2], Pascal Casari [2], Revathy Sankaran [3], Pau Loke Show [4,*], Aydin Berenjian [5,*] and Chyi-How Lay [6]

1 Faculty of Transportation Mechanical Engineering, the University of Da Nang—University of Science and Technology, 54 Nguyen Luong Bang, Da Nang City 550 000, Vietnam; Tvluan@dut.udn.vn
2 Institut de Recherche en Génie Civil et Mécanique (GeM) UMR CNRS 6183, Université de Nantes—Ecole Centrale Nantes, Equipe Etat Mécanique et Microstructure des Matériaux (E3M), 58 rue Michel Ange, BP 420, CEDEX 44606 Saint-Nazaire, France; vincent.legrand@univ-nantes.fr (V.L.); pascal.casari@univ-nantes.fr (P.C.)
3 Institute of Biological Sciences, Faculty of Science, University of Malaya, Kuala Lumpur 50603, Malaysia; revathy@um.edu.my
4 Department of Chemical Engineering, Faculty of Science and Engineering, University of Nottingham Malaysia, Jalan Broga, Semenyih, Selangor Darul Ehsan 43500, Malaysia
5 School of Engineering, Faculty of Science and Engineering, the University of Waikato, Hamilton 3240, New Zealand
6 Centre for General Education, Feng Chia University, Seatwen, Taichung 40724, Taiwan; chlay@fcu.edu.tw
* Correspondence: PauLoke.Show@nottingham.edu.my (P.L.S.); aydin.berenjian@waikato.ac.nz (A.B.)

Received: 15 December 2019; Accepted: 9 January 2020; Published: 13 January 2020

Abstract: In this study, the hygro–thermo–mechanical responses of balsa core sandwich structured composite was investigated by using experimental, analytical and numerical results. These investigations were performed on two types of specimen conditions: dry and moisture saturation sandwich composite specimens that are composed of E-glass/polyester skins bonded to a balsa core. The wet specimens were immersed in distilled water at 40 °C until saturated with water. The both dry and wet sandwich composite specimens were heated by fire. The mass loss kinetic and the mechanical properties were investigated by using a cone calorimeter following the ISO 5660 standard and three-point bending mechanical test device. Experimental data show that the permeability and fire resistance of the sandwich structure are controlled by two composite skins. Obtained results allow us to understand the Hygro–Thermo–Mechanical Responses of the sandwich structured composite under application conditions.

Keywords: sandwich composite fire; mechanical responses; moisture content; balsa core; mass loss kinetic; buckling failure

1. Introduction

The use of organic matrix composite materials has been continuously growing since the 1960s. As known to all, the material undergoes important physical and/or chemical modifications under extreme conditions, such as an appearance of metastable states or phase transitions [1,2]. Measurements in extreme conditions are facing scientific challenges to spot the properties of materials, and a technical challenge to apply new materials. Obviously, the increasing use of composites has reached a level that these materials compete with conventional materials such as steel and aluminium alloys in diverse areas, particularly aeronautics, aerospace and the shipbuilding industry due to their advantages in physical, chemical and mechanical properties [3–6].

Compared to other materials, organic matrix composites have low density, high specific stiffness and strength, good fatigue endurance and outstanding resistance to corrosion. However, there are several disadvantages compared to metals that include low impact tolerance, low fire performance and anisotropic properties [5–8]. Hence, the study on the effect of composite material properties under extreme conditions is needed to understand the behaviour and to optimize their properties. In the marine industry, the use of a sandwich structure consisting of a lightweight core made of polymer foam or balsa wood surrounded by thin stiff composite skins made of fiberglass and a major polymer as vinyl ester, epoxy or polyester, is common [5–8]. These combinations allowed a construction of an extraordinarily lightweight, durable and rigid structure. However, this type of material structure in the naval industry requires the special precaution of fire resistance. Composite sandwich materials are subject to strict regulation, and it is important to predict their thermomechanical properties as handling any applications [9–15]. The thermal degradation of materials as a composite sandwich have been widely described in detail [1,2,8,16–21]. There is limited study on the evaluation of the losses of the mechanical properties under the coupling effect of heat flux and moisture absorption. It is important to know the residual mechanical properties at room temperature of a burnt sandwich composite material in order to estimate the fire resistance of this structure after a fire exposure [9–13,15,21–25]. In this context, we were analysing the h hygro–thermo–mechanical responses of a sandwich structure composed of fiber-E-glass embedded in a polyester matrix, for the composition of the skins bonded to a balsa core. Previous studies focused on the thermal degradation of sandwich composite materials [15,23–25], and there are limited studies performed on the hygroscopically aged materials exploring the coupling hygro–thermo properties [20,22,24]. Officially, the water and temperature simultaneously could cause extreme degradation on the skins of this sandwich structure, thus the weak core would be exposed to application conditions. The resulting mechanical states of the core material can eventually induce the geometrical stability damage of such sandwich structure [13–15,23–25]. Balsa wood is widely used for cores of sandwich structures, especially in the shipbuilding industry due to its microstructure composing of long cells aligned in the axial direction which could provide the required axial strength and stiffness. However, only a few detailed studies were conducted on this subject [13–15,23–25]. Some research works focused specifically on the mechanical properties of balsa at high temperature and axial response failure under compression [1,7,8,16].

In the present work, we focus on determining the mass loss kinetic and flexural behaviour under fire of the two types of dry and wet composite sandwich samples by using a cone calorimeter and a Zwick universal testing machine. The dry sandwich samples that were obtained from the shipbuilding industry were immersed in distilled water at 40 °C until water saturation. Fire tests were processed with a heat flux of 50 kW/m^2 at different pyrolysis times. Additionally, a multi-layer analysis (skins and core) was conducted based on experimental results of the composite sandwich structure to estimate the hygro–thermo–mechanical properties of the global sandwich structure. This study enables the evaluation of the elastic modulus E and flexural load of the remaining sandwich structure material after enduring harsh working conditions such as exposure in water-fire.

2. Experiment Set-Up

2.1. Sandwich Composite Materials

The sandwich composite samples were made up of E-glass/polyester skins bonded to a balsa core by a direct infused process. These samples were cut from commercial plates that are used in the naval structure. Figure 1 indicated a studied E-glass/polyester/balsa sandwich specimen. The skins consist of E-glass fabric M450/QX868 made into a 2-plies layer surrounding the balsa core. The core was ordered to balsa wood pieces concocted in the form of about 50.0 × 30.0 × 16.0 mm^3 blocks. The wood fiber direction (D_3) of the core is perpendicular to the composite skins. The average value of the balsa wood's density was 126 ± 30 kg m^{-3}. The coefficients e_s = 1.2 (mm), l = 40.1 (mm), l* = 111.0 (mm) and e = 18.5 (mm), respectively, stand for the average skin thickness, average width, length and thickness of the sandwich specimens.

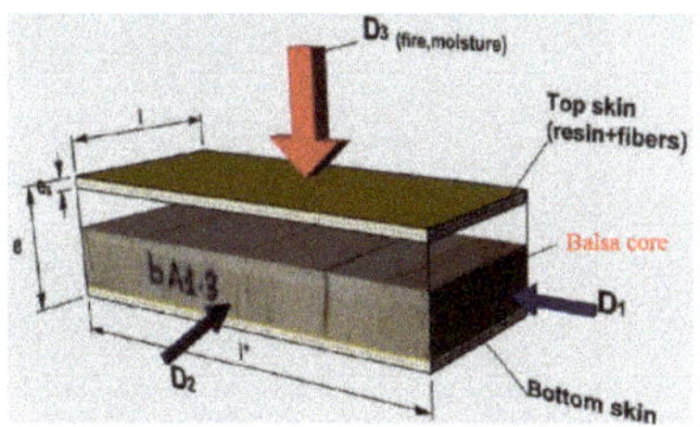

(a) (b)

Figure 1. The studied E-glass/polyester/balsa sandwich composite beam specimen (**a**). Dimensions were illustrated the chosen directions of the moisture and heat flux. D_1, D_2 and D_3 are, respectively, the transverse directions and the thickness direction (**b**).

2.2. Experimental Measurements

2.2.1. Water Absorption Measurements

The sandwich specimens for water uptake were dehydrated in an oven at 50 °C until the weight loss was stabilized, and they were then placed into a container at room temperature for 24 h. The second step involves the complete immersion of specimens in distilled water at 40 °C. In order to measure the quantity of the moisture absorption, the specimens were periodically removed out of the water bath one at a time, wiped off with an absorbent cloth, and immediately their mass was weighed. Sartorius, an MCBA 100 balance with precision ranging of (0–60) g ± 0.1 mg, (61–110) g ± 0.2 mg and (111–210) g ± 0.5 mg, was used. The moisture absorbed by the material at a specific time M (t) was experimentally determined by using the following equation:

$$M_t = \frac{W_t - W_0}{W_0} \times 100 \tag{1}$$

where, W_t is the weight of the specimen at the immersion time t and W_0 is the initial weight of the specimen. The weight measurements were taken initially at an interval of one quantification per day during the first 30 days, and later with a longer periodicity, since the mass fluctuations were not as large as during these days.

2.2.2. Mass Loss Kinetic Measurements

Fire-induced mass loss of the sandwich structured composite specimens were carried out using cone calorimetry (ISO 5660 standard). A radiative heat source was emitted from the cone constructed by winding an electrical resistance. The radiative source was kept at a uniform heat flux of 50 kW m^{-2}. Gas flux was diluted with fresh air and drawn into a chimney. The ignition of the sandwich composite material was caused by a pilot spark. The temperature of the flame during a fire exposure of the material was 750 °C. A surface of the test sample was positioned at a distance of 26 mm from the radiative source. Figure 2 shows the diagram of the position of the cone calorimetry heating setup. The direction of the heat flux was perpendicular to a surface of the sandwich sample test. During a fire resistance test time, the mass of the test sample was recorded as a function of the combustion time. For the purpose of quickly stopping the degradation of the combustion sample, we quickly removed a holder of the sample from the calorimeter cone and placed it into a chamber under nitrogen atmosphere. A residual mechanical property measurement was performed by using a Zwick universal testing machine. (Zwick Roell Group, Ulm, Germany). The mass loss kinetic measurement was conducted for the skin alone and the sandwich structured composite samples.

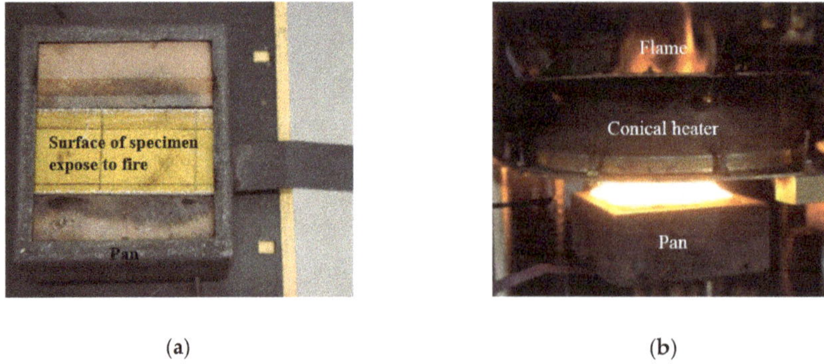

(a) (b)

Figure 2. Diagram of the ATLAS cone calorimetry: holder of the test specimen (**a**), conical heater (**b**).

2.2.3. Mechanical Property Measurement

In order to measure a remaining mechanical property of the material after an exposure time to fire, a three-point bending flexural test was performed on the samples that underwent the thermal degradation processes. Both unaged and thermally-aged sample types were measured for the maximum flexural force reached before failure as a function of degradation time. The Zwick universal testing machine with 15 mm radius supports was used with a displacement speed of 10 mm min^{-1}. During the test, both the flexural force and displacement of the test specimen were recorded. Figure 3 shows the supporting and the device for the three-point bending tests.

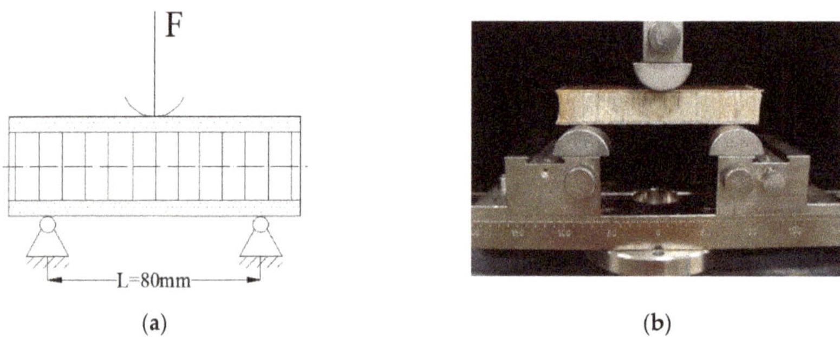

(a) (b)

Figure 3. Sketch of the supporting (**a**) and the device (**b**) for the three-point bending tests of sandwich beams.

3. Experimental Results

3.1. Moisture Diffusion

If balsa wood immersed in water, the moisture will absorb in two ways: the first water floods in free volumes (cracks, hollow fibers) and the second moisture diffuses the dense material. Thus, in the single balsa wood, the moisture is quickly absorbed, and the moisture content obtained is high [6,7]. In the case of balsa as the core materialin sandwich composite structures, the skins acted as waterproof barriers, and infusion resin made full an important amount of hollow fibers for the moisture's viable passage. Figure 4 showed the moisture absorption characterization in both the composite sandwich structure and the bi-blade (balsa core + skin) specimens. The polyester/E-glass fibers skins played as a boundary which limits the water diffusion in the longitudinal direction of the wood fibers. Thus, the sandwich structure absorbed the relative moisture slowly.

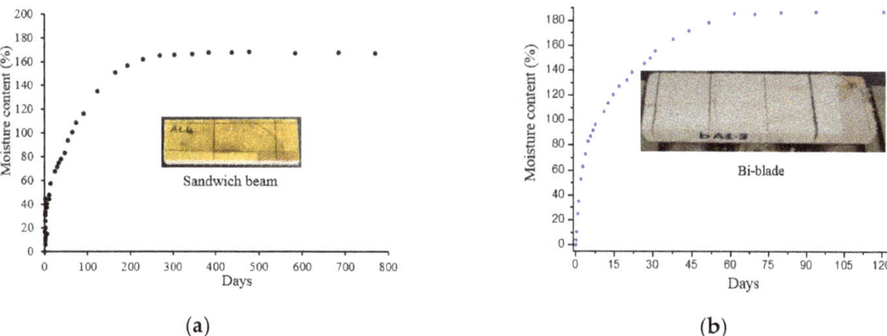

Figure 4. Diffusion kinetics of moisture at 40 °C in a sandwich beam (**a**) and a bi-blade (one skin and a core) (**b**).

The moisture content obtained in the sandwich material was lower than the bi-blade (balsa core + skin). This is due to the presence of one skin which limits moisture diffusion in one side of the sandwich structure. Figure 4a shows that the moisture diffusion was saturated in the sandwich structure after 400 days of immersion. For the bi-blade, the moisture saturation obtained after 75 days. This difference in the saturation time is due to the skins conducting a high restriction to the moisture penetration inside the material. Note that there is a large difference in the moisture content between both of the specimen types. The water absorption limitation corresponds to the value of the moisture content gain (Figure 4), which is approximately 180% for the bi-blade structure, and only 160% for the sandwich structure, even though the water diffusion takes place in three sides of the specimen. This measured value is consistent with values in the written works of the composite sandwich structure and single balsa [6,8,19,26].

3.2. Mass Loss Kinetic

In order to understand the hygro–thermo–mechanical responses of the sandwich structured composite, it is necessary to know the mass loss kinetic property of this structure. Thus, the two specimen types were measured: single skin and the entire sandwich structured composite. The skin consists of polyester resin and E-glass fibers, and the core is made of balsa wood impregnated with resin. The skin is practically insensitive to water uptake, whereas the water content in the core is very high, reaching more than 400% [6,7]. So, it is expected to know a fire response between the dry and the moisture saturation sandwich structured composite specimens.

In the first approximation for this paper, we consider that the fire response of the dry composite skin and the skin hygroscopically aged are similar because the skin absorbed a negligible amount of water. We therefore separated the skin from a dry sandwich composite sample (dimensions are indicated in Section 2.1), and measured its fire resistance property under 50 KW m^{-2} heat flux and 750 °C. The corresponding mass loss kinetic curve is shown in Figure 5. The entire combustion time was about 200 s, and it induced a mass loss of 60% (initial mass = 10.99 g). In the first 100 s, the single skin lost about 50% of its weight. This result provides an estimation of the average mass loss rate of the composite skin of about v_s = 0.05 g s^{-1}. At the start of the combustion, the skin burned and emitted a white smoke, and subsequently released a thick, black smoke. At the end of the combustion, only the fiberglass fabrics skin remained; the resin completely disappeared.

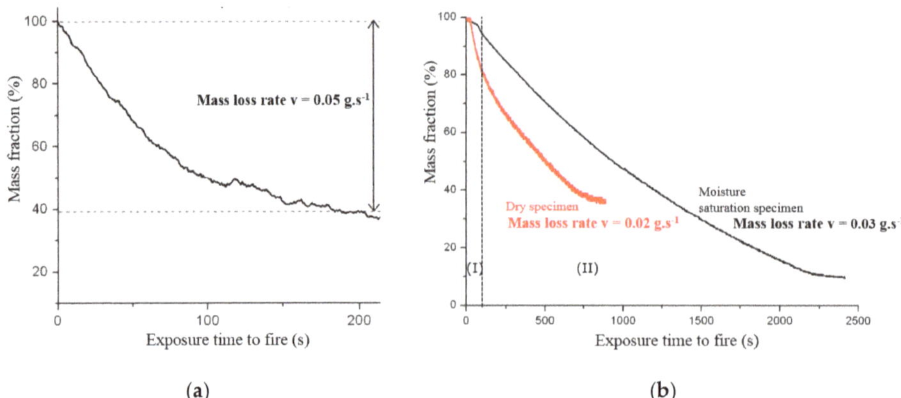

Figure 5. Mass loss curves as a function of fire exposure time for dry E-glass polyester single skin (**a**) and sandwich composite beam (**b**) from cone calorimetry measurements at 750 °C.

For the dry sandwich specimen as shown on the Figure 5b, (initial mass = 38.06 g), the material combustion curve was investigated in two successive portions: Portion I presents the pyrolyzed top skin, the exposure time to fire was 100 s, induced loss mass 20% and the mass loss rate was equal to v_{ts} = 0.08 g s^{-1}. Portion II respectively presents the char of the balsa core and the bottom skin, induced the loss mass about 45% during 775 s and the mass loss rate was about 0.02 g s^{-1}. The incombustible mass was about 35% of the initial mass. Combustion products of the sandwich structured composite at the end of the degradation were left with glass fiber fabric and wood charcoal. The glass fiber fabric and wood charcoal did not degrade due to the pyrolysis temperature that was conducted at 750 °C, which is lower than the melting temperature of the glass fibers and the wood charcoal [2,8,11]. The obtained results of the sandwich structured composite was pyrolyzed by the cone calorimetry, and it was found that the kinetic mass loss characteristic of three elements (top skin–core–bottom skin) was discontinuous. There was a superposition of the combustion at interface between top skin–core and core–bottom skin. During the first 100 s, the top skin was degraded and retracted by the heat flux, followed by the balsa core degradation. The top skin–core progressively induced delamination, an inflammation of the balsa core on the edges of the test specimen, and on resin between the balsa wood blocks. It made the balsa core lose its mass while the degradation in the top skin continues. The equivalent mass loss curve of the sandwich structure was obtained, which is the continuous curve in Figure 5b. The overview of the degradation processes on the cross section at different time intervals of the sandwich specimens during fire exposure is illustrated in Figure 6.

Finally, with the same measurements with above dry sandwich specimens, the determination of moisture saturation specimens (immersed in water at 40 °C) were performed, (initial mass = 78.93 g). Based on the observation, the fire behavior of the wet sandwich sample was similar to the dry sandwich sample. During the first 100 s, the combustion of the top skin induced a mass loss about 5% and the mass loss rate was equal to v_{ts} = 0.04 g s^{-1}. Then, the combustion of the equivalent bi-blade consisting of the balsa core and the bottom skin, induced a mass loss of 85% during 2150 s, and the mass loss rate was equal to 0.03 g s^{-1}. The remaining mass was about 10% of the initial mass of the test sample. Figure 5 presented that the evolution of the mass loss as a function of exposure time to fire of the moisture saturation sandwich sample. The degradation mechanism in the moisture saturation sandwich sample occurred slowly, but in the dry sandwich sample it occurred rapidly. This is because the water saturation sandwich sample contains a high water quantity which might significantly slow down the material pyrolysis. Thus, the complete deterioration due to fire of the water saturation sample was achieved in the 2250 s.

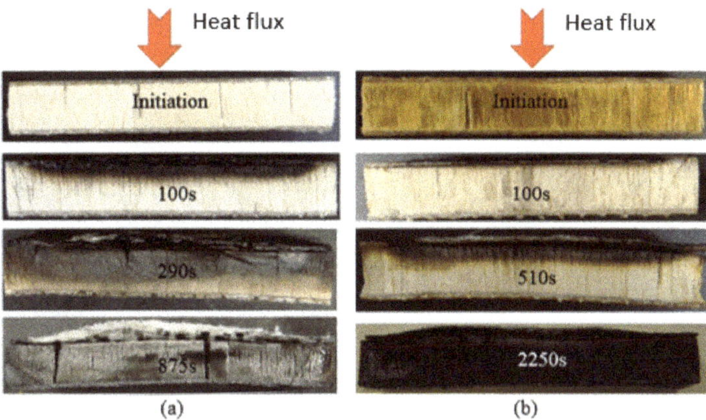

Figure 6. Cross-sectional view of the during fire exposure of the dry specimens (**a**) and the moisture saturation specimens (**b**).

3.3. Post-Hygro-Thermo-Mechanical Properties

Following mass loss measurements, the residual mechanical properties were determined to understand the hygro–thermo–mechanical response of these materials. Three-point bending measurements were performed on dry and water saturation samples at different times of the fire exposure. This analysis allows examining the residual mechanical strength of the sandwich structured composite after fire exposure.

Thus, the first step is to measure the ability of the sandwich structure to transmit a load up to the elastic limit. Figure 7 illustrated the evolution of the force as a function of displacement for dry and water saturation sandwich samples that did not undergo heat treatment. The mechanical behaviour consists of three periods: the first (period I) is similarly a behaviour of the porous materials. The displacement increases while the force is constant. This is supposed to be due to a change of the free volume of the top composite skins that were directly subjected to the action force. Thus, the residual deformation is maintained after discharge. A linear elastic behaviour of the sandwich structured composite material was observed. The force increases linearly with the displacement. Therefore, this linear portion of the curve was selected to make definite the Young's modulus. The final portion of the curve in period I corresponds to a nonlinear behaviour of the material until the abrupt rupture of the specimen. At the beginning of this phase, the force increases which corresponds to a small displacement. This clearly demonstrates that the flexural stiffness of the sandwich structure is improved by the shear stiffness of the balsa core. The end of this phase highlights a failure mode of the balsa core. The core is major enforced to shear, and failure takes place as the critical value (shear strength) of the core material is attained by the maximum shear stress. In the case of the moisture saturation specimen, after reaching critical load, the specimen did not break completely. The large amount of water contained in the core made it more elastic, so there was then a period of internal structure reorganization with constant force (period II). Finally, the force increases linearly with the increase of displacement (period III). The end of this period corresponds to a nonlinear behaviour of the moisture material until the rupture of the specimen. The critical load of the moisture specimen is lower than this one of the dry specimen, but the elastic deformation of the moisture specimen is larger by about 1.5 times. Figure 7b shows the failure mode of dry and water saturation sandwich specimens.

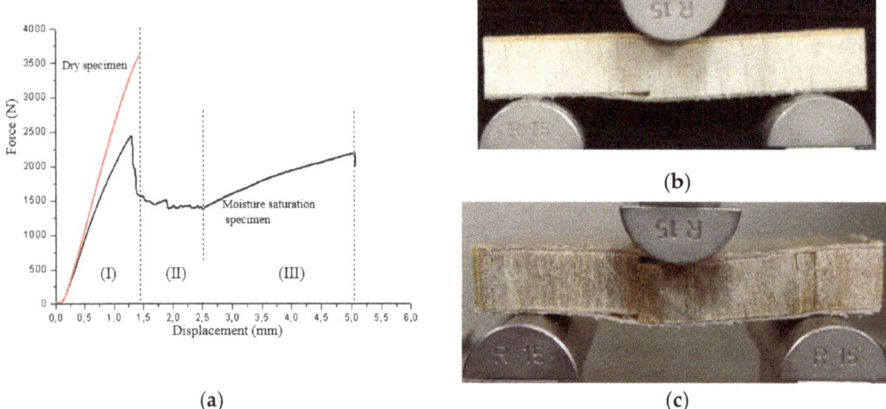

(a) (b) (c)

Figure 7. Three-point bending test: Force—displacement curve of the dry and water saturation sandwich specimens (**a**), failure mode of dry sandwich specimen (**b**) and water saturation sandwich specimen (**c**).

The second step was to measure the hygro–thermo–mechanical properties of the sandwich composite beam. The failure mode of the sandwich beam is characterized by two parts. The first is the core's shear failure and the second failure is due to global buckling of the skins. The displacement obtained by the three-point bending test is therefore composed of two parts: the deflection due to global buckling and the displacement caused by shear deformation of the core. The total displacement is expressed as follows [13,22,24]:

$$\frac{\delta}{FL} = \frac{L^2}{48D} + \frac{1}{4S} \tag{2}$$

where δ is the displacement (mm) measured at mid-span under load F (N). L is the span length (mm), $D = IE$ is the flexural stiffness of the sandwich beam (N·mm^2) within E (MPa) is the Young's modulus I is the inertia moment, and $S = Gel$ is the shear stiffness of the sandwich beam (N) with G being the shear modulus (MPa). The coefficients l and e are, respectively, the width and the thickness of the sandwich specimen. Equation (2) is used to determine the values of the Young's modulus E and the shear modulus G of the specimen by measuring three-point bending for five sandwich beam specimens with the different values of L. The corresponding curve is represented in the Figure 8 as well as the equation of the line obtained by linear regression analysis on the experimental points. The calculated values of the Young's modulus is $E = 7.8 \times 10^3$ MPa and of the shear modulus, it is $G = 85.9$ MPa for the dry balsa core sandwich beam specimen. Results are coincident with one found in the research in the particular case for pure balsa wood or E-glass-reinforced polyester resin [2,3,6,7,12].

It is supposed that the parameter $(1/(4S))$ of the Equation (2) is corresponding to zero (Ignoring the contribution of core to the flexural stiffness). The Equation (2) is also rewritten to identify the Young's modulus E as follows:

$$E = \frac{L^3}{4le^3} \times \frac{F}{\delta} \tag{3}$$

Identification of the elastic modulus E is performed by using the analytical resolving method, given by the Equation (3) and the experimental results F and δ, obtained by three-point bending test.

Finally, the residual flexural mechanical responses were investigated on the dry and moisture saturation sandwich beam specimens. The normalized evolutions of the maximum force and the Young's modulus as a function of the time exposed to fire are shown in Figure 9. These plots show the thermo–mechanical responses as the maximum force (F_m) and the flexural modulus (E) decreases followed exponential law. This law had also been observed for other composite materials [2–4,9,27].

The dry and water saturation sandwich structured composite materials lost their mechanical properties during 100 s of exposure time to fire. This time characteristic corresponds to the thermal degradation of the top skin. When the top skin is degraded to weaken the sandwich structure, the maximum force and the flexural modulus decreases like the exponential law. The dependence of the thermo–mechanical properties on the moisture content is visible in Figure 9. These obtained experimental results indicate that the sandwich structured composite specimen and the thermo–mechanical responses significantly related to the degradation of the top skin, and they are not assured of the moisture content in the balsa core. The moisture concentration only strongly influences the duration of the fire exposure. The normalized force curve expresses a light disparity between dry and moisture materials and the normalized Young's modulus indicated the similar trend. Thus, it allows us to express that the relative elasticity of the sandwich material is independent of the internal water concentration, but it is significantly influenced by the duration of the fire exposure time. Figure 10 represents the cross-section of the sandwich structured composite beam after 100 s exposed to fire for the dry and the moisture saturation specimens to confirm well the weakened sandwich structure due to the degradation of its top skin.

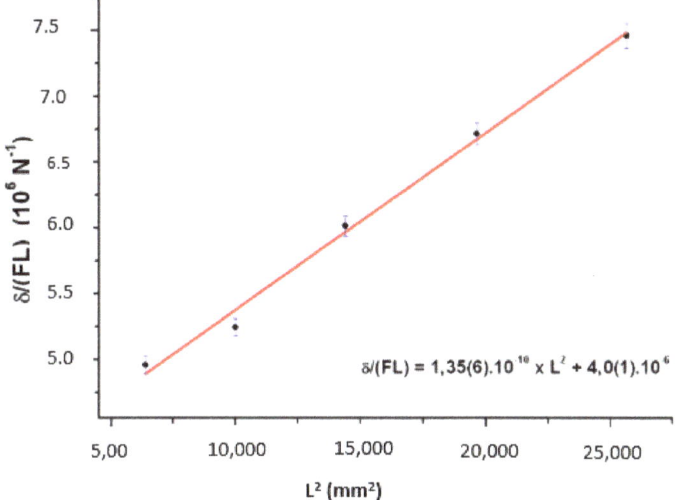

Figure 8. The curve of the $\delta/(FL) = L^2$ to determine the Young's modulus (E) and the shear modulus (G).

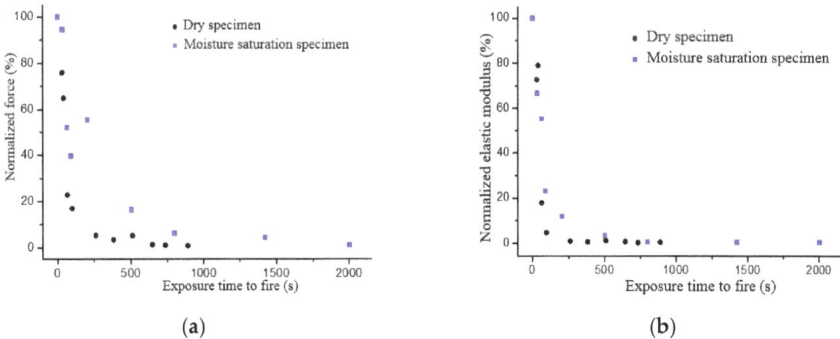

Figure 9. Post-residual flexural mechanical responses as a function of fire exposure time: normalized residual force (**a**) and normalized residual modulus (**b**).

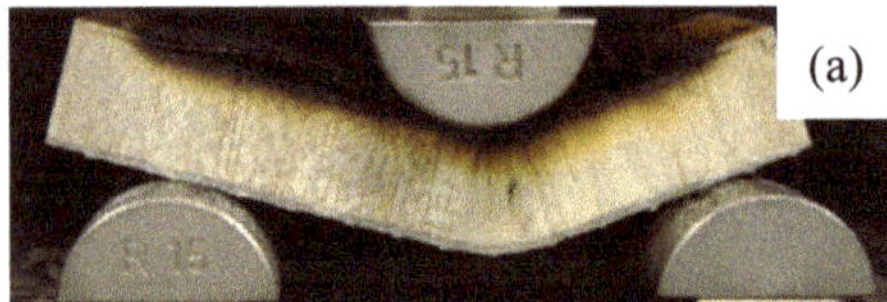

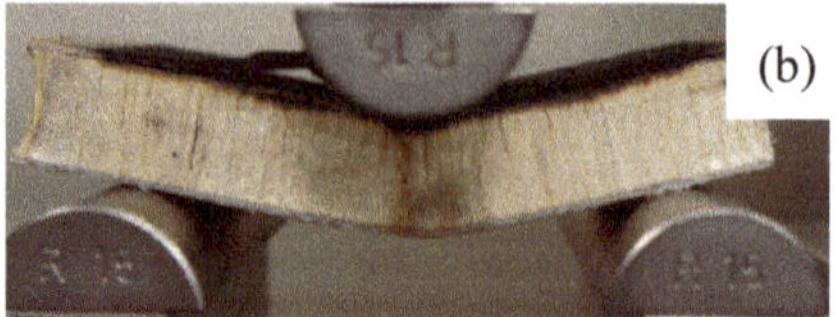

Figure 10. Cross-section of the sandwich structured composite beam after 100 s exposed to fire for the dry specimen (**a**) and the moisture saturation specimen (**b**).

4. Conclusions

The diffusive behaviour of the sandwich specimen and the bi-blade specimen (one skin and core) were investigated to understand the moisture impervious barrier significance of the skin. Thermal degradation rates under fire were identified for each skin and bi-blade sandwich specimens. The residual flexural mechanical responses as a function of the fire exposure time under a heat flux of 50 kW m^{-2} and 750 °C were analysed for the dry and moisture saturation sandwich specimens. The obtained results show that the sandwich structure undergoes a rapid thermo–mechanical degradation in the first 100 s of fire exposure, and this degradation is strongly influenced by the degradation of the top skin. Although the relative elasticity modulus of the sandwich structure composite material is independent of the moisture content, it is strongly influenced by the fire exposure time.

Author Contributions: Conceptualization, L.T.; Data curation, V.L.; Methodology, Software and Resources: P.C.; Original Draft Writing: L.T. and R.S.; Review & Editing: R.S.; Supervision: P.L.S. and A.B.; Project Administration: P.L.S. and C.-H.L. All authors have read and agreed to the published version of the manuscript.

Funding: This research received no external funding.

Acknowledgments: Luan TranVan is indebted to the Ministry of Education and Training (MOET) for funding under grant number KYTH-74 (B2017.DNA.11). We would like to thank Didier Andeler and Philippe Frétaud (IUT Saint-Nazaire, France) for technical support.

Conflicts of Interest: The authors declare no conflict of interest in any choice of research project funding sponsors; design of the study; in the collection, analyses or interpretation of data; in the writing of the manuscript; or in the decision to publish the results.

References

1. Lua, J. Hybrid Progressive Damage Prediction Model for Loaded Marine Sandwich Composite Structures Subjected to a Fire. *Fire Technol.* **2011**, *47*, 851–885. [CrossRef]
2. Mouritz, A.; Mathys, Z.; Gardiner, C. Thermomechanical modelling the fire properties of fibre–polymer composites. *Compos. Part B Eng.* **2004**, *35*, 467–474. [CrossRef]
3. Ulven, C.A.; Vaidya, U.K. Post-fire low velocity impact response of marine grade sandwich composites. *Compos. Part A Appl. Sci. Manuf.* **2006**, *37*, 997–1004. [CrossRef]
4. Mouritz, A.P.; Feih, S.; Kandare, E.; Mathys, Z.; Gibson, A.G.; Des Jardin, P.E.; Case, S.W.; Lattimer, B.Y. Review of fire structural modelling of polymer composites. *Compos. Part A Appl. Sci. Manuf.* **2009**, *40*, 1800–1814. [CrossRef]
5. Joshi, N.; Muliana, A. Deformation in viscoelastic sandwich composites subject to moisture diffusion. *Compos. Struct.* **2010**, *92*, 254–264. [CrossRef]
6. Legrand, V.; TranVan, L.; Jacquemin, F.; Casari, P. Moisture-uptake induced internal stresses in balsa core sandwich composite plate: Modeling and experimental. *Compos. Struct.* **2014**, *119*, 355–364. [CrossRef]
7. Tranvan, L.; Legrand, V.; Jacquemin, F. Thermal decomposition kinetics of balsa wood: Kinetics and degradation mechanisms comparison between dry and moisturized materials. *Polym. Degrad. Stab.* **2014**, *110*, 208–215. [CrossRef]
8. Mouritz, A.P. Fire resistance of aircraft composite laminates. *J. Mater. Sci. Lett.* **2003**, *22*, 1507–1509. [CrossRef]
9. Sereir, Z.; Adda-bedia, E.A.; Boualem, N. The evolution of transverse stresses in hybrid composites under hygrothermal loading. *Mater. Des.* **2011**, *32*, 3120–3126. [CrossRef]

10. Sinmazçelik, T.; Avcu, E.; Bora, M.Ö.; Çoban, O. A review: Fibre metal laminates, background, bonding types and applied test methods. *Mater. Des.* **2011**, *32*, 3671–3685. [CrossRef]
11. Frostig, Y. Hygothermal (environmental) effects in high-order bending of sandwich beams with a flexible core and a discontinuous skin. *Compos. Struct.* **1997**, *37*, 205–221. [CrossRef]
12. Le Duigou, A.; Deux, J.M.; Davies, P.; Baley, C. PLLA/flax mat/balsa bio-sandwich manufacture and mechanical properties. *Appl. Compos. Mater.* **2011**, *18*, 421–438. [CrossRef]
13. Steeves, C.A.; Fleck, N.A. Collapse mechanisms of sandwich beams with composite faces and a foam core, loaded in three-point bending. Part II: Experimental investigation and numerical modelling. *Int. J. Mech. Sci.* **2004**, *46*, 585–608. [CrossRef]
14. Tagarielli, V.L.; Fleck, N.A.; Deshpande, V.S. Collapse of clamped and simply supported composite sandwich beams in three-point bending. *Compos. Part B Eng.* **2004**, *35*, 523–534. [CrossRef]
15. Manalo, A.C.; Aravinthan, T.; Karunasena, W. Flexural behaviour of glue-laminated fibre composite sandwich beams. *Compos. Struct.* **2010**, *92*, 2703–2711. [CrossRef]
16. Zhu, F.; Li, K. Numerical Modeling of Heat and Moisture Through Wet Cotton Fabric Using the Method of Chemical Thermodynamic Law Under Simulated Fire. *Fire Technol.* **2011**, *47*, 801–819. [CrossRef]
17. Voss, M.; Wannicke, N.; Deutsch, B.; Bronk, D.; Sipler, R.; Purvaja, R.; Ramesh, R.; Rixen, T. Internal Cycling of Nitrogen and Nitrogen Transformations. In *Treatise on Estuarine and Coastal Science*; Academic Press: Cambridge, MA, USA, 2012; ISBN 9780080878850.
18. Gigliotti, M.; Jacquemin, F.; Vautrin, A. Assessment of approximate models to evaluate transient and cyclical hygrothermoelastic stress in composite plates. *Int. J. Solids Struct.* **2007**, *44*, 733–759. [CrossRef]
19. Choi, H.S.; Jang, Y.H. Bondline strength evaluation of cocure/precured honeycomb sandwich structures under aircraft hygro and repair environments. *Compos. Part A Appl. Sci. Manuf.* **2010**, *41*, 1138–1147. [CrossRef]
20. Abot, J.L.; Yasmin, A.; Daniel, I.M. Hygroscopic behavior of woven fabric carbon-Epoxy composites. *J. Reinf. Plast. Compos.* **2005**, *24*, 195–207. [CrossRef]
21. Fiore, V.; Di Bella, G.; Valenza, A. Glass-basalt/epoxy hybrid composites for marine applications. *Mater. Des.* **2011**, *32*, 2091–2099. [CrossRef]
22. Steeves, C.A.; Fleck, N.A. Material selection in sandwich beam construction. *Scr. Mater.* **2004**, *50*, 1335–1339. [CrossRef]
23. Bažant, Z.P.; Beghini, A. Sandwich buckling formulas and applicability of standard computational algorithm for finite strain. *Compos. Part B Eng.* **2004**, *35*, 573–581. [CrossRef]
24. Reyes, G.; Rangaraj, S. Fracture properties of high performance carbon foam sandwich structures. *Compos. Part A Appl. Sci. Manuf.* **2011**, *42*, 1–7. [CrossRef]
25. Chen, C.; Harte, A.M.; Fleck, N.A. Plastic collapse of sandwich beams with a metallic foam core. *Int. J. Mech. Sci.* **2001**, *43*, 1483–1506. [CrossRef]
26. Chen, H.; Miao, M.; Ding, X. Influence of moisture absorption on the interfacial strength of bamboo/vinyl ester composites. *Compos. Part A Appl. Sci. Manuf.* **2009**, *40*, 2013–2019. [CrossRef]
27. Le Lay, F.; Gutierrez, J. Improvement of the fire behaviour of composite materials for naval application. *Polym. Degrad. Stab.* **1999**, *64*, 397–401. [CrossRef]

Article

Multivariate Analysis and Machine Learning for Ripeness Classification of Cape Gooseberry Fruits

Miguel De-la-Torre [1], Omar Zatarain [1], Himer Avila-George [1,*], Mirna Muñoz [2], Jimy Oblitas [3], Russel Lozada [4], Jezreel Mejía [2] and Wilson Castro [3,5]

1 Departamento de Ciencias Computacionales e Ingenierías, Universidad de Guadalajara, Ameca 46600, Jalisco, Mexico; miguel.dgomora@academicos.udg.mx (M.D.-l.-T.); omar.zatarain@academicos.udg.mx (O.Z.)
2 Centro de Investigación en Matemáticas, Zacatecas 98160, Mexico; mirna.munoz@cimat.mx (M.M.); jmejia@cimat.mx (J.M.)
3 Facultad de Ingeniería, Universidad Privada del Norte, Cajamarca 06001, Peru; jimy.oblitas@upn.edu.pe
4 Escuela Profesional de Ingeniería Electrónica, Facultad de Producción y Servicios, Universidad Nacional de San Agustín, Arequipa 04000, Peru; rlozadav@unsa.edu.pe
5 Facultad de Ingeniería de Industrias Alimentarias, Universidad Nacional de Frontera, Sullana 20100, Peru; wcastro@unfs.edu.pe
* Correspondence: himer.avila@academicos.udg.mx
† This paper is an extended version of our paper published in Proceedings of the 8th International Conference on Software Process Improvement, León, Guanajuato, México, 23–25 October 2019.

Received: 1 November 2019; Accepted: 3 December 2019; Published: 5 December 2019

Abstract: This paper explores five multivariate techniques for information fusion on sorting the visual ripeness of Cape gooseberry fruits (principal component analysis, linear discriminant analysis, independent component analysis, eigenvector centrality feature selection, and multi-cluster feature selection.) These techniques are applied to the concatenated channels corresponding to red, green, and blue (RGB), hue, saturation, value (HSV), and lightness, red/green value, and blue/yellow value (L*a*b) color spaces (9 features in total). Machine learning techniques have been reported for sorting the Cape gooseberry fruits' ripeness. Classifiers such as neural networks, support vector machines, and nearest neighbors discriminate on fruit samples using different color spaces. Despite the color spaces being equivalent up to a transformation, a few classifiers enable better performances due to differences in the pixel distribution of samples. Experimental results show that selection and combination of color channels allow classifiers to reach similar levels of accuracy; however, combination methods still require higher computational complexity. The highest level of accuracy was obtained using the seven-dimensional principal component analysis feature space.

Keywords: Cape gooseberry; color space selection; color space combination; food engineering

---

1. Introduction

In the advent of the fourth industrial revolution, the growing tendency of automation of human activities encourages the use of robotic systems in the food industry [1]. In this context, the automation of food packing processes is essential to accelerating the production rate, and reducing human contact and possible contamination of food products. Moreover, machine vision techniques allow robotic systems to retrieve information from food products, using different sensors that depend on the particular characteristics to be measured, and each sensor represents an additional cost to construct an information retrieval system. For instance, an application that requires such automation systems is the classification of Cape gooseberry (*Physalis peruviana* L.) fruits according to their level of ripeness. Current industry practices address this repetitive task through visual inspection of color, size, and shape parameters [2]. While automated sorting systems based on computer vision techniques have been proposed to improve production methods and provide high-quality products, their operation relies on classification algorithms that consider either different color spaces or a combination of them [3,4].

The most common representation of color images employed by computer vision systems is a combination of the three primary colors: Red, green, and blue (RGB). The triplet with the values for each primary color is typically considered as a coordinate system with either Euclidean, Mahalanobis, Hamming, or a different metric of distance. In such a three-dimensional coordinate system, each point (e.g., 3D vector) represents a different color in the visible spectrum. Other color spaces different than RGB are commonly employed, providing different three-dimensional representations, and can be classified into three categories according to [5]: Hardware-orientated spaces, human-orientated spaces, and instrumental spaces. In the first category, hardware-orientated spaces (e.g., RGB, YIQ, and CMYK) are defined based on the properties of the hardware devices used to display images. On the other hand, human-orientated spaces (e.g., HSI, HSL, HSV, and HSB) are based on hue and saturation, following the principles of an artist and based on inherent color characteristics. Finally, instrumental spaces (e.g., XYZ, L*a*b*, and L*u*v*) are those used for color instruments, where the color coordinates of an instrumental space are the same on all output media.

As will be considered in Section 2, the color spaces that are most commonly employed in the classification of fruits are RGB, L*a*b*, and HSV. However, the accuracy of the same classifier on the same dataset may vary from one color space to the other. Some authors have investigated these differences in classification accuracy due to the variation of the distribution of pixels in distinct color spaces or the use of different segmentation techniques. According to [6,7], the practice of color measurement in food engineering, the L*a*b* color space, is the most commonly used. The main reasons are related to the uniform distribution of colors and because the L*a*b* is perceptually uniform (i.e., equal changes in data are visually perceived as equal changes in the color space). However, it is known that color spaces like RGB, L*a*b*, and HSV are equivalent up to a transformation.

Regardless of the color space used for classification, the objective of classifiers applied to fruit sorting consists of finding a criterion to separate samples from one or other ripeness levels in the so-called feature space. The goal is to establish a decision boundary that may be applied as a fixed borderline between categories. Supervised classifiers employ labeled samples to learn a model that is used to predict a category in new, never seen unlabeled samples. Supervised classifiers commonly employed in the food industry include support vector machines (SVM), k-nearest neighbors (KNN), artificial neural networks (ANN), and decision tree (DT) [8,9].

In practice, any pattern classifier may be employed, presenting a trade-off between accuracy and complexity. While the equivalence between color spaces is well-known [10], it has been found that different color spaces allow the same classifier to produce distinct classification rates, due to variations in the distribution of color samples [3,11]. Moreover, the combination of color spaces using multivariate analysis may provide a feature space where an increase in classification accuracy is possible. For instance, in [3], a methodology to compare different combinations of machine learning techniques and color spaces (RGB, HSV, and L*a*b*) was proposed in order to evaluate their ability to classify Cape gooseberry fruits. The results showed that the classification of Cape gooseberry fruits by their ripeness level was sensitive to both the color space and the classification technique used. The models based on the L*a*b* color space and the support vector machine (SVM) classifier showed the highest performance regardless of the color space. An improvement was obtained by employing principal component analysis (PCA) for the combination of the three-color spaces at the expense of increased complexity. An extension of such a study was proposed in [4], where a supervised multivariate analysis method was compared with previous results (linear discriminant analysis, LDA).

In this paper, an extension of previous work described in [3,4] is proposed to compare multivariate analysis methods and machine learning techniques for ripeness classification. The color channels from RGB, HSV, and L*a*b* color spaces were concatenated to spam a nine-dimensional feature space. The five multivariate methods employed to combine information from the nine color channels include PCA, LDA, independent component analysis (ICA), multi-cluster feature selection (MCFS), and eigenvector centrality feature selection (ECFS). In the last case, selection methods applied to find the most relevant features for classification were MCFS and ECFS. The main contribution of this paper is the use of multivariate techniques to find an appropriate feature space for classification.

The manuscript is organized as follows. Section 2 summarizes the most recent works published on ripeness classification, including diverse approaches and methodologies. Some of the most popular methods were selected for this comparison, and Section 3 describes the material and methods employed in experiments to compare the distinct approaches. Section 4 presents the results and a discussion on the relevant findings. Finally, Section 5 presents conclusions and future work.

2. Ripeness Classification

As reported in the literature, different color spaces have been used to create automated fruit classification systems, presenting different levels of accuracy that are related to both, the color space and the sorting algorithm. Table 1 shows common methods and color spaces reported in the literature used to classify distinct fruits according to their ripeness level.

Table 1. Color spaces and classification approaches for fruit classification in literature. NA stands for non-available information MDA stands for multiple discriminant analysis, QDA for quadratic discriminant analysis, PLSR for partial least squares regression, RF for the random forest, and CNN for the convolutional neural network. The table was taken from [3] and updated with new findings.

Item	Colorspace	Classification Method	Accuracy	Ref
Apple	HSI	SVM	95	[12]
Apple	L*a*b*	MDA	100	[13]
Avocado	RGB	K-Means	82.22	[14]
Banana	L*a*b*	LDA	98	[15]
Banana	RGB	ANN	96	[16]
Blueberry	RGB	KNN and SK-Means	85-98	[17]
Date	RGB	K-Means	99.6	[18]
Lime	RGB	ANN	100	[19]
Mango	RGB	SVM	96	[5]
Mango	L*a*b*	MDA	90	[20]
Mango	L*a*b*	LS-SVM	88	[21]
Oil palm	L*a*b*	ANN	91.67	[22]
Pepper	HSV	SVM	93.89	[23]
Persimmon	RGB + L*a*b*	QDA	90.24	[24]
Tomato	HSV	SVM	90.8	[25]
Tomato	RGB	DT	94.29	[26]
Tomato	RGB	LDA	81	[27]
Tomato	L*a*b*	ANN	96	[28]
Watermelon	YCbCr	ANN	86.51	[29]
Soya	HSI	ANN	95.7	[8]
Banana	RGB	Fuzzy logic	NA	[9]
Banana	RGB	CNN	87	[9]
Watermelon	VIS/NIR	ANN	80	[30]
Watermelon	RGB	ANN	73.33	[31]
Tomato	FTIR	SVM	99	[32]
Kiwi	Chemometrics MOS E-nose	PLSR, SVM, RF	99.4	[33]
Coffee	RGB + L*a*b* + Luv + YCbCr + HSV	SVM	92	[34]
Cape Gooseberry	RGB + HSV + L*a*b*	ANN, DT, SVM and KNN	93.02	[3,4]

According to Table 1, the most common color space used for classification is RGB, with 50% of the works, followed by L*a*b* with 32%, and HSV with 14%. Similarly, the most common classifiers are ANN and SVM, with 32% of the experiments reporting results using color spaces that include RGB,

L*a*b*, and HSV. The accuracy obtained by each approach depends on the experimental settings and are not comparable at this point. However, reported accuracy ranges between 73 and 100 percent.

2.1. Methods for Color Selection and Extraction

The distribution of samples in the feature space depends on the measurements obtained from sensors, and in this case, the color channels for the distinct color spaces. The search for the color channels that are most relevant for classification is important to help classifiers to find the decision frontier between classes. Features that are noisy or not relevant may difficult classification problems and may conduce to a low performance even by the most sophisticated classifiers. Finding a subset of the k most relevant features, either by selecting them or applying feature extraction techniques, favors the reduction of redundant and irrelevant information. The so obtained k-dimensional feature space employed for classification instead of the original d-dimensional feature space is suitable to facilitate finding a separation criterion between classes. Whereas feature extraction algorithms find a mapping to a new feature space, feature selection methods aim to select a subset of vectors that spans a feature subspace that facilitates classification.

Feature extraction approaches can be categorized according to the use of data labels in supervised and unsupervised. Unsupervised feature extraction techniques consider the underlying distribution of data solely, and aim to find a mapping to a new feature space with desired characteristics. An example of unsupervised methods is PCA, which employs the Eigenvectors of the covariance matrix of samples to maximizes their spread in each new axis. Additionally, supervised approaches employ the information from class-labels to find the mapping. A representative supervised approach is the linear discriminant analysis (LDA), that aims to maximize the spread of samples distinct classes, and minimize the within-class spread.

Analogously, supervised feature selection considers class labels to find the most relevant features, and unsupervised feature selection strategies are based exclusively on the underlying distribution of samples. The selection of the subset of the most relevant features is a computational expensive combinatorial problem. The optimally of an algorithm to find a good enough feature subset may depend on the strategy followed for ranking or selection of features.

In feature selection and extraction, the problem can be stated by considering a set of points (sample tuples or feature vectors) $X = \{x_1, x_2, ..., x_N\}$, where $x_i \in R^d$. The algorithms for feature extraction and selection, find a new set $X' = \{x'_1, x'_2, ..., x'_k\}$, where $x'_i \in R^k$, and $k \leq d$ is the new dimension of the feature vectors.

2.2. Principal Component Analysis (PCA)

The PCA method is applied to find a linear transformation that finds the directions of maximum variance data. Sample patterns are projected onto a new feature space, and the axes with more explained variance provide a distribution that facilitates the separation between classes. The algorithm is shown in the Figure 1 depicts the general procedure to transform data samples from X to their new representation in the k-dimensional feature space X′. The new k-dimensional feature space corresponds to the k eigenvectors of the covariance matrix C. The axis with the highest eigenvalues expresses a higher explained variance.

Input: X <- Set of d-dimensional data samples
Output: Samples X' mapped to the new k-dimensional feature space
Normalize samples in the input d-dimensional data
Compute covariance matrix $C_{d \times d}$
Decompose the covariance matrix C into its eigenvectors
Select the k eigenvectors with largest eigenvalues ($k \leq d$)
Generate a $d \times k$ projection matrix W from the highest k eigenvectors
Map the samples to the new k-dimensional feature space using $X' = X \cdot W$

Figure 1. The procedure followed by principal component analysis (PCA) to map the input data samples to the new k-dimensional feature space.

2.3. Linear Discriminant Analysis (LDA)

The LDA method allows obtaining and applying a linear transformation that finds the directions of maximum variance input data samples. The main difference with PCA is that LDA aims to minimize intraclass variability, whereas it maximizes interclass variability employing class labels. The main limitation is that the number of classes bounds the number of features in the new k-dimensional space (e.g., $1 < k < c$, where c is the number of classes). This limitation makes this approach advantageous only with a high number of classes, and unpractical for data with a few classes (e.g., $c << d$). The procedure followed in computing the mapping and transforming the data is shown in Figure 2.

Input: X <- Set of d-dimensional data samples
L <- Set of labels for the N data samples in X
Output: Samples X' mapped to the new k-dimensional feature space
Standardize the samples in the d-dimensional dataset
Compute the d-dimensional mean vector corresponding to each class
Construct the between-class scatter matrix S_B, and the within-class scatter matrix S_w
Compute the Eigenvectors and corresponding Eigenvalues of the matrix $S_w^{-1} S_B$
Generate a ($d \times k$)-dimensional transformation matrix W_{LDA} with the k eigenvectors with the k largest eigenvalues as columns
Map the samples onto the new feature subspace using W_{LDA}

Figure 2. The procedure followed by linear discriminant analysis (LDA) to map the data samples to the new k-dimensional feature space.

2.4. Independent Component Analysis (ICA)

The ICA method finds underlying components from multivariate statistical data, where data is decomposed into components that are maximally independent in an appropriate sense (e.g., kurtosis and negentropy). The difference between PCA and LDA is that low-dimensional signals do not necessarily correspond to the directions of maximum variance; rather, the ICA components have maximal statistical independence and are nongaussian. In practice, ICA can be used to find disjoint underlying trends in multi-dimensional data [35].

The algorithm is shown in the Figure 3 depicts the procedure followed by the FastICA algorithm to obtain the independent components from X, using kurtosis as a measure of non-gaussianity. In this case, dimensionality reduction is not obvious, given that there is no measure of how important a particular independent component is. The relevance of the individual features obtained with PCA and LDA is given by the algorithms shown in Figures 1 and 2, respectively. In the case of ICA, feature selection techniques may be employed to provide a relevance level for each of the features that result from the transformation, as described in Sections 2.5 and 2.6.

Input: X <- Set of d-dimensional data samples
Output: Samples X' mapped to the new k-dimensional feature space ($k = d$)
1 Center data samples in $X_c = center(X)$
2 Whiten data and obtain $X_w = D^{1/2}E^T X_c$, where E is the matrix of Eigenvectors, and D is the diagonal matrix of Eigenvalues of the covariance matrix of X_c
3 Set a random initial weight N-dimensional vectors $\mathbf{w} = rand(N)$
4 Estimate nongaussianity according to kurtosis; $G = 4s$, and $Gp = 12S_k^2$
5 Update weight $\mathbf{w} = GX_{pw} - Gp\mathbf{w}$
6 If $\Delta = max(1 - |W, W_{prev}) > \gamma$, for a given threshold γ, go back to line 4
7 Repeat for each vector $\mathbf{w}$ to produce the new mapping W
8 $X' = WX$ is the new representation of data in the new feature space

Figure 3. The procedure followed by FastICA to map data samples in X to the new feature space that respects nongaussianity using kurtosis.

2.5. Eigenvector Centrality Feature Selection (ECFS)

The feature selection via eigenvector centrality is a supervised method that ranks features by identifying the most important ones. It maps the selection problem to an affinity graph with features as nodes and assesses the rank features according to the eigenvector centrality (EC) [36].

The algorithm shown in the Figure 4 presents the method to rank and select the most relevant features from the data samples X. While this does not constitute a proper transformation in terms of linear algebra, every sample is represented in a new k-dimensional feature space with the highest-ranked features.

Input: X <- Set of d-dimensional data samples
L <- Set of labels for the N data samples in X
Output: v_0 ranking scores for each of the d features
Build the weighted graph $G = (V, E)$; where vertices V correspond to data samples and edges E among features:
Compute the Fisher score : $f_i = \frac{\sum_{c=1}^{C}(\mu_{i,C} - \mu_i)^2}{\sigma_i^2}$, where μ_{ij} represents the mean, and σ_{ij} is the standard deviation for the whole dataset corresponding to feature i
Compute mutual information: $m_i = \sum_{y \in Y} \sum_{z \in x^{(i)}} p(z, y) \log \left(\frac{p(z,y)}{p(z)p(y)} \right)$, where $p(\cdot, \cdot)$ is the joint probability distribution
Obtain the adjacency matrix $A = \alpha k + (1 - \alpha)\Sigma$, where $k = f \cdot m^T$ and $\Sigma(i, j) = \max(\sigma^{(i)}, \sigma^{(j)})$ Ranking: Compute the eigenvalues $\{\Lambda\}$and eigenvectors $\{V\}$ of A, where $\lambda_0 = \max_{\lambda \in \Lambda}(|\lambda|)$
v_0 is the eigenvector associated to λ_0

Figure 4. The procedure followed by eigenvector centrality feature selection to select the variables that constitute the new feature space.

2.6. Multi Cluster Feature Selection (MCFS)

Multi-class feature selection (MCFS) is an unsupervised technique that aims to find those features that preserve the multi-cluster underlying structure of the samples used for training [37]. Given that the number of clusters is unknown a priori, it is a good practice to explore distinct values to find a good feature subspace. The most relevant features are found following the procedure shown in the algorithm shown in the Figure 5.

Input: X <- Set of N d-dimensional data samples
K <- Number of clusters, default $K = 5$
Output: The top k features according to their MCFS scores
1 Construct a p-nearest neighbor graph Laplacian (e.g., $W_{ij} = 1$ `iif` i and j are connected by an edge; $D_{ii} = \sum_j W_{ij}$; and $L = D - W$)
2 Solve the generalized eigenproblem $Ly = \lambda Dy$, where $Y = y_1, ..., y_K$ are the top K eigenvectors with respect to the smallest eigenvalues.
3 Solve the regression problem: $\min_{a_k} ||y_k - X^T a_k||^2$, with a user-defined cardinality to control the sparseness of a_k
4 Compute: $MCFS(j) = \max_k |a_{k,j}|$

Figure 5. The procedure followed by eigenvector centrality feature selection to select the variables that constitute the new feature space.

While the simplest method to choose W was presented in Step 1, other methods exist that range between accuracy and complexity (See [37]). According to the authors, the default number of nearest neighbors is $p = 5$, and the default number of eigenfunctions is $K = 5$. This last parameter K usually influences the accuracy of the algorithm and should be optimized before usage.

2.7. Classification for Fruit Sorting

According to Table 1, some of the most popular supervised classifiers in fruit sorting are the artificial neural networks (ANN), decision trees (DT), support vector machines (SVM), and k-nearest neighbor (KNN). These classifiers were used in this paper for the experimental settings. While these techniques have been present in the literature for many years now, see [3,38], their usage in practice increased due to their capacity to address diverse real-world problems.

ANN is a non-linear classifier that simulates biological neural networks. A common implementation of ANN corresponds to the probabilistic ANN, which produces an estimated posterior probability for each input sample to belong to any of the classes, and the max function allows to select the most likely class. In this research, the Matlab's Neural Network Toolbox was used to implement the probabilistic ANN classifier, byways of the newpnn function, tunned to optimize hyperparameters using linear search.

DT is a tree-based example of the knowledge used to represent the classification rules. Internal nodes are representations of tests of an attribute; each branch represents the outcome of the test, and leaf nodes represent class labels. In this paper, the Matlab's Machine Learning Toolbox (MLT) was used the train and simulate DTs, using the Classification & Regression Trees (CART) algorithm to create decision trees, with the fitctree and predict functions. The function fitctree employes standard classification and regression trees algorithm to create DTs.

SVM is a non-parametric statistical learning classifier that constructs a separating hyperplane (or a set of hyperplanes) in a high-dimensional feature space. Some versions use the so-called kernel trick to map data to higher dimensional feature space and find the separating hyperplane there. The functions fitcecoc and predict functions were used for simulations, both implemented in Matlab's MLT. The fitcecoc function was tunned to use a linear kernel and Bayesian hyperparameter optimization.

KNN is a non-parametric classifier that keeps all training samples, and prediction is based on the number of closest neighbors belonging to a class. Given an input sample, the distance to all stored samples is computed and compared to all pre-stored samples, presenting a high computational complexity at prediction. For simulations, the fitcknn and predict functions from Matlab's MLT were used. This function employs Bayesian hyperparameter optimization.

3. Materials and Methods

For experiments, a set of 925 samples of Cape gooseberry fruits were collected from a plantation located in the village of El Faro, Celendin Province, Cajamarca, Peru (UTM: −6.906469, −78.257071). Fruit samples were homogeneously disposed on a conveyor belt used in a production line (160 × 25 cm, and 80 cm high). A Halion-HA-411 VGA webcam was used for image capture, which provides RGB raw images in JPG format. The resolution of the resulting images is 1280 × 1720 pixels. The camera was fixed 35 cm over the conveyor belt, and the captured scene was covered with black matte panels to reduce variations in light, as implemented by Pedreschi et al. [39]. The illumination system included two long fluorescent tubes (Philips TL-D Super, cold daylight, 80 cm, 36 W) that were placed on both sides of the conveyor belt. Additionally, a circular fluorescent tube (Philips GX23 PH-T9, cold daylight, 21.6 cm, 22 W) was located at the top of the setting. Images captured with the camera were stored on a portable computer running Matlab to control image acquisition and data analysis.

Seven levels of ripeness were employed for visual classification, following the standard for Cape gooseberry, and the visual scale proposed in [40] and shown in Figure 6. The process followed for evaluation is depicted in Figure 7. Images captured from the conveyor belt (step 1) were employed to find the regions of interest corresponding to Cape gooseberry fruits in the image, employing standard segmentation techniques (steps 2 and 3); the size of resulting regions depends on the size of the actual fruit. Color versions of segmented fruits were labeled by five experts according to the categorization proposed by Fischer et al. in [40] (step 4). One-color sample was selected for each fruit region in each of the RGB, HSV and L*a*b* color spaces, by computing the average for each color channel; and a nine-dimensional feature vector was built through concatenation: $x = [R, G, B, H, S, V, L*, a*, b*]^T$ (step 5). Then, multivariate analysis techniques for feature extraction/selection were applied to the set of feature vectors (step 6), and the resulting samples were organized for five-fold cross-validation. Four standard classifiers were trained (step 7) and performance evaluation computed (step 8).

Figure 6. Levels of ripeness employed for supervised visual classification.

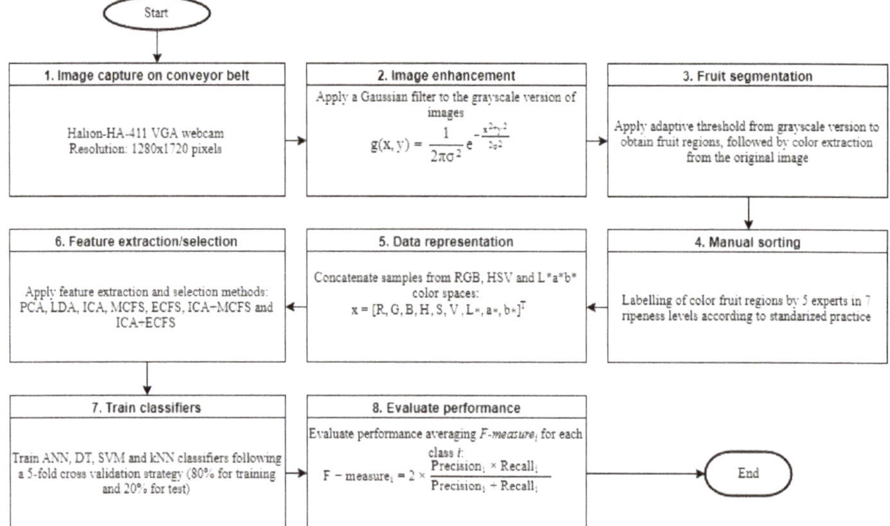

Figure 7. Experimental process followed to evaluate the system with distinct feature extraction/selection methods and different classifiers.

The performance of the seven-class classifiers was evaluated using the F-measure, as defined in [3]. First, the confusion matrix is computed according to the responses of each classifier, and true positives (TP_i), false positives (FP_i), true negatives (TN_i), and false negatives (FN_i) are obtained for each class i, using the elements N_{ij} of the confusion matrix. Class-specific precision and recall are computed using Equations (1) and (2), respectively.

$$Precision_i = \frac{TP_i}{TP_i + FP_i} \tag{1}$$

$$Recall_i = \frac{TP_i}{TP_i + FN_i} \tag{2}$$

Finally, the multiclass F-measure was used for comparison along with the experimental results, due to its representativeness of the classification performance on target classes (Equation (3)).

$$F - measure_i = 2 \times \frac{Precision_i \times Recall_i}{Precision_i + Recall_i} \tag{3}$$

The three analyses followed to characterize the performance of the system started by fixing the classifier (e.g., SVM). First, the k (number of clusters) was explored in order to find the k that allows the highest classification performance. Then, the size of the feature space was explored in terms of average F-measure. The last analysis explores the performance using the parameters found in previous steps, and the four classifiers presented in Section 2 ANN, SVM, DT, and KNN.

4. Experimental Results

As explained in Section 2.6, MCFS needs a search to find the number of clusters that maximizes the classification performance. The number of characteristics was set to seven, to make it comparable with previous results using PCA [3].

Figure 8 shows the box plots that summarize the distribution of performance for the SVM classifier trained with seven color channels (features) selected with the MCFS algorithm. The parameter that controlled the number of clusters was moved from 1 to 9 (i.e., the maximum number of possible features). In most cases, the median of the F-measure was maintained around 71.75, and only two cases were different: 2 and 9. Using nine clusters appears to provide lower performance related to the creation of an excessive number of clusters. On the other hand, using only two clusters for feature selection seems to provide a higher level of accuracy. However, and regardless of the median accuracy, the variability between cases shows a difference that makes no significant difference in using a different number of clusters. Therefore, in the following experiments, the number of clusters is fixed at 2, and it explored other variables.

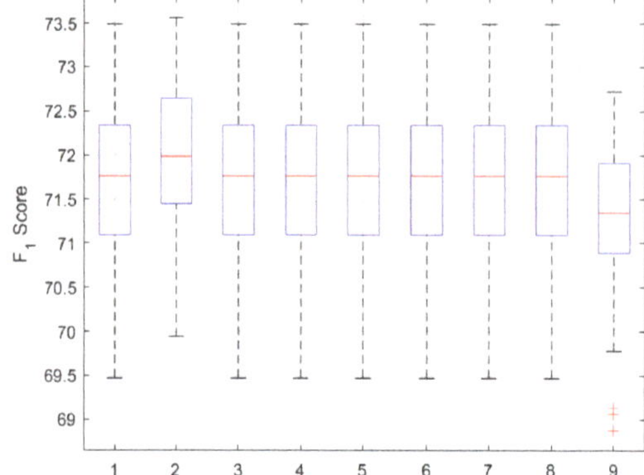

Figure 8. Boxplots corresponding to the F-measure for nine distinct values of the parameter controlling the number of clusters in multi-cluster feature selection (MCFS). The number of characteristics was fixed at 7, and the experimentation follows a five-fold cross-validation strategy.

4.1. Analysis of Feature Spaces

Table 2 shows the average F-measure and standard deviation corresponding to the outputs generated by the SVM classifier after training on feature spaces selected or extracted with the different methods explained in Section 2. The feature space was varied from $d \in \{1...9\}$ features, generating nine d-dimensional feature spaces for classification. The performance was estimated using a five-fold cross-validation strategy to obtain a measure of dispersion.

Table 2. Average F-measure of the support vector machine (SVM) classifier applied to distinct feature spaces obtained with the four methods for feature extraction/selection. IC stands for the independent component, bold numbers symbolize the highest F-measure obtained for each method, and numbers in parenthesis symbolize standard deviation.

Method	1D	2D	3D	4D	5D	6D	7D	8D	9D
PCA	40.89	68.56	69.48	71.23	71.83	71.69	71.99	71.70	71.65
	(0.34)	(0.91)	(0.95)	(0.82)	(0.92)	(0.70)	(0.81)	(0.92)	(0.91)
LDA	52.43	69.10	69.48	70.05	70.02	71.48	-	-	-
	(0.81)	(1.24)	(1.17)	(1.00)	(1.05)	(0.74)	-	-	-
ICA	8.12	25.21	53.89	58.93	62.18	63.74	68.10	70.38	71.67
	(0.40)	(0.45)	(1.16)	(1.12)	(1.16)	(1.02)	(0.91)	(0.87)	(0.90)
MCFS − 2 clusters	64.74	65.67	70.04	70.72	71.02	71.92	71.99	71.83	71.66
	(0.70)	(0.68)	(1.13)	(1.04)	(0.96)	(0.76)	(0.79)	(0.89)	(0.87)
Color channel	L*(7)	V(6)	H(4)	b*(9)	R(1)	G(2)	B(3)	S(5)	a*(8)
ECFS	40.93	68.81	69.55	71.33	71.89	71.76	71.86	71.84	71.66
	(0.32)	(1.18)	(1.2)	(0.72)	(0.72)	(0.79)	(0.83)	(0.79)	(0.87)
Color channel	G(2)	R(1)	a*(9)	b*(8)	H(4)	L*(7)	S(5)	V(6)	B(3)
ICA + ECFS	23.21	25.18	28.82	36.14	51.84	51.70	61.22	62.71	71.67
	(0.41)	(0.46)	(0.54)	(0.71)	(0.79)	(0.74)	(0.73)	(0.75)	(0.90)
IC	2	1	9	8	4	7	5	6	3
ICA + MCFS	26.61	44.68	57.24	61.13	61.81	63.07	65.14	68.57	71.67
	(0.57)	(1.07)	(1.08)	(1.09)	(1.15)	(1.26)	(1.14)	(0.84)	(0.90)
IC	3	2	4	8	9	1	6	5	7

Results showed in Table 2 evidence that was using seven features provide a level of performance that is similar either using MCFS or PCA. On the other hand, ECFS and LDA present the highest level of performance using five and six features, respectively, with a slightly lower average performance compared to PCA and MCFS. Moreover, in all cases, using more than three features allows classifiers to obtain a significantly higher performance with a lower standard deviation. In that sense, when a feature space with more color channels—or features—is employed, the SVM classifier presents a higher and more stable classification performance, at the expense of the evident increase in computational complexity. This is evident either if features are selected (e.g., MCFS, ECFS) or extracted (e.g., PCA, LDA). Different behavior is presented when ICA is employed for feature extraction, due to the strategy to find the independent components instead of the axis of maximum spread.

In the hypothetic case that only three-color channels were allowed, and these channels could be arbitrarily chosen from the nine provided by our three-color spaces, in this case, a selection method should be used. Then, a quick look at the 3D column of Table 2 evidences that the MCFS provided a better channel selection, achieving the highest performance level with a feature space composed of channels $[L*, V, H]$.

4.2. Performance across Classifiers

The comparative of performance in terms of F-measure, between the ANN, DT, SVM, and KNN classifiers, evaluated on the best d-dimensional feature space found in Section 4.1, is presented in Figure 9. The distinct feature spaces provided a different optimal number of characteristics, and those features were employed in each case. In particular, seven features were selected for PCA and MCFS, six features for LDA, and five features for ECFS. In the case of ICA, all nine features were employed to obtain the highest level of performance.

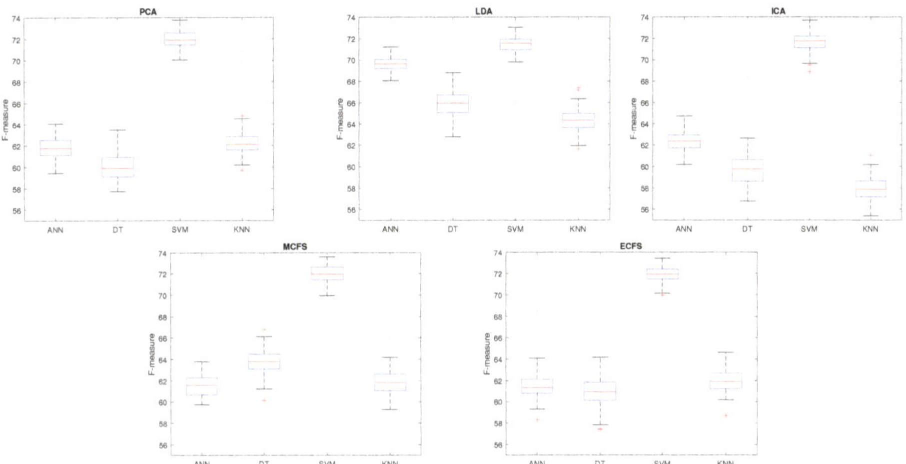

Figure 9. Boxplots representing the distribution of F-measure performance for the feature selection/extraction approaches, using four different classifiers.

As shown in Figure 9, the SVM classifier F-measure outperforms the rest of the approaches, and only ANN performance is close to SVM performance on the six-dimensional LDA feature space. The highest level of F-measure achieved by ANN is shown in the space extracted with LDA. In general, in these settings, the performance of all classifiers presents its highest level on the seven-dimensional PCA space. The settings suggest that PCA provides a feature space that facilitates the work of a classifier after combining information from multiple color spaces. On the other hand, focusing on the two feature selection methods, a similar level of performance is provided by all classifiers, without significant difference. The only case where KNN presents a lower performance is on the nine-dimensional ICA feature space. However, if a minimum number of features is required for a given application (e.g., to reduce computational complexity and cost), a feature selection method may provide the means to select a few color channels (sensors), at expenses of a reduction in performance.

5. Conclusions and Future Work

In this paper, an extension of a food packaging process was proposed for Cape gooseberry fruit sorting, according to its ripeness level. As a difference from previous works, five techniques from the multivariate analysis were employed to find the feature spaces that facilitate classification. Whereas PCA, LDA, and ICA provided mapping to a new feature space, the two selection methods (MCFS and ECFS) provided the most relevant features for classification. The configuration of the experiments provided a realistic scenario, including accommodating Cape gooseberry fruits with distinct levels of ripeness on a conveyor belt, that were captured with a VGA camera located on top. Segmentation and manual sorting were performed before feature extraction/selection and classification. Four classifiers commonly employed in literature for ripeness classification were compared, including ANN, SVM, DT, and kNN.

Results reveal that selection and extraction methods allow classifiers to reach similar levels of accuracy, but feature extraction methods require an increased computational complexity. This evidence must be considered in a final implementation, and the real-time performance of the whole system should be observed once running with the distinct algorithms on the selected computational platform.

Considering the classifiers, the SVM classifier outperformed the rest in terms of F-measure regardless of the feature space. This may respond to the organization of samples in the feature space, and the capacity of the Bayesian optimization on SVM to find a good separating hyperplane. Moreover, the four classifiers employed in the test presented the highest level of accuracy on the seven-dimensional PCA feature space. This combination of 7-D PCA feature space and the SVM classifier should be considered when a final implementation. However, the hyperparameters for this (and other classifiers) were fixed before training, and some optimization may allow finding a higher level of performance for the distinct classifiers used in experiments.

On the other hand, the lowest level of accuracy was achieved on the one-dimensional feature space, employing the ICA feature extraction technique without a feature selection method. This evidences the need for a feature selection method when ICA is employed for finding new feature spaces with independent spanning vectors. On the opposite, the highest level of accuracy for a one-dimensional feature space was obtained with the MCFS channel selection, obtaining an F-measure of 64.74 (0.07), with a single feature (L*) using the SVM classifier. This result suggests that the L* color channel from the L*a*b* perceptually uniform color space is the feature that provides the highest degree of separation between classes. The L* feature, combined with the other six features (R, G, B, b*, H, and V), allows it to obtain a performance that is similar to PCA with the same seven features.

The results obtained in the experiments suggest some paths for further research. Future work may include the use of distinct and more sophisticated algorithms for feature selection and extraction that may be explored and combined to find the best combination for a particular application. Similarly, other algorithms for classification may be tested in this configuration, such as those employing deep learning and large data sets. Additionally, optimization techniques like particle swarm optimization or evolutionary algorithms may be employed to find the best hyperparameters for the application.

On the other hand, information fusion techniques like classifier combination strategies may also enhance the establishment of the decision borderlines between classes, with the inherent performance increase. Finally, another kind of problem may benefit from feature extraction and selection techniques in food engineering, like using multi- or hyperspectral sensors to measure the level of ripeness of Cape gooseberry or any different type of fruit.

Author Contributions: Individual contributions for the authors are as follows: conceptualization, W.C., H.A.-G. and M.D.-l.-T.; methodology, J.O., M.M. and J.M.; software, M.D.-l.-T. and J.O.; validation, W.C. and H.A.-G.; formal analysis, M.D.-l.-T. and O.Z.; resources, R.L., M.M. and J.M.; data curation, H.A.-G. and M.D.-l.-T.;

writing–original draft preparation, M.D.-l.-T.; writing–review and editing, O.Z., H.A.-G. and W.C.; visualization, M.D.-l.-T. and H.A.-G.; supervision, M.M. and J.M.; project administration, H.A.-G.; funding acquisition, W.C.

Funding: This research received no external funding.

Conflicts of Interest: The authors declare no conflicts of interest.

References

1. Bader, F.; Rahimifard, S. Challenges for Industrial Robot Applications in Food Manufacturing. In Proceedings of the 2nd International Symposium on Computer Science and Intelligent Control, Stockholm, Sweden, 21–23 September 2018; ACM: New York, NY, USA, 2018; pp. 37:1–37:8.
2. Zhang, B.; Huang, W.; Li, J.; Zhao, C.; Fan, S.; Wu, J.; Liu, C. Principles, developments and applications of computer vision for external quality inspection of fruits and vegetables: A review. *Food Res. Int.* **2014**. [CrossRef]
3. Castro, W.; Oblitas, J.; De-La-Torre, M.; Cotrina, C.; Bazan, K.; Avila-George, H. Classification of Cape Gooseberry Fruit According to its Level of Ripeness Using Machine Learning Techniques and Different Color Spaces. *IEEE Access* **2019**. [CrossRef]
4. De-la Torre, M.; Avila-George, H.; Oblitas, J.; Castro, W. Selection and Fusion of Color Channels for Ripeness Classification of Cape Gooseberry Fruits. In *Trends and Applications in Software Engineering*; Mejia, J., Muñoz, M., Rocha, Á., A. Calvo-Manzano, J., Eds.; Springer International Publishing: Berlin/Heidelberg, Germany, 2019; pp. 219–233.
5. Nandi, C.; Tudu, B.; Koley, C. A machine vision-based maturity prediction system for sorting of harvested mangoes. *IEEE Trans. Instrum. Meas.* **2014**. [CrossRef]
6. Du, C.; Sun, D. Multi-classification of pizza using computer vision and support vector machine. *J. Food Eng.* **2008**. [CrossRef]
7. Taghadomi-Saberi, S.; Omid, M.; Emam-Djomeh, Z.; Faraji-Mahyari, K. Determination of cherry color parameters during ripening by artificial neural network assisted image processing technique. *J. Agric. Sci. Technol.* **2015**, *17*, 589–600.
8. Abdulhamid, U.; Aminu, M.; Daniel, S. Detection of Soya Beans Ripeness Using Image Processing Techniques and Artificial Neural Network. *Asian J. Phys. Chem. Sci.* **2018**. [CrossRef]
9. Hadfi, I.; Yusoh, Z. Banana ripeness detection and servings recommendation system using artificial intelligence techniques. *J. Telecommun. Electron. Comput. Eng.* **2018**, *10*, 83–87.
10. Schwarz, M.; Cowan, W.; Beatty, J. An Experimental Comparison of RGB, YIQ, LAB, HSV, and Opponent Color Models. *ACM Trans. Graph. (TOG)* **1987**. [CrossRef]
11. Bora, D.; Gupta, A.; Khan, F. Comparing the Performance of L*A*B* and HSV Color Spaces with Respect to Color Image Segmentation. *Int. J. Emerg. Technol. Adv. Eng.* **2015**, *5*, 192–203.
12. Zou, X.; Zhao , J.; Li, Y. Apple color grading based on organization feature parameters. *Pattern Recognit. Lett.* **2007**, *28*, 2046–2053. [CrossRef]
13. Cárdenas-Pérez, S.; Chanona-Pérez, J.; Méndez-Méndez, J.; Calderón-Domínguez, G.; López-Santiago, R.; Perea-Flores, M.; Arzate-Vázquez, I. Evaluation of the ripening stages of apple (Golden Delicious) by means of computer vision system. *Biosyst. Eng.* **2017**, *159*, 46–58. [CrossRef]
14. Guerrero, E.; Benavides, G. Automated system for classifying Hass avocados based on image processing techniques. In Proceedings of the 2014 IEEE Colombian Conference on Communications and Computing, COLCOM 2014—Conference Proceedings, Bogota, Colombia, 4–6 June 2014. [CrossRef]
15. Mendoza, F.; Aguilera, J. Application of Image Analysis for Classification of Ripening Bananas. *J. Food Sci.* **2004**, *69*, E471–E477. [CrossRef]
16. Paulraj, M.; Hema, C.; Sofiah, S.; Radzi, M. Color recognition algorithm using a neural network model in determining the ripeness of a banana. In Proceedings of the International Conference on Man-Machine Systems, Penang, Malaysia, 26–27 August 2009; Universiti Malaysia Perlis: Perlis, Malaysia, 2009; pp. 2B71–2B74.
17. Li, H.; Lee, W.; Wang, K. Identifying blueberry fruit of different growth stages using natural outdoor color images. *Comput. Electron. Agric.* **2014**, *106*, 91–101. [CrossRef]
18. Pourdarbani, R.; Ghassemzadeh, H.; Seyedarabi, H.; Nahandi, F.; Vahed, M. Study on an automatic sorting system for Date fruits. *J. Saudi Soc. Agric. Sci.* **2015**, *14*, 83–90. [CrossRef]

19. Damiri, D.; Slamet, C. Application of Image Processing and Artificial Neural Networks to Identify Ripeness and Maturity of the Lime (citrus medica). *Int. J. Basic Appl. Sci.* **2012**, *1*, 175–179. [CrossRef]
20. Vélez-Rivera, N.; Blasco, J.; Chanona-Pérez, J.; Calderón-Domínguez, G.; de Jesús Perea-Flores, M.; Arzate-Vázquez, I.; Cubero, S.; Farrera-Rebollo, R. Computer Vision System Applied to Classification of "Manila" Mangoes During Ripening Process. *Food Bioprocess Technol.* **2014**, *7*, 1183–1194. [CrossRef]
21. Zheng, H.; Lu, H. A least-squares support vector machine (LS-SVM) based on fractal analysis and CIELab parameters for the detection of browning degree on mango (*Mangifera indica* L.). *Comput. Electron. Agric.* **2012**, *83*, 47–51. [CrossRef]
22. Fadilah, N.; Mohamad-Saleh, J.; Halim, Z.; Ibrahim, H.; Ali, S. Intelligent color vision system for ripeness classification of oil palm fresh fruit bunch. *Sensors* **2012**, *12*, 14179–14195. [CrossRef]
23. Elhariri, E.; El-Bendary, N.; Hussein, A.; Hassanien, A.; Badr, A. Bell pepper ripeness classification based on support vector machine. In Proceedings of the 2nd International Conference on Engineering and Technology, Cairo, Egypt, 19–20 April 2014; pp. 1–6. [CrossRef]
24. Mohammadi, V.; Kheiralipour, K.; Ghasemi-Varnamkhasti, M. Detecting maturity of persimmon fruit based on image processing technique. *Sci. Hortic.* **2015**, *184*, 123–128. [CrossRef]
25. El-Bendary, N.; El Hariri, E.; Hassanien, A.; Badr, A. Using machine learning techniques for evaluating tomato ripeness. *Expert Syst. Appl.* **2015**, *42*, 1892–1905. [CrossRef]
26. Goel, N.; Sehgal, P. Fuzzy classification of pre-harvest tomatoes for ripeness estimation–An approach based on automatic rule learning using decision tree. *Appl. Soft Comput.* **2015**, *36*, 45–56. [CrossRef]
27. Polder, G.; Van der Heijden, G. Measuring ripening of tomatoes using imaging spectrometry. In *Hyperspectral Imaging for Food Quality Analysis and Control*; Elsevier: Amsterdam, The Netherlands, 2010; pp. 369–402. [CrossRef]
28. Rafiq, A.; Makroo, H.; Hazarika, M. Neural Network-Based Image Analysis for Evaluation of Quality Attributes of Agricultural Produce. *J. Food Process. Preserv.* **2016**, *40*, 1010–1019. [CrossRef]
29. Shah Rizam, M.S.; Farah Yasmin, A.R.; Ahmad Ihsan, M.Y.; Shazana, K. Non-destructive watermelon ripeness determination using image processing and artificial neural network (ANN). *Int. J. Electr. Comput. Eng.* **2009**, *3*, 332–336.
30. Abdullah, N.; Madzhi, N.; Yahya, A.; Rahim, A.; Rosli, A. ANN Diagnostic System for Various Grades of Yellow Flesh Watermelon based on the Visible light and NIR properties. In Proceedings of the 2018 4th International Conference on Electrical, Electronics and System Engineering (ICEESE), Kuala Lumpur, Malaysia, 8–9 November 2018; pp. 70–75. [CrossRef]
31. Syazwan, N.; Rizam, M.; Nooritawati, M. Categorization of watermelon maturity level based on rind features. *Procedia Eng.* **2012**, *41*, 1398–1404. [CrossRef]
32. Skolik, P.; Morais, C.; Martin, F.; McAinsh, M. Determination of developmental and ripening stages of whole tomato fruit using portable infrared spectroscopy and Chemometrics. *BMC Plant Biol.* **2019**, *19*, 236. [CrossRef] [PubMed]
33. Du, D.; Wang, J.; Wang, B.; Zhu, L.; Hong, X. Ripeness Prediction of Postharvest Kiwifruit Using a MOS E-Nose Combined with Chemometrics. *Sensors* **2019**, *19*, 419. [CrossRef]
34. Ramos, P.; Avendaño, J.; Prieto, F. Measurement of the ripening rate on coffee branches by using 3d images in outdoor environments. *Comput. Ind.* **2018**, *99*, 83–95. [CrossRef]
35. Hyvärinen, A.; Erkki, O. Independent component analysis: Algorithms and applications. *Neural Netw.* **2000**, *13*, 411–430. [CrossRef]
36. Roffo, G.; Melzi, S. Ranking to learn. In Proceedings of the International Workshop on New Frontiers in Mining Complex Patterns, Riva del Garda, Italy, 19 September 2016; Springer: Cham, Switzerland, 2016; pp. 19–35. [CrossRef]
37. Cai, D.; Zhang, C.; He, X. Unsupervised feature selection for multi-cluster data. In Proceedings of the 16th ACM SIGKDD International Conference on Knowledge Discovery and Data Mining, Washington, DC, USA, 25–28 July 2010; ACM: New York, NY, USA, 2010; pp. 333–342. [CrossRef]
38. Duda, R.; Hart, P.; Stork, D. *Pattern Classification*; Wiley & Sons: New York, NY, USA, 2001.

39. Pedreschi, F.; Leon, J.; Mery, D.; Moyano, P. Development of a computer vision system to measure the color of potato chips. *Food Res. Int.* **2006**, *39*, 1092–1098. [CrossRef]
40. Fischer, G.; Miranda, D.; Piedrahita, W.; Romero, J. *Avances en Cultivo, Poscosecha y Exportación de la Uchuva Physalis peruviana L.*; Universidad Nacional de Colombia: Bogotá, Colombia, 2005.

Article

Recovery of Protein from Dairy Milk Waste Product Using Alcohol-Salt Liquid Biphasic Flotation

Pei En Tham [1], Yan Jer Ng [1], Revathy Sankaran [2], Kuan Shiong Khoo [1], Kit Wayne Chew [3,*], Yee Jiun Yap [3], Masnindah Malahubban [4], Fitri Abdul Aziz Zakry [4] and Pau Loke Show [1,*]

1 Department of Chemical and Environmental Engineering, Faculty of Science and Engineering, University of Nottingham Malaysia, Jalan Broga, Semenyih 43500, Selangor Darul Ehsan, Malaysia; peien0405@gmail.com (P.E.T.); yanjer98@hotmail.com (Y.J.N.); kuanshiong.khoo@hotmail.com (K.S.K.)
2 Institute of Biological Sciences, Faculty of Science, University of Malaya, Jalan Universiti, Kuala Lumpur 50603, Malaysia; revathysankaran@ymail.com
3 School of Mathematical Sciences, Faculty of Science and Engineering, University of Nottingham Malaysia, Jalan Broga, Semenyih 43500, Selangor Darul Ehsan, Malaysia; yeejiun.yap@nottingham.edu.my
4 Faculty of Agriculture and Food Sciences, Universiti Putra Malaysia Sarawak Campus, Bintulu 97008, Sarawak, Malaysia; masnindah@upm.edu.my (M.M.); zakryfitri@upm.edu.my (F.A.A.Z.)
* Correspondence: KitWayne.Chew@nottingham.edu.my or kitwayne.chew@gmail.com (K.W.C.); PauLoke.Show@nottingham.edu.my or showpauloke@gmail.com (P.L.S.); Tel.: +603-8924-8605 (P.L.S.)

Received: 25 October 2019; Accepted: 19 November 2019; Published: 21 November 2019

Abstract: Expired dairy products are often disposed of due to the potential health hazard they pose to living organisms. Lack of methods to recover valuable components from them are also a reason for manufactures to dispose of the expired dairy products. Milk encompasses several different components with their own functional properties that can be applied in production of food and non-food technical products. This study aims to investigate the novel approach of using liquid biphasic flotation (LBF) method for protein extraction from expired milk products and obtaining the optimal operating conditions for protein extraction. The optimized conditions were found at 80% concentration ethanol as top phase, 150 g/L dipotassium hydrogen phosphate along with 10% (*w*/*v*) milk as bottom phase, and a flotation time of 7.5 min. The protein recovery yield and separation efficiency after optimization were 94.97% and 86.289%, respectively. The experiment has been scaled up by 40 times to ensure it can be commercialized, and the protein recovery yield and separation efficiency were found to be 78.92% and 85.62%, respectively. This novel approach gives a chance for expired milk products to be changed from waste to raw materials which is beneficial for the environment and the economy.

Keywords: milk; protein; liquid biphasic flotation; dairy waste; recovery

---

1. Introduction

A large quantity of dairy waste is produced per annum in every country. Taking UK as an example, a total of 330,000 tons of milk waste is produced annually with approximately 90% of the total waste produced from homes. This is equivalent to 490 million pints nationwide or 18.5 pints per household. Milk should be kept at the right temperature to prevent it from spoiling before the expiry date. However, the typical household UK fridge operates at a temperature that is 2 °C warmer than the recommended storage temperature of milk, which is between 0 and 5 °C [1]. This amount of milk waste creates an environmental problem as it creates greenhouse gas emissions equivalent to approximately 20,000 cars annually [2].

Milk contains approximately 87.4% of water and 12.6% of milk solids. Fats makes up 3.7% of the 12.6% milk solids while the remaining 8.9% is 3.4% of proteins, 4.8% lactose, and 0.7% minerals.

Of the proteins in milk, 80% is Casein and the remaining 20% is Whey protein [3]. Casein is chiefly phosphate-conjugated and mainly consists of calcium phosphate-micelle complex. Whey protein is a collection of a globular proteins with a high level of α-helix structure and the acidic-basic and hydrophobic-hydrophilic amino acids are distributed in a fairly balanced form. Whey proteins have substantial levels of secondary, tertiary, and quaternary structure. They are heat-labile stabilizing their protein structure through intermolecular disulfide linkage [4]. The proteins in milk are considered to be complete as they contain all types of essential amino acids in amounts that match the amino acid requirements. They are used as a standard reference for proteins to compare with other food proteins due to their high quality. Branched-chain amino acids contents such as valine, isoleucine, and leucine in milk are also higher than many other foods [4].

Since the conventional technique for extracting bioactive compounds need longer extraction time yet cost-consuming with complex scale-up, the liquid biphasic flotation (LBF) method was proposed [5]. LBF system is an integration of the adsorptive bubbles floatation system, where the biphasic system is supported with air bubbling to transport the biomolecules from one phase to another. The surface-active compound of biomolecules present will be absorbed onto the surface of ascending gas bubble and be brought from the bottom phase to the top organic phase [6]. LBF is formed by combining an immiscible polymer and a salt solution. Addition of salt to water will cause segregation of ions into their preferred water structuring [7]. Aqueous biphasic systems will occur when certain solutes cause an aqueous solution to fully separate into two aqueous phases. The basic aqueous two-phase system (ATPS) phasing strategy is based on the separation of proteins into one phase with the contaminants being present in the other phase. The smaller biomolecules will be present mostly at the bottom phase which also can be known as the salt-rich phase. Whereas proteins will be brought up to the top phase [8]. Polarity is believed to play a role in the separation; molecules with lower polarity will be partitioned to the top phase, while molecules with higher polarity will be partitioned to the bottom phase. This aqueous liquid–liquid two-phase system is more widely used in the extractive separation of labile biomolecules such as proteins. This system operates under mild conditions due to the low interfacial tension between the two phases, achieving small droplet size, large interfacial areas, and efficient mixing under very gentle stirring and rapid partition.

LBF is a well-known method for the separation, concentration, and purification of biological material, particularly for protein, enzyme, and DNA [9]. Extraction using LBF is based on the separation of biomolecules between the two aqueous phases [10]. Much work has been done by using LBF to exploit and study the behavior of the aqueous rich phase and driving forces which will affect the partitioning of biomolecules in the separation process. These systems were based on aqueous mixtures of two incompatible polymers, such as polyethylene glycol (PEG), dextran, and/or maltodextrin [11]. Since then, many immiscible aqueous systems were found by using hydrophilic solutions. However, other types of LBF, with components of different phase, had been focused on to increase the mass transfer rates and the selectivity of certain biomolecules. Ionic liquids [12], inorganic salts, and carbohydrates are three examples of solutes used in LBF [13]. These molecules were applied in the separation or purification of a wide range of compounds, including proteins, enzymes, antibiotics, organic acids, and many other bio- or synthetic molecules [13].

LBF is a very promising method, and it indicates a great potential for a wider usage in partitioning, concentrating, and purification of labile biology products from natural sources or fermentation broths, as well as in enzyme technology during industrial or laboratory production of enzymes. LBF is an integration of the principles of ATPS but with additional bubbling action to enhance the separation of biomolecules. This integration will utilize the adsorptive gas bubble separation technique in which the biomolecules with surface-active sites in the bottom aqueous phase are selectively adsorbed onto the surface of the ascending gas bubbles which are then collected in the immiscible top aqueous phase. With this, water soluble biomolecules can be separated from their crude aqueous extracts [9].

A detailed study was made with aims to obtain optimal operating conditions for the extraction of protein from expired milk using alcohol/salt LBF. The effect of milk concentration, type of salt, type of alcohol, concentration of salt solution, concentration of alcohol solution, pH, flotation time along with a scaled-up LBF system were studied. Up to current date, no study has been made on recovery of protein from expired milk using alcohol/salt LBF. Partitioning of the milk protein into the alcohol phase through LBF using low-cost and recyclable phase forming components would lead to a cost-efficient protein recovery process. Additionally, alcohol-salt LBF has a low viscosity, easier constituent for recovery and short settling time. As such, this approach would enable the production of milk protein to be economically feasible and sustainable. This study has led to a novel discovery of liquid biphasic flotation application for protein extraction from milk waste with economic processes that will be beneficial at the industrial scale.

2. Materials and Methods

2.1. Materials

Food grade alcohols of ethanol, 1-propanol, 2-propanol (R&M Chemicals, Selangor, Malaysia) were used as the extraction solvents of proteins. Salts for the bottom phase that are utilized in this study were ammonium sulphate [$(NH_4)_2SO_4$], di-potassium hydrogen phosphate (K_2HPO_4), sodium sulphate (Na_2SO_4), di-sodium hydrogen phosphate (Na_2HPO_4) and magnesium sulphate ($MgSO_4$) purchased from R&M Chemicals (Selangor, Malaysia). Bradford reagent was used to quantify protein concentration in the two solutions (top phase and bottom phase) after the flotation. Bradford reagent is also purchased from R&M Chemicals (Selangor, Malaysia).

2.2. Apparatus

Liquid biphasic flotation unit of 50 mL volume capacity was used as the separation system, and it was obtained from Donewell Resources (Puchong, Selangor, Malaysia). A 50 mL glass tube was connected from the bottom to a gas compressor. The bottom of the glass tube was drilled and fitted with a rubber tube to be connected to the gas compressor. A sintered glass disk (Grade 4 (G4) porosity) was fitted at the bottom of the glass tube so that air bubbles will be produced when compressed air is passed through it. The flowrate of air supplied to the LBF system is controlled by using a flowmeter (model: RMA-26-SSV) with a range of 50 to 200 cc/min (Dwyer, Michigan, IN, USA). The air compressor is powered by plugging it into a wall socket. Figure 1 shows the schematic diagram of the LBF system.

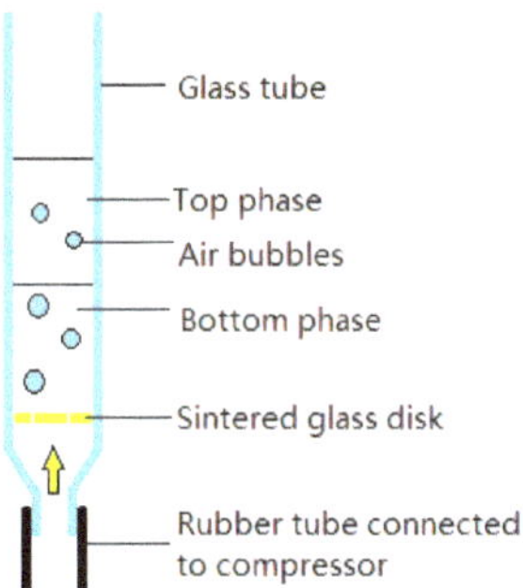

Figure 1. Figure illustrating schematic diagram of LBF system.

2.3. Preparation of Milk Samples

Expired milk was supplied by local producers (Dutch Lady) and was stored under room temperature. The concentration of proteins in the milk was tested to be 13.64 mg/mL. The milk was stored into a refrigerator at a temperature of 2 °C to reduce the effects of bacteria activity.

2.4. Protein Assay

The protein is determined by using Bradford Reagent. The dilution has been prepared with 10× dilution with adding 2 mL reagent with 0.2 mL of either top or bottom solution and was incubated for 10 min before the reading was tested using UV-Vis spectrophotometer at a reading wavelength of 595 nm. The absorbance of the protein concentration was based on the calibration between BSA concentration and OD_{595}.

2.5. Protein Extraction Using LBF

A mixture of salt solution and expired milk was mixed and top up to 15 mL to be used as the bottom phase of the experiment. 15 mL of pure alcohol was used as the top phase of the experiment. After pouring the two solutions into the LBF system, the mixture was allowed to settle for 30 s so that two layers of liquid can be formed inside the system. The flowmeter was set to 25 cc/min to allow compressed air to flow into the system. Air was passed through the system for 10 min before closing the flowmeter and allowing the system to settle for 5 min again. The top and bottom layer was pipetted out from the glass tube and tested for their respective protein concentrations.

2.6. Optimization of LBF Operating Parameters

The operating parameters of LBF such as type of salt/alcohol, concentration of salt/alcohol/milk, pH of the bottom phase, and the flotation time were investigated by one factor at a time (OFAT) approach to maximize protein extraction and recovery. The optimization started with the initial operating conditions which is 100% of 2-propanol, 20 g/L salt solution, 15% (*w*/*v*) milk solution, flotation time of 10 min and the initial pH of the solution. The initial volume for both top phase and bottom phase was kept at 15 mL each, and the experiment was carried out at room temperature. Table 1 shows the parameters and variables tested for this experiment.

Table 1. Parameters and variables.

No.	Condition	Initial Setting	Variables	Unit	Justification
1.	Type of salt	-	Ammonium sulphate, Magnesium sulphate, Sodium sulphate, Dipotassium hydrogen phosphate, Disodium hydrogen phosphate	-	The salt that would result in the best yield was chosen from the five salts used
2.	Type of alcohol	2-Propanol	Ethanol, 1-Propanol, 2-Propanol	-	After the selection of salt was completed, the type of alcohol that would result in the best yield was determined
3.	Concentration of salt	20	150, 200, 250, 300, 350	g/L	The percentage of salt was set according the Separation and Purification Technology Recovery of lipase derived from Burkholderia cenocepacia ST8 using sustainable aqueous two-phase flotation composed of recycling hydrophilic organic solvent and inorganic salt.
4.	Concentration of alcohol	100	60, 70, 80, 90, 100	%	A total of 15 mL of top phase solution is added into the system and the concentration of alcohol adjusted by using deionized water.
5.	Concentration of milk	15	5, 10, 15, 20, 25	% (*w*/*v*)	After the few parameters are stable, alteration of milk concentration begins.
6.	pH	9.15	6.5, 7, 7.5, 8, 9.15	-	Condition of milk has been altered to set the right pH for the whole system.
7.	Operation time	10	5, 7.5, 10, 12.5, 15	min	Initial setting without alcohol; after 10-min alcohol is added because no two-phase forming.

2.7. Calculations of Recovery Yield and Separation Efficiency

Recovery yield (R) of proteins in the hydrophilic organic solvent phase was measured using the following equation:

$$R = \frac{C_T V_T}{M} \times 100\%$$

where,

R is the recovery yield
C_T is the concentration of proteins in the hydrophilic organic solvent phase
V_T is the volume of the hydrophilic organic solvent phase
M is the total mass of proteins in the initial milk used

The separation efficiency from the milk to the top phase after LBF is calculated by the following equation:

$$E = \frac{C_T V_T}{C_T V_T + C_B V_B} \times 100\%$$

where,

E is the separation efficiency
C_T is the concentration of proteins in the hydrophilic organic solvent phase
V_T is the volume of the hydrophilic organic solvent phase
C_B is the concentration of proteins at the bottom phase
V_B is the volume of the bottom phase

3. Results and Discussion

3.1. Effects of Different Types of Inorganic Salt in Protein Recovery Using LBF

The type of salts used for LBF is a key for protein extraction in this system as different salts induce different interactions with the protein, causing the separation efficiency of the proteins to alter. This is because salt solutions at the bottom phase are responsible of manipulating the surface tension of water thus changing the hydrophobic interactions between proteins and water at bottom phase [14]. As a result, when the protein solubility is reduced, proteins will start to migrate to the top phase when aided by flotation. It is reported that the Gibbs free energy of hydration of salt was the key to the formation of a biphasic system between salt and alcohol solution [15].

This experiment was conducted by firstly determining the most suitable type of salt to be used to obtain the best results and then changing the type of alcohol used. The types of salts used include ammonium sulphate, dipotassium hydrogen phosphate, disodium hydrogen phosphate, sodium sulphate, and magnesium sulphate. The results are illustrated in Figure 2. Dipotassium hydrogen phosphate and disodium hydrogen phosphate show a higher efficiency followed by ammonium sulphate and sodium sulphate in the descending order of 85.18%, 68.99%, 58.85%, and 53.81%, respectively. Magnesium sulphate showed the lowest separation efficiency which is 50.10%. For disodium hydrogen phosphate, the bottom phase forms salt crystals after flotation has been completed. This is due to the volume of the bottom phase being reduced; thus, the salt is unable to be fully dissolved in the remaining solution. Higher maintenance cost of the system is required if disodium hydrogen phosphate were to be used as a medium to create the biphasic conditions of this system.

As for the protein recovery yield, dipotassium hydrogen phosphate exhibits a higher protein recovery yield than all other salts, with a recovery yield of 29.997%. The lowest recovery yield of all salt tested was ammonium sulphate, which has a recovery yield of 7.74%. The other salts, which are disodium hydrogen phosphate, magnesium sulphate, and sodium sulphate, each has a recovery yield of 20.33%, 16.04%, and 10.81% respectively. The recovery yield for dipotassium hydrogen phosphate is significantly higher than that of disodium hydrogen phosphate. Taking into consideration the fact that when disodium hydrogen phosphate is used, the bottom phase will form salt crystals after flotation, and the two results obtained, dipotassium hydrogen phosphate was chosen to be the inorganic salt used in the following tests.

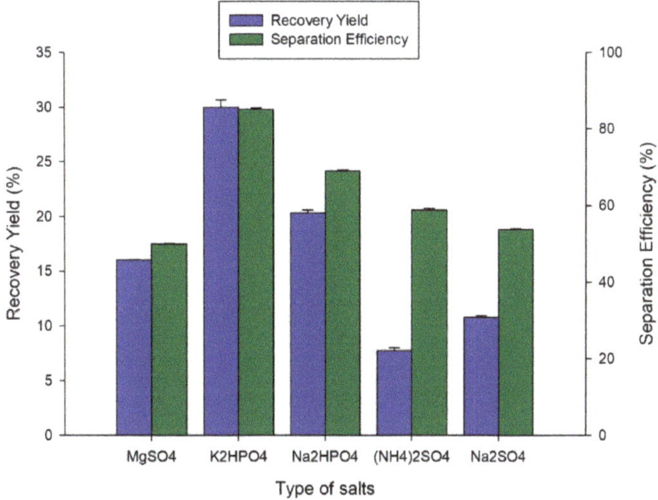

Figure 2. Figure showing the effect of different types of salts on the protein recovery yield and separation efficiency using LBF system.

3.2. *Effect of Different Types of Alcohols for Protein Recovery*

The type of alcohol used for LBF plays an important role in the system as different types of alcohol have different levels of interactions with the proteins, which will determine how much protein can the system extract. The type of alcohol used also affects the formation of biphasic system with salt solution. Many proteins are found to be non-compatible with the alcohol-rich top phase in the LBF process [16]. Some alcohols such as methanol will form triphasic solutions rather than biphasic solutions when mixed with salt solution. The selection of the type of alcohol used is very crucial as it will affect the overall performance of the system. In this study, ethanol, 1-propanol and 2-propanol of 100% were selected to form a biphasic system with dipotassium hydrogen phosphate at a concentration of 20 g/L.

All three alcohols were found to successfully form a biphasic system with the dipotassium hydrogen phosphate solution and the protein recovery yield for ethanol, 1-propanol and 2-propanol found to be 42.00%, 13.23%, and 23.35%, respectively. Based on Figure 3, ethanol has outperformed both 1-propanol and 2-propanol in terms of recovery yield, being almost two times the yield of 2-propanol and almost three times the yield of 1-propanol. In terms of separation efficiency, ethanol also outperforms both 1-propanol and 2-propanol. The separation efficiency of ethanol, 1-propanol, and 2-propanol is found to be 92.20%, 87.54%, and 90.56%, respectively. Alcohols usually contain a carbon chain and a functional group (-OH); the difference between ethanol and the other two alcohols is that ethanol has a shorter carbon chain, resulting in it having more ethanol molecules at the same volume. This can be proven by dividing the density of the respective alcohol with its molar mass. For example, the density of ethanol is 0.789 g/cm^3 at 20 °C [17] and ethanol has a molar mass of 46.07 g/mol [18], this will result in ethanol a volume of 0.01715 mol/cm^3. This is higher than 1-propanol and 2-propanol as they have volumes of 0.01336 mol/cm^3 and 0.01306 mol/cm^{3,} respectively. The density of 1-propanol and 2-propanol is 0.803 g/cm^3 [19] and 0.785 g/cm^3 [20], respectively at 20 °C while their molar mass is 60.096 g/mol [21] and 60.1 g/mol [22], respectively. This difference will result in the protein molecules being able to interact more with the alcohol molecules and not settle back to the bottom phase after flotation. Additionally, the high recovery yield and separation efficiency in ethanol could be because of high polarity of the alcohol compared to the other two alcohols used. High hydroxyl group in ethanol could allow more protein to be accumulated at the top phase thus, giving high recovery yield [23]. Due to the above reasons, ethanol is chosen to carry out the following tests.

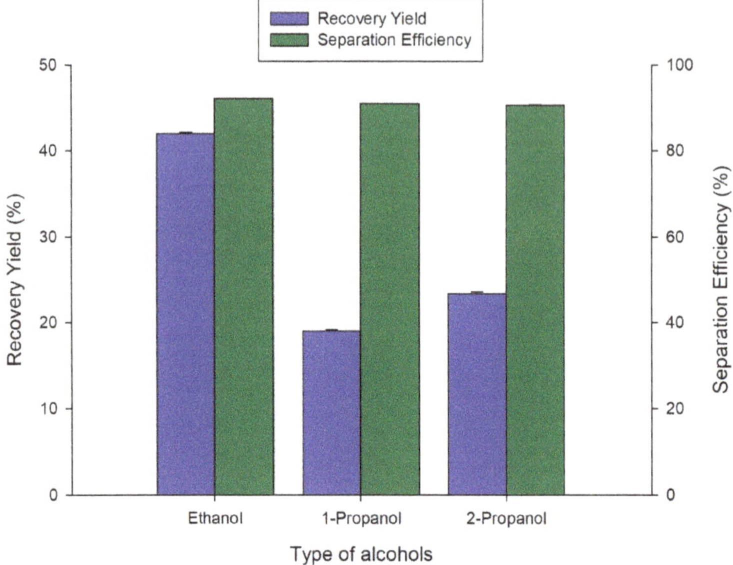

Figure 3. Effect of various types of alcohols on protein recovery yield and separation efficiency using LBF system.

3.3. *Effect of Different Concentration of K_2HPO_4 Salt on the Recovery of Proteins*

The concentration of salt used for the bottom phase is also optimized in this study. Varying salt concentrations have been used in the separation of proteins. When dipotassium hydrogen phosphate concentration is increased from 150 g/L to 350 g/L, the volume of top phase showed a decreasing trend while the volume of bottom phase showed an increasing trend. More proteins are retained in the lower phase.

From Figure 4, the highest protein recovery yield is exhibited by a salt concentration of 150 g/L, with a yield of 46.83%, while the lowest yield was obtained from a salt concentration of 350 g/L, with a yield of 37.14%. At salt concentration of 200 g/L, 250 g/L, and 300 g/L, the recovery yield of proteins is 45.01%, 39.36%, and 44.96% respectively. When increasing the salt concentration, the yield shows a decreasing trend due to more proteins being denatured when exposed to higher concentrations of salts. Thus, the lowest concentration of salt to form a biphasic solution should be obtained [24]. As for the separation efficiency when salt concentration is altered, the separation efficiency shows a decreasing trend when the salt concentration is increased. This is also due to the proteins in the milk being denatured by the salts when the solution is mixed together. Due to the above reasons, a salt concentration of 150 g/L was used for the following tests.

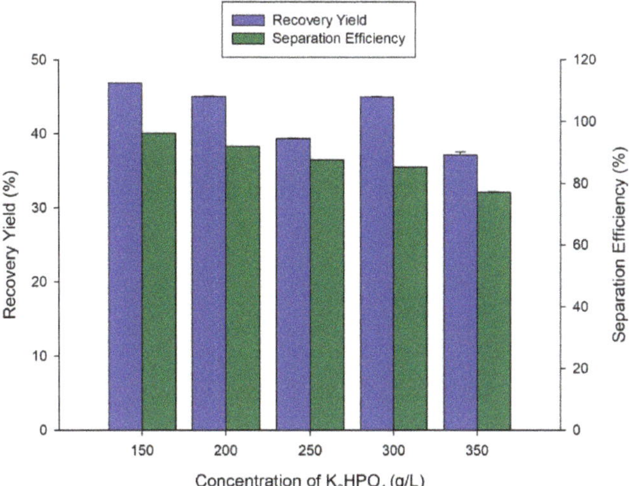

Figure 4. Figure illustrating effect of various K_2HPO_4 concentrations on protein recovery yield and separation efficiency using LBF system.

3.4. *Effect of Different Concentrations of Ethanol on the Recovery of Proteins*

The concentration of alcohol used will also affect the overall performance of the LBF system. Thus, the next parameter to be optimized is the concentration of alcohol. Various concentrations from 60% to 100% of ethanol were tested by using 15% (g/L) dipotassium hydrogen phosphate. As shown in Figure 5, 80% (*W*/*V*) shows the best recovery yield of 77.30%, followed by 70% of ethanol concentration which shows a yield of 54.50%. Concentrations of 60%, 90%, and 100% show 45.88%, 46.30%, and 41.48% yield, respectively. The protein recovery yield increases when the ethanol concentration increases, however when the alcohol concentration exceeds a certain point, the protein recovery yield starts to reduce. This is because the formation of the biphasic layers is weak when the concentration of alcohol is low. This result follows the trend in the previous study on protein recovery of wet microalgae using LBF where the highest protein recovery was obtained at 60% of 1-propanol, and the recovery yield decreased when the concentration of alcohol reduced below 40% [25]. These phenomena are due to the concentration of alcohol decreasing; more hydrophilic proteins can be dissolved into the top phase when the proteins are brought up by flotation air bubbles. However, when the concentration of alcohol decreases, the water at the top phase tends to migrate down to the bottom phase of the LBF system [16].

In terms of separation efficiency, ethanol solution with 80% concentration also has the highest separation efficiency, being at a value of 93.82%. The separation efficiency is followed by a pure ethanol solution which has a value of 90.59%. Values for 60%, 70%, and 90% are 69.16%, 54.17%, and 84.60%, respectively. The high value obtained at 80% of ethanol could be contributed due to cluster formation of ethanol. The number and size of the clusters strongly depend on the number of hydrogen bonds, and at higher concentration of ethanol, the cluster size is higher, which contributed to higher recovery of protein. Generally, ethanol has maximum viscosity of 75% to 80%, thus, this supports the high recovery of protein and separation efficiency. This indicates that at 80% of ethanol concentration, more proteins can be separated from the bottom phase which is the main point of this study. Therefore, 80% concentration ethanol solutions will be used in the following optimizations.

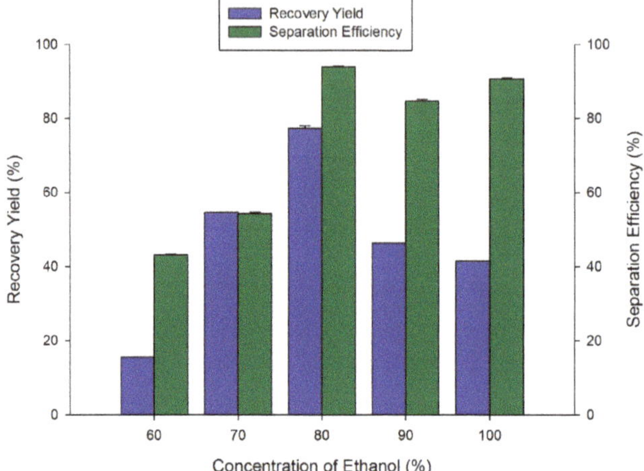

Figure 5. Figure showing effect of various concentrations of ethanol on protein recovery yield and separation efficiency using LBF system.

3.5. *Effect of Various Concentrations of Milk*

In this section, the effect of milk concentration of the bottom phase of the LBF system was studied. The concentration of milk may pose a potential effect on protein extraction by affecting the formation of the biphasic system, indicating that the concentration of milk used for extraction will have an impact on the yield of proteins recovered [26]. Milk with concentrations of 5% (*w*/*v*), 10% (*w*/*v*), 15% (*w*/*v*), 20% (*w*/*v*), and 25% (*w*/*v*) mixed along with dipotassium hydrogen phosphate solution with 150 g/L concentration were tested to investigate the effects of milk concentration on protein recovery. The highest protein yield obtained is at 10% (*w*/*v*), closely followed by 5% (*w*/*v*), which has the values of 93.96% and 93.75%, respectively. At 25% (*w*/*v*), the recovery yield of proteins is the lowest, having a value of 50.98%. At 15% (*w*/*v*) and 20% (*w*/*v*), it has values of 89.77% and 64.74%, respectively. According to Figure 6, it was observed that the concentration of milk has increased from 5% (*w*/*v*) to 10% (*w*/*v*), the recovery yield has only increased slightly; however, when the concentration is further increased, the yield starts to drop significantly, especially from 15% (*w*/*v*) to 20% (*w*/*v*), with a total drop of more than 25% protein yield. This is due to the fact that when a high concentration of milk is used, the salt solution mixture tends to form a liquid of which its high viscosity will result in the formation of flotation bubbles to be too difficult to control. By increasing the concentration of milk used, the performance of the LBF would be reduced as the level of impurities within the solution will also increase. The overall composition of the bottom phase will be altered as there is a lot of impurities in the milk [16,27]. In terms of separation efficiency, however, the highest value obtained is 94.02%, which is achieved by 25% (*w*/*v*). The separation efficiency gradually increases as the concentration is increased, starting from 5% (*w*/*v*), the separation efficiency was found to be 72.62%, 87.14%, 90.96%, and 92.59%, respectively. Due to the higher separation efficiency, 10% (*w*/*v*) of milk concentration was selected to carry on the following tests.

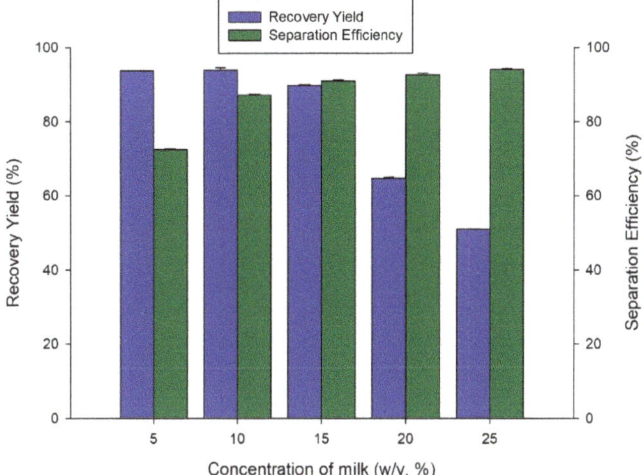

Figure 6. The effect of different concentrations of milk on the protein recovery yield and separation efficiency utilizing LBF system.

3.6. Effect of pH on the Recovery of Proteins

Impact of pH partitioning of proteins and enzymes to the phases in the LBF system depends on their isoelectric points. The pH of the system, however, affects the charge of target protein molecules and ionic composition, as well as introduces differential partitioning into the two phases. Most of the biomolecules, especially proteins and enzymes, are stable at neutral pH, which is a favorable condition to conduct the LBF partitioning. However, an increase in pH of the LBF from 7.0 to 9.0 reduced the protein recovery and activity recovered. Enzyme stability is slightly reduced in the acidic area except at the lowest pH and was dramatically lost at pH above 9.0. This dependence on pH for optimal protein recovery can be explained in terms of the charge in the protein. The protein in the LBF is predominantly casein. From the literature, isoelectric point of casein is 4.6, and since the pH of milk is 6.6, casein molecules are positively charged due to the protons provided by the milk medium. Given that the formula for pH is $pH = -log[H^+]$, where H^+ is concentration of hydrogen ions, a pH of 7 to 8 in the system (see Figure 7 corresponds to a negatively charged medium. With a positive charge, the casein molecules are thus hydrophobic, making them less soluble in water. Given that bubbles are used to push the particles up to the top phase and that the casein molecules are positively charged, the surface charge on the bubbles plays a vital role in the protein extraction efficiency. In particular, the charge of the casein and the surface charge of the bubble will be responsible for the adsorption of the protein molecule to the bubble surface. In order for the protein molecules to be attached to the bubble surface, the charge of the bubble surface must therefore be negative.

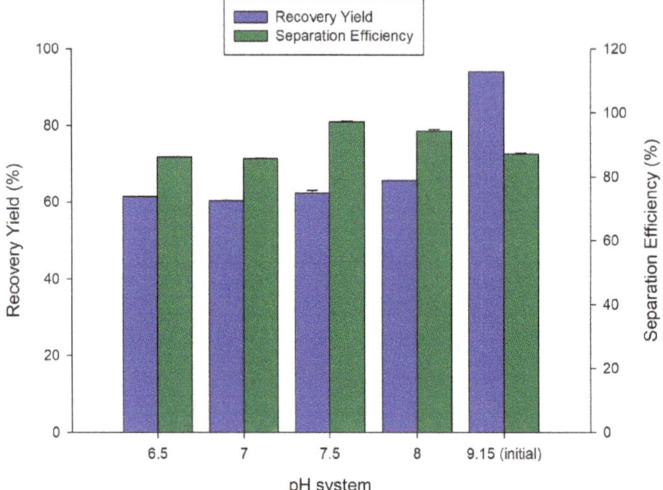

Figure 7. Figure showing the effect of pH system on protein recovery yield and separation efficiency.

To understand the mechanics behind the adsorption of the bubbles, we have developed a mathematical model with the use of partial differential equations (PDEs).

The region of interaction in our model is constructed using the same approach as the Lamm equations (Lamm O., 1929), which is to divide the volume of the container into sector-shaped cells. With reference to Figure 8 consider a region R in a sector-shaped cell within the chamber of the flotation system. Let M, M_{in}, and M_{out} be the mass of solute inside R, the mass flow into R, and the mass flow out of R, respectively. By the principle of conservation of mass,

$$\frac{\partial M}{\partial t} = \frac{\partial M_{in}}{\partial t} - \frac{\partial M_{out}}{\partial t} \quad (1)$$

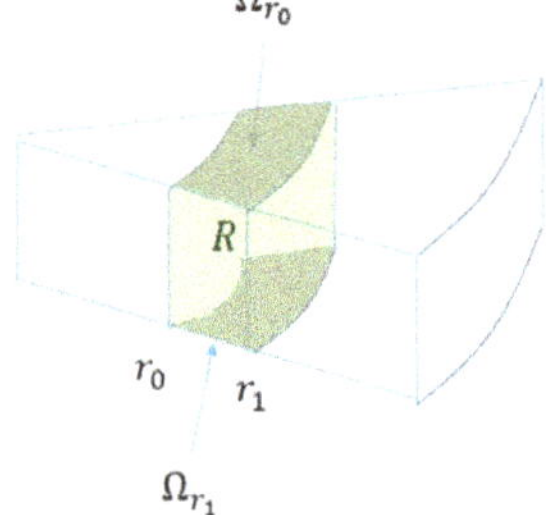

Figure 8. Region of collision of the protein particles against the bubbles, R.

Flux $\underline{j}$ is defined as the number of bubbles passing though an area A per unit time. Assuming no diffusion, i.e., only convection,

$$\underline{j} = \sigma \underline{s} \quad (2)$$

where σ and $\underline{s}$ are the density and velocity of the bubbles respectively.

Putting Equation (2) into (1) gives

$$\frac{\partial \sigma}{\partial t} = -\nabla \cdot \underline{J}. \quad (3)$$

The bubbles are assumed to move into the region R, resulting in a negative $\nabla \cdot \underline{j}$, and the negative sign is to make it positive.

Integrating,

$$\frac{\partial}{\partial t}\int_{h_0}^{h_1} \sigma(h,t)A(\Omega_h)dh = -\int_{h_0}^{h_1} \frac{\partial}{\partial h}[j(h,t)A(\Omega_h)]dh \tag{4}$$

where $\sigma(h,t)$ and $j(h,t)$ are density and magnitude of flux of bubbles respectively, $A(\Omega_h)$ is the area in the top and bottom surfaces of the region R (see Figure 1), and cylindrical coordinates are used: $r = \sqrt{x^2+y^2}$, $\phi = \frac{y}{x}$, $h = z$.

From Sminov and Berry (Smirnov et al., 2015), the velocity of the bubble is given by

$$s = \frac{2dg{r_b}^2}{9v} \tag{5}$$

where g is the free fall acceleration, d is the density difference for liquid and air inside bubbles or the liquid density, r_b is the radius of bubble, v is the liquid viscosity.

From previous work (Lin YK, 2015), the number of bubbles that can be adsorbed to the surface of a bubble, N_p, is given by $N_p = \pi\left(\frac{r_b}{r_p}\right)^2$ where r_p is the radius of the particle.

For maximum adsorption, the number of particles in the region R must be $\pi\left(\frac{r_b}{r_p}\right)^2$ times the number of bubbles in the same region R, and assuming a very small R, all bubbles that go into the R will collide with all particles that go into R.

From Equation (4), substituting $A(\Omega_r) = h\phi r$ and rearranging gives

$$\int_{h_o}^{h_1} h\phi r\frac{\partial\sigma}{\partial t} + \frac{\partial\sigma}{\partial r}(h\phi rj)dh = 0 \tag{6}$$

From Equation (4),

$$\int_{z_0}^{z_1} \frac{\partial}{\partial t}[\sigma(z,t)A(\Omega_z)] + \frac{\partial}{\partial z}[j(z,t)A(\Omega_z)]dz = 0. \tag{7}$$

Putting $Area = r\phi(r_1 - r_0)$ into Equation (7) and integrated w.r.t. z gives

$$r\phi(r_1 - r_0)\frac{\partial}{\partial t}\sigma(z,t) + r\phi(r_1 - r_0)\frac{\partial}{\partial z}j(z,t) = 0$$

Rearranging, we have

$$\frac{\partial}{\partial t}\sigma(z,t) + \frac{\partial}{\partial z}j(z,t) = 0 \tag{8}$$

Putting $j(z,t) = \sigma s$ into Equation (8) yields

$$\frac{\partial}{\partial t}\sigma(z,t) + \frac{\partial}{\partial z}(\sigma s) = 0$$

$$\frac{\partial}{\partial t}\sigma(z,t) + s\frac{\partial}{\partial z}\sigma(z,t) + \sigma(z,t)\frac{\partial}{\partial z}s = 0 \tag{9}$$

Given that $s = \frac{2dg{r_b}^2}{9v}$, we can take $d \approx \rho$. We know that

$$P = \rho g h \tag{10}$$

where P is pressure of fluid, which is equal to the pressure in the bubble, ρ is density of fluid, g is acceleration due to gravity, and h is height. From Sminov and Berry,

$$P = P_{ex} + \frac{2\alpha}{r_b} \tag{11}$$

where P_{ex} is exteral pressure acting on liquid, α is the surface tension and is a constant, and r_b is the radius of the bubble. Combining Equations (10) and (11) gives:

$$r_b = \frac{2\alpha}{\rho g h - P_{ex}} \tag{12}$$

where P_{ex} is the pressure of air in the room (most likely atmospheric pressure, 1 atm).

Putting Equation (12) into Equation (6) gives:

$$s = \frac{2\rho g\left(\frac{2\alpha}{\rho g h - P_{ex}}\right)^2}{9v} = \frac{8\rho g}{9v}\left(\frac{\alpha}{\rho g z - P_{ex}}\right)^2 = f(z) \tag{13}$$

where, $h = z$.

Putting (13) into (9) gives:

$$\frac{\partial}{\partial t}\sigma(z,t) + \frac{8\rho g}{9v}\left(\frac{\alpha}{\rho g z - P_{ex}}\right)^2 \frac{\partial}{\partial z}\sigma(z,t) - \frac{16(\rho g)^2\alpha^2}{9v(\rho g z - P_{ex})}\sigma(z,t) = 0 \tag{14}$$

$$\frac{\partial}{\partial t}\sigma(z,t) + f(z)\frac{\partial}{\partial z}\sigma(z,t) + \sigma(z,t)\frac{d}{dz}f(z) = 0 \tag{15}$$

$$f(z)\sigma(z,t)_z + \sigma(z,t)_t + F(z)\sigma(z,t) = 0 \tag{16}$$

where, $F(z) \equiv \frac{d}{dz}f(z)$.

To solve Equation (16), Laplace transform is applied to Equation (16), giving

$$f(z)\underline{\sigma}_z(z,s) + s\underline{\sigma}(z,s) - \sigma(z,0) + F(z)\underline{\sigma}(z,s) = 0$$

$$f(z)\underline{\sigma}_z(z,s) + [s + F(z)]\underline{\sigma}(z,s) - \sigma(z,0) = 0 \tag{17}$$

Since $\sigma(z,0) = 0$, Equation (17) becomes

$$f(z)\underline{\sigma}_z(z,s) + [s + F(z)]\underline{\sigma}(z,s) = 0 \tag{18}$$

To solve Equation (18), we divide Equation (18) by $f(z)$ to give

$$\underline{\sigma}_z(z,s) + \frac{s + F(z)}{f(z)}\underline{\sigma}(z,s) = 0 \tag{19}$$

Multiplying Equation (19) with integrating factor $e^{\int \frac{s+F(z)}{f(z)}dz}$ yields

$$\frac{d}{dz}\left[e^{q(z,s)}\underline{\sigma}(z,s)\right] = 0 \tag{20}$$

where,

$$q(z,s) = \frac{9sv}{8\rho g\alpha^2}\left(\frac{\rho^2 g^2 z^3}{3} - \rho g z^2 P_{ex} + (P_{ex})^2 z\right) + ln ln\left|\frac{8\rho g\alpha^2}{9v(\rho g z - P_{ex})^2}\right| \tag{21}$$

Integrating Equation (20) gives

$$\underline{\sigma}_z(z,s) = Ce^{Q(z,s)} \tag{22}$$

where C is a constant, and $Q(z,s) = q(z,s)^{-1}$. Transforming Equation (22) from the s domain back to the t domain yields

$$\begin{aligned} L^{-1}\{\underline{\sigma}_z(z,s)\} = & \quad \sigma_z(z,t) \\ & = \delta\left[t - \frac{9v}{8\rho g\alpha^2}\left(\rho g z^2 P_{ex} - (P_{ex})^2 z - \frac{\rho^2 g^2 z^3}{3}\right) - ln ln\left|\frac{8\rho g\alpha^2}{9v(\rho g z - P_{ex})^2}\right|\right] \end{aligned} \tag{23}$$

From Equation (23), for the bubbles to hit all protein molecules, a sufficient amount of time is required. In particular, the Dirac delta function in Equation (23) indicates that for maximum protein extraction, $t \sim O(z^2)$. Thus, with a uniform distribution of bubble holes at the bottom of the chamber, the time it takes to collect all molecules increases as the square of the height of the chamber. This explains the increasing amount of time required for higher protein collection in Figure 9, which shows both the theoretical and experimental data. It can be seen from this figure that the experimental data is in agreement with the theoretical data for flotation time less than 8 min. The reduction in yield after 8 min can be attributed to the lack of protein in the system after prolonged extraction.

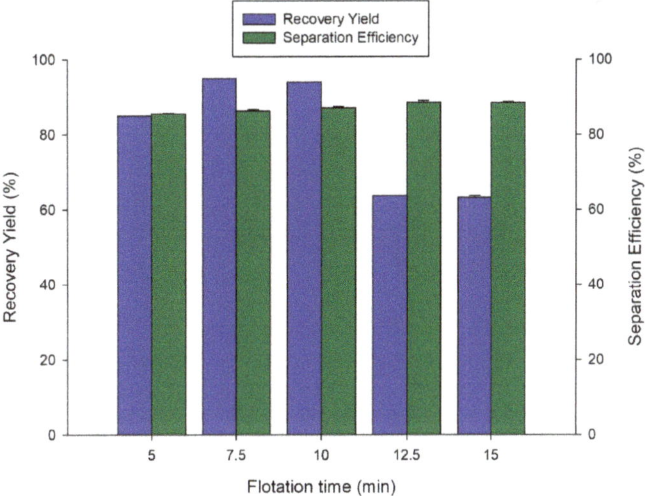

Figure 9. Figure illustrating the effect of flotation time on protein recovery yield and separation efficiency.

Besides allowing for a better understanding of the mechanism behind the system, this model also serves to provide some insights on design of efficient flotation systems at industrial scale. In particular, since it has been mathematically shown that an increase in height of the chamber drastically increases the flotation time, future chambers of flotation should preferably be as low as industrially viable.

The effect of pH has been studied by altering the pH of the bottom phase by using 1M hydrochloric acid. The initial pH of the bottom phase is 9.15, the tested pH is 6.5, 7.0, 7.5, and 8.0. The acid is added drop by drop until the bottom phase reached the desired pH with the aid of a pH meter. The highest recovery yield of proteins is obtained by the solution to which hydrochloric acid has not been added, which is the solution with a pH of 9.15. The recovery yield of the solution with this pH is at 93.96%, surpassing the second highest, which is a pH of 8.0 with a value of 65.71%, by more than 28%. The other three pH, being 6.5, 7.0, and 7.5, each has a recovery yield of 61.45%, 60.37%, and 62.33%, respectively. The big difference in yield is caused by the hydrochloric acid used to alter the pH having denatured the proteins, thus greatly reducing the yield of the LBF [24]. As the acid was added drop by drop, each time a drop of acid hits the surface of the bottom phase, the extreme pH of the acid will denature some of the proteins at the bottom phase before being diluted by the rest of the solution at this phase. This results in the big difference in the recovery yield when comparing between a bottom phase of pH 9.15 and 8.0; this is also why the difference in recovery yield between pH 8 and pH 6.5 does not show a big difference when compared with 9.15 and 8.0. The separation efficiency increases when the pH is changed from neutral to 6.5, which is from 85.59% to 86.21%. At a pH of 7.5, the separation efficiency is the highest which is at 97.18%. The separation efficiency of pH 8.0 and 9.15 is 94.33% and 87.14%, respectively. Due to the above reasons, the following test was conducted without altering the pH system.

3.7. Effect of Flotation Time on the Recovery of Proteins

The duration of the flotation process being conducted is very important as it could cause a major impact on the area of air–water interface per unit volume of aqueous solution in time [27]. Flotation times of 5, 7.5, 10, 12.5, and 15 min are tested for this study. The highest recovery yield of proteins is obtained when the system is run for 7.5 min, having a yield of 94.97%, closely followed by a flotation time of 10 min having a yield of 93.96%. With a flotation time of 5, 12 and 15 min, the yield is 85.03%, 63.76%, and 63.24%, respectively. The yield of proteins obtained shows an increasing trend when the flotation time is increased; however, it starts to decrease when the flotation time exceeds 10 min. This is because molecules other than proteins is being blown to the top phase, causing the overall concentration of the proteins at the top phase to be reduced. This is proven as the volume of the bottom phase shows a decreasing trend when the flotation time is increased, decreasing from 4 mL for 5 min flotation time to 2.5 mL for 15 min flotation time. In terms of separation efficiency, the values show a general increasing trend, but the increase in efficiency is only by a small amount, starting from 5 min flotation time, the separation efficiency is 85.58%, 86.29%, 87.20%, 88.64%, and 88.46%, respectively. Due to the reasons stated above, a flotation time of 7.5 min is taken to be optimum for this study. In general, the recovery of protein from the beginning of the experiment is low, however; it increases as the optimization is carried out. The low in recovery may be due to high contamination of bacteria that could affect the recovery of protein.

3.8. Effect of Scaling up LBF for Industrial Application Purposes

For industrial reasons, this experiment was tested again under large scale conditions. The experiment was scaled up by 40 times (from 30 mL to 1.2 L), and the results show that this experiment is suitable to be scaled up, with a recovery yield of more than 70% and a separation efficiency of more than 80%. This indicates that LBF is suitable to be commercialized as a method to separate proteins from expired milk wastes.

In this study, several parameters and their impacts on the LBF system have been studied by using the one-factor-at-a-time approach. Given that there might be a possibility that there will be interaction effects between several parameters, further studies on interaction factors such as milk concentration and flotation time could have a high chance of further optimizing this process. There is also a need to further improve this method so that protein recovery rate may be increased. An air compressor that can achieve higher and more accurate flowrates can be used in future LBF experiments. Furthermore, different types of gases can be used in replacement of atmospheric air for flotation. Different types of gases may help to bring up proteins or have other interactions with proteins that can improve the protein recovery yield. Thus, the effects of different types of gases such as pure oxygen or pure nitrogen can be tested to improve the recovery yield of the protein. Besides, the liquid used as top phase can be changed to different materials as using large amount of alcohol in industrial scale is a safety hazard. Alternative materials such as other organic solvents can be considered as an alternative to alcohol. Another aspect that is worth mentioning is the brand of milk used. As each brand of milk has a different formula for the milk they produce, changing the brand of milk used may also improve the protein recovery yield as there might be some components in milk from other brands that help in protein separation. Moreover, milk of different expiry dates can also be tested as well, as the level of microorganisms inside the milk may differ as time progresses. Further studies can be carried out by comparing milk of different expiry dates and some milk that is close to their expiry date so that the effects are clear.

4. Conclusions

The parameters for protein extraction from expired dairy products were optimized in this study. The effects of the type of inorganic salt used, the type of alcohol used, the concentration of salt used, the concentration of alcohol used, the concentration of raw material (milk) used, pH of the bottom phase, and the flotation time of the LBF system were discussed. The optimum conditions for protein extraction from dairy wastes tested in this study were found to be 150 g/L dipotassium hydrogen phosphate, 80% of ethanol, 10% (*w*/*v*) milk, a pH system of 9.15 (initial pH), and a flotation time of 7.5 min. The final protein recovery yield and separation efficiency after optimization were 94.97% and 86.29%, respectively. A scaling up of the LBF system was also performed at a factor of 40 times, and the protein recovery yield and separation efficiency for this test were 78.92% and 85.62% respectively. This study showed that proteins can be extracted from dairy waste effectively. The advantages of this novel approach include providing a use for expired milk products, reducing wastes being thrown away and benefiting the environment, and turning waste that once needed money to be disposed of into a raw material that can provide profit. Additionally, the utilization of high concentration of alcohol and salt will help inhibit further contamination by bacteria as they cannot survive in high alcohol and salt concentration environments. The concern with environmental impact due to high salt and alcohol concentration can be avoided by studying the recycling ability of the phase components. Studies have shown that there is a great potential to reuse recycling phase components in the subsequent extraction of LBF. Future studies on the interaction of parameters and methods for recycling the top and bottom phases after separation will provide a great opportunity for future industries to apply this method as a waste treatment process.

Author Contributions: Conceptualization, K.W.C.; data curation, P.E.T., Y.J.N., and K.W.C.; formal analysis, P.E.T. and Y.J.N.; funding acquisition, P.L.S.; methodology, R.S. and K.W.C.; supervision, K.W.C. and P.L.S.; visualization, K.S.K.; writing—original draft, P.E.T. and Y.J.N.; writing—review and editing, R.S., K.S.K., Y.J.Y., M.M., F.A.A.Z., and P.L.S.

Funding: This research was supported financially by SATU Joint Research Scheme (ST014-2018, ST022-2019), Geran Penyelidikan Universiti Malaya (UMRG Programme)—SUS (Sustainability Science) (RP025B-18SUS) and Fundamental Research Grant Scheme (Malaysia FRGS/1/2019/STG05/UNIM/02/2).

Acknowledgments: This study is supported by the Fundamental Research Grant Scheme (Malaysia, FRGS/1/2019/STG05/UNIM/02/2).

Conflicts of Interest: The authors declare that they have no conflicts of interests.

References

1. Milk Waste Produced. Available online: http://www.wrap.org.uk/content/wrap-comes-winning-formula-tackle-milk-waste (accessed on 9 July 2019).
2. Wasted Milk Is a Drain on Resources. Available online: https://www.ed.ac.uk/news/all-news/140512-climate (accessed on 9 July 2019).
3. American Association of Cereal Chemist. *Properties of Milk and Its Components*; American Association of Cereal Chemists: St. Paul, MN, USA, 1997; Volume 1, pp. 1–10.
4. Davoodi, S.H.; Shahbazi, R.; Esmaeili, S.; Sohrabvandi, S.; Mortazavian, A.; Jazayeri, S.; Taslimi, A. Health-Related Aspects of Milk Proteins. *Iran. J. Pharm. Res.* **2016**, *15*, 573–591. [PubMed]
5. Reen, C.S.; Wayne, C.K.; Loke, S.P.; Manickam, S.; Chuan, L.T.; Yang, T.A.O. Isolation of protein from Chlorella sorokiniana CY1 using liquid biphasic fl otation assisted with sonication through sugaring-out effect. *J. Oceanol. Limnol.* **2019**, *37*, 898–908.
6. Wayne, C.K.; Reen, S.; Krishnamoorthy, R.; Tao, Y.; Chu, D. Liquid biphasic flotation for the purification of C-phycocyanin from Spirulina platensis microalga. *Bioresour. Technol.* **2019**, *288*, 121519.
7. Bian, H.; Li, J.; Zhang, Q.; Chen, H.; Zhuang, W.; Gao, Y.Q.; Zheng, J. Ion Segregation in Aqueous Solutions. *J. Phys. Chem. B* **2012**, *116*, 14426–14432. [CrossRef] [PubMed]

8. Yuzugullu, Y.; Duman, Y.A. Aqueous Two-Phase (PEG4000/Na2SO4) Extraction and Characterization of an Acid Invertase from Potato Tuber(Solanum tuberosum). *Prep. Biochem. Biotechnol.* **2015**, *45*, 696–711. [CrossRef] [PubMed]
9. Show, P.L.; Ooi, C.W.; Anuar, M.S.; Ariff, A.; Yusof, Y.A.; Chen, S.K.; Annuar, M.S.; Ling, T.C. Recovery of lipase derived from Burkholderia cenocepacia ST8 using sustainable aqueous two-phase flotation composed of recycling hydrophilic organic solvent and inorganic slat. *Sep. Purif. Technol.* **2013**, *110*, 112–118. [CrossRef]
10. Lee, S.Y.; Khoiroh, I.; Ling, T.C.; Show, P.L. Aqueous Two-Phase Flotation for the Recovery of Biomolecules Aqueous Two-Phase Flotation for the Recovery of Biomolecules. *Sep. Purif. Rev.* **2015**, *45*, 81–92. [CrossRef]
11. Albertsson, P. Partition of cell particles and macromolecules in polymer two-phase systems. *Adv. Protein Chem.* **1970**, *24*, 309–341. [PubMed]
12. Ying, S.; Loke, P.; Chuan, T.; Chang, J. Single-step disruption and protein recovery from Chlorella vulgaris using ultradonication and ionic liquid buffer aqueous solutions as extractive solvents. *Biochem. Eng. J.* **2017**, *124*, 26–35.
13. Asenjo, J.A.; Andrews, B.A. Aqueous two-phase systems for protein separation. Phase separation and applications. *J. Chromatogr. A* **2012**, *1238*, 1–10. [CrossRef] [PubMed]
14. Wingfield, P.T. Protein precipitation using ammonium sulfate. *Curr. Protoc. Protein Sci.* **1998**, *13*, A.3F.1–A.3F.8.
15. Lu, Y.; Yu, M.; Tan, Z.; Yan, Y. Phase Equilibria and Salt Effect on the Aqueous Two-Phase System of Polyoxyethylene Cetyl Ether and Sulfate Salt at Three Temperatures. *J. Chem. Eng. Data* **2016**, *61*, 2135–2143. [CrossRef]
16. Sankaran, R.; Show, P.L.; Cheng, Y.S.; Tao, Y.; Ao, X.; Nguyen, T.D.; Van Quyen, D. Integration Process for Protein Extraction from Microalgae Using Liquid Biphasic Electric Flotation (LBEF) System. *Mol. Biotechnol.* **2018**, *60*, 749–761. [CrossRef] [PubMed]
17. Properties of Ethanol. Available online: https://pubchem.ncbi.nlm.nih.gov/compound/Ethanol#section=Density (accessed on 12 July 2019).
18. Properties of Ethanol. Available online: https://pubchem.ncbi.nlm.nih.gov/compound/Ethanol#section=Computed-Properties (accessed on 12 July 2019).
19. Properties of 1-Propanol. Available online: https://pubchem.ncbi.nlm.nih.gov/compound/1031#section=Density (accessed on 17 July 2019).
20. Properties of Isopropyl. Available online: https://pubchem.ncbi.nlm.nih.gov/compound/Isopropyl-alcohol#section=Density (accessed on 17 July 2019).
21. Properties of 1-Propanol. Available online: https://pubchem.ncbi.nlm.nih.gov/compound/1031#section=Computed-Properties (accessed on 17 July 2019).
22. Properties of 2-Propanol. Available online: https://pubchem.ncbi.nlm.nih.gov/compound/Isopropyl-alcohol#section=Computed-Properties (accessed on 12 July 2019).
23. Mao, Y.; Sheng, X.; Pan, X. The effects of NaCl concentration and PH on the stability of hyperthermophilic protein Ssh10b. *BMC Biochem.* **2007**, *8*, 1–8. [CrossRef] [PubMed]
24. Phong, W.N.; Show, P.L.; Teh, W.H.; Teh, T.X.; Lim, H.M.; binti Nazri, N.S.; Tan, C.H.; Chang, J.S.; Ling, T.C. Proteins recovery from wet microalgae using liquid biphasic flotation. *Bioresour. Technol.* **2017**, *244*, 1329–1336. [CrossRef] [PubMed]
25. Ling, P.; Suffian, M.; Annuar, M.; Loke, P.; Chuan, T. Extractive bioconversion of poly-caprolactone by Burkholderia cepacia lipase in an aqueous two-phase system. *Biochem. Eng. J.* **2015**, *101*, 9–17.
26. Ooi, C.W.; Tey, B.T.; Hii, S.L.; Ariff, A.; Wu, H.S.; Lan, J.C.W.; Juang, R.S.; Kamal, S.M.M.; Ling, T.C. Direct purification of Burkholderia Pseudomallei lipase from fermentation broth using aqueous two-phase systems. *Biotechnol. Bioprocess. Eng.* **2009**, *101*, 811–818. [CrossRef]
27. Khoo, K.S.; Chew, K.W.; Ooi, C.W.; Ong, H.C.; Ling, T.C.; Show, P.L. Extraction of natural astaxanthin from Haematococcus pluvialis using liquid biphasic flotation system. *Bioresour. Technol.* **2019**, *290*, 121794. [CrossRef] [PubMed]

Article

Characterization and Modelling Studies of Activated Carbon Produced from Rubber-Seed Shell Using KOH for CO_2 Adsorption

Azry Borhan [1,*], Suzana Yusup [1], Jun Wei Lim [2] and Pau Loke Show [3]

1 Department of Chemical Engineering, Institute of Sustainable Building, Centre for Biofuel and Biochemical Research, Universiti Teknologi PETRONAS, 32610 Seri Iskandar, Perak, Malaysia
2 Department of Fundamental and Applied Sciences, Institute of Sustainable Building, Centre for Biofuel and Biochemical Research, Universiti Teknologi PETRONAS, 32610 Seri Iskandar, Perak, Malaysia
3 Department of Chemical and Environmental Engineering, Faculty of Science and Engineering, University of Nottingham Malaysia, Jalan Broga, 43500 Semenyih, Selangor, Malaysia
* Correspondence: azrybo@utp.edu.my; Tel.: +605-368-7576; Fax: +605-365-6176

Received: 22 October 2019; Accepted: 11 November 2019; Published: 14 November 2019

Abstract: Global warming due to the emission of carbon dioxide (CO_2) has become a serious problem in recent times. Although diverse methods have been offered, adsorption using activated carbon (AC) from agriculture waste is regarded to be the most applicable one due to numerous advantages. In this paper, the preparation of AC from rubber-seed shell (RSS), an agriculture residue through chemical activation using potassium hydroxide (KOH), was investigated. The prepared AC was characterized by nitrogen adsorption–desorption isotherms measured in Micrometrices ASAP 2020 and FESEM. The optimal activation conditions were found at an impregnation ratio of 1:2 and carbonized at a temperature of 700 °C for 120 min. Sample A6 is found to yield the largest surface area of 1129.68 m^2/g with a mesoporous pore diameter of 3.46 nm, respectively. Using the static volumetric technique evaluated at 25 °C and 1.25 bar, the maximum CO_2 adsorption capacity is 43.509 cm^3/g. The experimental data were analyzed using several isotherm and kinetic models. Owing to the closeness of regression coefficient (R^2) to unity, the Freundlich isotherm and pseudo-second kinetic model provide the best fit to the experimental data suggesting that the RSS AC prepared is an attractive source for CO_2 adsorption applications.

Keywords: rubber-seed shell; activated carbon; CO_2 adsorption; isotherms; kinetics modeling

1. Introduction

With the fast escalation of the overall industrialization and population in many countries, the utilization of energy is exclusively expending. Presently over 85% of the international energy requirement is being financed by the burning of fossil fuels [1]. The reasons for this dependence on energy sources are a result of instinctive energy density, supply, and dependency of modern society on the procurement and exchange of these resources. Fossil fuels will still dominate in the predictable future, primarily in power production and industrial manufacturing. The utilization of these fossil fuels, especially in electricity generation, residentiary, transportation, and industrial area discharges massive amounts of carbon dioxide (CO_2) into the atmosphere, and thus upsets the carbon balance of our planet, which has been stable over millions of years. Although anthropogenic emissions of CO_2 can be considered comparatively limited related to the natural carbon changes such as photosynthetic fluxes, its escalation has distinct impacts on the global climate over a very short duration of time. The concentration of CO_2 in the atmosphere recorded in December 2018 has increased from 280 to 408 ppm since the beginning of the industrial revolution [2]. The rise of CO_2 concentration dominates the balance of incoming and outgoing energy in the earth's atmosphere. Hence, CO_2 has often been pointed out as the main anthropogenic greenhouse gas (GHG) as well as the leading offender in climate change. Due to this serious environmental problem, there is pressure all over the world to address this issue. International agreement known as the Kyoto protocol was reached and signed by 192 nations under the United Nations Framework Convention on Climate Change (UNFCC) to limit GHG emissions of industrialized nations by 5.2% [3]. Despite Malaysia contributing 0.2% in global greenhouse gases emissions [4], the increment of CO_2 emission is not to be taken lightly, because the continuous increase of the CO_2 rate to the surroundings can cause global warming as well as affect human health due to long term exposure to high concentrations of CO_2 in the future.

Many CO_2 capture technologies are proposed and being investigated these years, including chemical and physical adsorption, cryogenic separation, and membrane separation [5,6]. At the time being, chemical absorption (scrubbing), utilizing amine-based solvents, is the most considered technology used in industries. Although it is viewed as the most practical technology for CO_2 capture in post-combustion processes and has been used for more than 60 years, its application brings some negative impact, such as equipment corrosion, it requires high absorber volume, and is harmful to human health. In addition, this method of CO_2 separation is energy consuming and expensive as it requires large amounts of low-pressure steam for adsorbent regeneration [6,7]. Among the viable technologies for CO_2 capture, adsorption using solid material is chosen due to low energy prerequisites, low essential and running cost, including controlled waste generation. Due to its diverse benefits, such as economical, accessible for regeneration, indifferent towards the moisture, high CO_2 adsorption uptake at ambient situation, high specific surface area, high mechanical durability, sufficient pore size distribution, as well as low in energy needed, [8] activated carbon (AC) is one of the up-and-coming solid adsorbents that can be employed to capture CO_2. Based on the advantages by AC, it has been extensively engaged in different utilizations including in gas and liquid phases. The capability of AC in CO_2 occupation also depends on several criteria such as the nature of the activating method and the quality of starting materials, which in turn alters the surface chemistry and porosity and of the synthesized AC [8–10].

Many researchers have investigated the manufacture of low-cost adsorbents from inexhaustible and economical precursors, which are mainly industrial and agricultural derivatives ranging from palm kernel [11], banana peel [12], coconut shell [13], doum seed coat [14], walnut shell [15], etc., and their findings are quite conclusive. Meanwhile, rubber plantations in Malaysia have increased ever since the year 2010. The increase of rubber plantations have led to the increase in rubber production. Apart from producing latex, rubber seeds with hard shells were produced at the same time. It is estimated that about 800 to 1200 kg of rubber seed per ha per year is produced in a rubber plantation. The increase of rubber seed is causing significant environment and disposal problems [16]. To reduce the waste disposal issue, the rubber-seed shell (RSS) is proposed to be used as AC for CO_2 capture, since there is limited research on its preparation, and to address the problems.

In this study, a chemical activation method by potassium hydroxide (KOH) was selected due to its numerous advantages over physical activation methods and favorable conditions compared to other chemical activating agents. The research study was set in the direction towards evaluating the potential of using RSS as AC material in the removal of CO_2. Since chemical activation was adopted in this study, impregnation ratio (IR), temperature (T_{act}), and activation time (t_{act}) would be the main factors affecting the extent of reaction. Therefore, these parameters were also investigated to evaluate the effects of operation conditions on pore advancement of AC prepared from RSS. Samples with the highest surface area are preferred to examine on the performance of CO_2 adsorption at ambient pressure and temperature. In addition, the CO_2 adsorption isotherm is evaluated through several models, such as Langmuir, Freundlich, and Temkin. Lastly, the kinetic property of the sample with the highest adsorption capacity of purified CO_2 is evaluated using the pseudo-first order, pseudo-second order, and Elovich kinetic models.

2. Materials and Methods

2.1. Materials and Pre-Treatment

The raw RSS is collected from the Rubber Industry Smallholders Development Authority (RISDA) rubber plantation located at Chemor, Perak. It is then washed using distilled water to remove impurities and dried overnight in an oven at 110 °C to remove surplus water content. Once dried, the RSS is crushed and pulverized and then sieved into a particle size of 250 μm and stored in airtight plastic containers for further use. All chemicals were of industrial reagent grade and acquired from R & M Chemicals supplier located at Semenyih, Selangor.

2.2. Activation and Carbonization

The RSS prepared was impregnated by mixing it with a desired ratio (1:1, 1:2, and 1:3) of KOH based on the weight of the dry sample. The mixing and impregnation processes were allowed to sit for overnight to ensure complete reaction takes place between the chemical reagent and raw material. The impregnated material was then carbonized in a fixed bed activation unit with heating temperature ranging from 400, 500, 600, 700, 800 to 900 °C and activation time between 60 and 120 min. The one-factor-at-a-time (OFAT) method was adopted in this study so that reduction of sample is achievable; it is a method of designing experiments involving the testing of factors one at a time instead of all simultaneously. Throughout the process, N_2 gas, which acts as the carrier gas and promotes the pore formation in RSS [17], was allowed to flow in the rotary kiln. After the heat treatment, the material was left to cool to room temperature and subsequently washed with distilled water to discard the excess KOH solution and ash. The AC sample produced was preserved in an oven for overnight at 90 °C and kept in a desiccator to prevent moisture.

2.3. Characterization

To investigate the best operating parameter for producing AC from RSS, several analytical equipments were employed. To investigate the morphology surface structure, a Zeiss EVO-50 Field Emission Scanning Electron Microscope (FESEM), model Supra 55 VP acquired from Zeiss Jena, Germany, is used to compare the structural images of RSS before and after activation by generating real space enhanced images of its surface. In addition to the standard electron microscope detectors, the instrument is also furnished with Energy Dispersive X-ray (EDX) Spectroscopy for elemental analysis investigation.

For specific surface and porosity analysis, Micrometrics ASAP 2020 is used to determine the surface area (S_{BET}), pore size distribution (D), and the total pore volume (V_T) through nitrogen adsorption–desorption isotherms analysis. N_2 (99.9% purity) gas is applied as adsorbate and the condition was allowed to flow at 350 °C for 2.5 h [16]. The specific surface area of the AC samples is determined using the Brunauer–Emmett–Teller (BET) method using nitrogen adsorption isotherm data tabulated from a computer.

2.4. CO_2 Adsorption Capacity Analysis

CO_2 adsorption analysis were carried via purified CO_2 (99.98% purity and supplied by Linde Malaysia Sdn. Bhd.) volumetric adsorption method using High Pressure Volumetric Analyzer (HPVA II) supplied by Particulate System. To ensure all impurities were removed from the samples, about 0.5 g of AC was inserted inside a 5 cm^3 sample glass cylinder and degassed at 170 °C for 8 h under vacuum. A 60 μm filter gasket was then planted on top of the sample cylinder, to avoid the fine particles from entering the valve [17]. After completion of the degassing step, the samples were cooled to ambient temperature and prepared for adsorption studies. The CO_2 adsorption process was started by introducing the gas adsorbate (CO_2) into the system. This is accomplished by granting the valve between the loading and sample cylinder to open and allow the CO_2 to interact with the AC material. To ensure equilibrium of the adsorption process, the holding time at each pressure interval was fixed at

45 min. By taking the differences between the amounts of dosed gas into and the amounts of gas staying in the system upon adsorption process, the volumetric CO_2 sorption capacity during experimental run was calculated [18]. For isotherm studies, using the resulting points of volumes adsorbed at equilibrium pressures, several isotherm models such as Langmuir, Freundlich, and Temkin [19] were plotted and fitted to experimental data to find suitable representation of adsorption process of RSS based AC. The suitability of the above-mentioned models is assessed by R^2 values that are close to unity. Table 1 outlines the non-linear and linear equations of these three models.

Table 1. Model isotherm equations.

Isotherm	Non-Linear Equation	Linear Equation
Langmuir	$q_e = \frac{q_m k_L P_e}{1+k_L P_E}$	$\frac{P_e}{q_e} = \frac{1}{q_m} P_e + \frac{1}{k_L q_m}$
Freundlich	$q_e = k_F P^{\frac{1}{n}}$	$\log q_e = \frac{1}{n} \log P_e + \log k_F$
Temkin	$q_e = B\,(\ln k_T P_e)$	$q_e = B \ln P_e + B \ln k_T$

P_e is equilibrium pressure (bar); q_e and q_m are the amount capacity of CO_2 adsorbed at equilibrium and at maximum, respectively (cm^3/g); k_L is the Langmuir constant (1/bar); k_F (cm^3/g·bar$^{1/n}$) and n is the Freundlich constant; $B = RT/b_T$; b_T (J/mol); and k_T (cm^3/g·bar) is the Temkin constant.

2.5. Kinetic Studies

CO_2 adsorption kinetics of KOH impregnated RSS AC are desirable to evaluate the accomplishment of sorbents and to understand the overall mass transfer in the CO_2 adsorption process. In addition, it served as baseline to predict CO_2 adsorption/desorption kinetics for the rational simulation and design of gas-treating systems. In this study, the usefulness of Lagergen's pseudo-first order model, pseudo-second order model, and Elovich model approaching the experimental values using purified CO_2 adsorption at 25 °C is examined. The compliance of the predicted adsorption capacity was evaluated by the magnitude of coefficient regression R^2 closeness towards unity.

2.5.1. Pseudo-First Order Kinetic Model

The linearized Lagergen's pseudo-first order model was the first adsorption rate equation depicted for sorption of a liquid/solid system and one of the most frequently used adsorption rate models. It is expressed by the equation below [19,20]:

$$\log(q_{eq} - q_t) = \log q_e - \frac{k1}{2.303} t, \tag{1}$$

where q_{eq} and q_t are the amount of adsorption at equilibrium and at that particular time t, respectively. It has the unit of mg/g. k_1 represents the rate constant for pseudo-first order adsorption. A linear plot will be obtained from the graph when this model is applicable. In addition, the slope and interception point of the plot can be used to determine pseudo-first order parameters.

2.5.2. Pseudo-Second Order Kinetic Model

The pseudo-second order model is expressed by the equation as follows [19,20]:

$$\frac{t}{q_t} = \frac{1}{k2qe^2} + \frac{1}{q_e} t, \tag{2}$$

where q_e and q_t are the amount of adsorption at equilibrium and at that particular time t, in mg/g, respectively. k_2 represents the overall rate constant for pseudo-second order adsorption with the unit of g/mg min. Similar to first order, a linear plot will be achieved from the graph of t/q_t versus time, t when it is applicable.

2.5.3. Elovich Kinetic Model

Elovich kinetic model is usually used in a gas–solid system and is expressed by [19,20]:

$$\frac{dq_t}{dt} = \alpha \ \exp(-\beta q_t) \tag{3}$$

where q_t is the amount of CO_2 adsorbed in mg/g at a particular time, t. α represents the initial adsorption rate in mg/g min, while β is the extent of surface coverage in g/mg and the process activation energy.

3. Results and Discussion

3.1. Elemental Composition Analysis

Aside from examining the surface morphology of the samples, FESEM is also furnished with Energy-Dispersive X-ray (EDX) spectroscopy for detecting the elemental composition. Table 2 below shows the corresponding elemental content before and after activation. Two main elements were detected on the samples before activation, which are carbon and oxygen. It was described in specific that the acceptable range of carbon presence should lie between 40 and 80% [17]. The result shows that RSS fulfills the criteria of producing AC. It was found that the percentage of carbon content has escalated after carbonization and activation due to the release of more volatile matter during the process. An additional element identified as potassium was detected in the sample after activation. Presence of potassium element is as a result the use of KOH as chemical activating agent during the activation process. Repetitive washing can greater reduce the potassium element, but complete elimination is hardly possible. Calcium content in RSS is expected and considered normal as the RSS is rich in protein, minerals, and amino acid. Eka et al. [21] has mentioned this in a report that rubber seed has a low content of calcium.

Table 2. Comparison of elemental composition.

Elements	Before Carbonization (RAW)		After Carbonization (A6)	
	Weight %	Atomic %	Weight %	Atomic %
Carbon	50.8	59.2	56.6	65.0
Oxygen	44.4	39.4	27.3	21.7
Potassium	-	-	12.5	8.3
Calcium	4.8	1.4	3.6	5.0
Total	100	100	100	100

3.2. Characterization Study

The preparation conditions and results of AC samples produced from KOH impregnated RSS are exhibited in Table 3. The result demonstrated that sample A6, which is arranged at an impregnation ratio (IR) of 1:2, T_{act} of 700 °C, and t_{act} of 120 min, yields the highest values of specific surface area, S_{BET} of 1129.60 m^2/g, average pore diameter, D of 3.46 nm, and total pore volume, V_T as high as 0.412 cm^3/g. This is followed by samples A5 at 826.31 m^2/g, 3.21 nm, and 0.376 cm^3/g, and A8 at 731.06 m^2/g, 3.21 nm, and 0.301 cm^3/g, respectively. In comparison, samples A1 and A2, which are developed at an IR of 1:1, all exhibit reduced S_{BET} and V_T values in comparison to A3. The reason is due to the IR of KOH to RSS which plays a critical role in the pores formation. High IR supposedly helps to increase the amount of potassium metal that can be intercalated and thus develops more pore formation [22]. Nevertheless, the outcome of the analysis confirms that there is an utmost number of ions that can be accepted above which would reduce pore progression. The reason behind this is because additional or excess activating agents probably form an insulating layer (or skin) coating the AC particles, and therefore lowering the activation process and the influence with the surrounding atmosphere [23]. This phenomenon is particularly observed in sample A4 where most likely activation is hindered, resulting in a lower surface area (S_{BET} of 492.55 m^2/g) and pore volume formation (V_T of 0.182 cm^3/g) as compared to A3.

Table 3. Surface area and porosity results.

Sample	IR	Act. Temp (°C)	Act. Time (min)	Specific Surface Area, S_{BET} (m^2/g)	Total Pore Volume, V_T (cm^3/g)	Average Pore Diameter, D (nm)	Percentage Micropores (%)
Raw	-	-	-	1.11	0.007	1.26	19.82
A1	1:1	400	60	203.81	0.113	1.81	78.71
A2	1:1	500	60	481.19	0.191	2.06	78.90
A3	1:2	500	60	571.86	0.203	2.17	85.29
A4	1:3	500	60	492.12	0.182	2.26	53.73
A5	1:2	600	120	826.31	0.376	3.21	34.24
A6	1:2	700	120	1129.60	0.412	3.46	31.67
A7	1:2	700	180	618.13	0.406	3.37	28.34
A8	1:2	800	120	731.06	0.301	3.21	24.27
A9	1:2	900	120	701.45	0.235	3.72	15.13

As for activation temperature, the optimum temperatures have been reported to be between 500–800 °C by most of the earlier researchers [24]. Hence, the experiments were conducted by varying temperatures from 400 to 900 °C, and it turned out that the recommended temperature is reliable due to verification provided by samples A1 to A3 and A5 to A9. Sample A1, which is carbonized at 400 °C, yields the lowest S_{BET} (203.81 m^2/g) and V_T (0.113 cm^3/g) than any additional samples. It was reported that 400 °C is the starting carbonization temperature in the development rudimentary of pores of AC material. Increasing the activation temperature to 700 °C will intensify the expulsion of molecular weight of unstable compounds and further generating new pores, resulting in the hastening of porosity growth of the AC. Sample A5 and A6 have the highest S_{BET} (826.31 m^2/g and 1129.60 m^2/g) and V_T (0.376 cm^3/g and 0.412 cm^3/g) values, respectively. Nevertheless, when the activation temperature is elevated to 900 °C, the excessive heat energy supplied to the carbon will result in the collapsing and knocking of some porous wall [17]. The outcome is the decrease quantity of S_{BET} and V_T. This can be interpreted by comparing sample A9 with other samples' characterization result.

Extended activation time during the carbonization process may cause in over-activation, where surface erosion is accelerated more quickly than pore formation. Sample A7 shows rapid decreasing in S_{BET}, V_T, and D after an activation time of 180 min. Considering the well-developed porous structure at a temperature of 700 °C, any increment in activation time will causes the carbon structures to break between its cross-links, resulting in pore collapsing [25,26]. Although sample A6 has the highest S_{BET}, V_T, and D, only 31.67% of its volume existed as micropores. Sample A3 has the highest micropore volume, with 85.29% of its total pore, while sample A9 has the lowest micropore volume with 15.13%. According to The International Union of Pure and Applied Chemistry (IUPAC) classification pore size are categorized as macro if the size >50 nm, meso if the size is between 2–50 nm, and micropore when <2 nm [26]. All samples, except raw and A1, show pore diameter in the range between 2–4 nm, and thus distinctly categorize that the pores belong to the mesopores classification.

3.3. Morphology

Figure 1 shows the microscopic morphology structure of the raw RSS and some selected samples prepared at different operating parameters. By using FESEM, structural images with magnification up to 300 times are taken. The structural image of fresh RSS in Figure 1a indicates that the raw material before undergoing activation shows no noticeable pores. However, the image clearly shows the existence of fine pores on the surface, which is one of the important aspects for manufacturing AC. After activation, a greater distribution of pores emerged to become active sites for adsorption to take place more readily. The difference in pore structure before activation and after activation is clearly illustrated through comparison made between Figure 1a with Figure 1b,c. The canal structure on the surface of sample A6 has been partially broken, which indicates carbonization process occurred and the surface was eroded by longer activation temperature and time than sample A1. Therefore, the S_{BET} of sample A6 is much higher. As shown in Figure 1c, sample A6 has the most well-developed structure compared the other three AC. This justified that sample A6 has the highest S_{BET}, V_T, and D compared to other samples. On the other hand, collapse of porous wall due to excessive heat exposure is observed in Figure 1d with sample A9 being activated at highest temperature. Two promising samples, A5 and A6, are selected for further analysis.

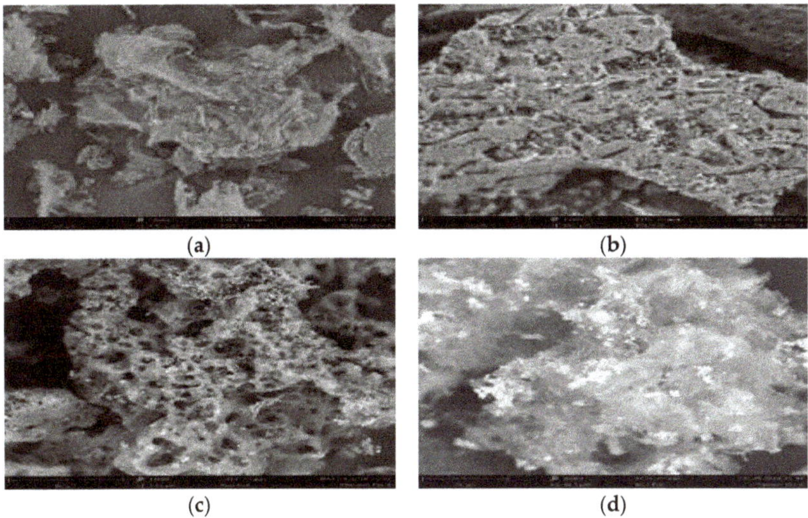

Figure 1. FESEM images of selected samples, (**a**) fresh RSS, (**b**) sample A1, (**c**) sample A6, and (**d**) sample A9.

3.4. Nitrogen Adsorption–Desorption Isotherms Study

Figure 2 shows the N_2 adsorption–desorption isotherms of the selected three AC. The quantity of N_2 adsorbed is projected against the relative pressure p/p_o where p = pressure at given condition and p_o = saturated vapor pressure of N_2. Based on IUPAC classification of adsorption isotherm [25], it can be seen that for raw AC the isotherm follows Type II classification, which signify the presence of microporous pores existed and within the micropores the surface resides almost exclusively. Once it was fully occupied by N_2 adsorbate, very few or no exterior surface left for further adsorption. Samples A5 and A6, however, show a combination of Type I and Type III classification where the trend line was initially following Type I with Type III trend line appearing at the end of high relative pressure. This combination is associated with a combination of microporous and mesoporous structures with mesopores as the dominant species [26]. The result is consistent with the D of 3.21 and 3.46 nm, respectively, as shown in Table 3, where it complies with the IUPAC classification of mesoporous material.

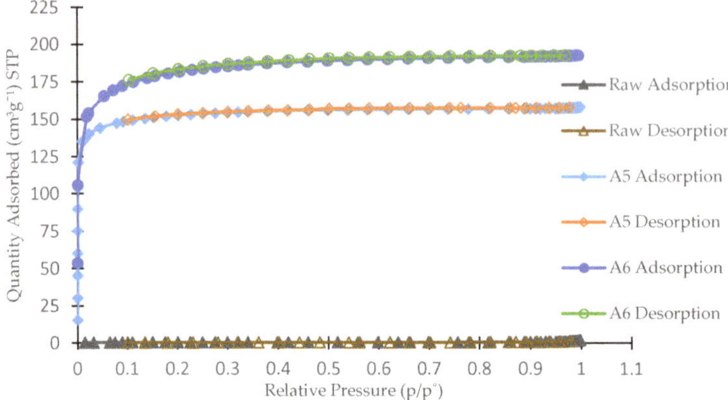

Figure 2. N_2 adsorption–desorption isotherms.

3.5. CO_2 Adsorption and Isotherm Modeling

Figure 3a depicts the CO_2 adsorption capacity for samples A5 and A6 at an ambient temperature and pressure. According to Estaves et al. [27], it is ideal to adsorb CO_2 at lower a temperature (25 °C in this research) as CO_2 adsorption process onto AC is exothermic due to the physical adsorption (physisorption) process. Weak van der Waals forces are basically involved in physisorption. At high temperature condition, these weak forces are easily broken and result in the decrease of the adsorption capacity. Instability of the CO_2 adsorbate on the carbon surface will result in a desorption process, owing to higher surface adsorption energy and molecular diffusion at high temperature. According to Hauchhum et al. [28], the rise in bed temperature of AC will expedite the internal energy of the adsorbent, and therefore CO_2 molecules are released from the surface. To summarize, exothermic process during adsorption is controlled by physisorption when there is reduction in the adsorption capacity with respect to the temperature [29]. Based from Figure 3a, higher CO_2 adsorption capacity for sample A6 is recorded compared to A5. This finding is parallel to the nitrogen adsorption analysis carried out earlier. This verifies that higher surface area does lead to higher adsorption capacity. The highest CO_2 adsorption capacity for sample A6 is 43.5094 cm^3/g at a pressure of 1.2523 bar, while for the A5 sample it is 39.2496 cm^3/g at a pressure of 1.2525 bar, respectively.

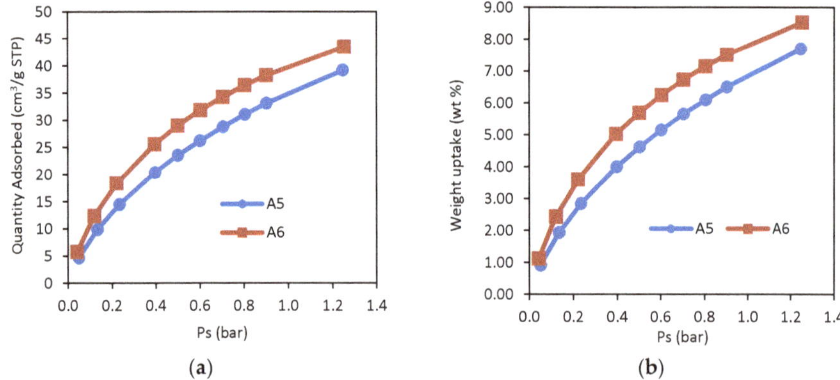

Figure 3. (**a**) CO_2 adsorption capacity at 25 °C; (**b**) weight uptake of CO_2 adsorption at 25 °C.

The effect of pressure on CO_2 capacity is also observed where higher adsorption occurs at higher pressure. This phenomenon happens because high pressure tends to push CO_2 molecules onto adsorption site within the pore. In Figure 3b, the highest CO_2 sorption quantity that is exhibited by samples A6 and A5 is 8.5431 wt% at a pressure of 1.2523 bar and 7.7067 wt% at a pressure of 1.2525 bar, respectively. The disparity in CO_2 weight uptake of these solid AC as a result of the surface area value is shown in Table 3. Sample A6, which has the largest S_{BET}, is noticed to adsorb more CO_2 compared to other samples, which implies that there are more available surface sites for CO_2 adsorption processes to take place.

Table 4 shows the calculated isotherm constants and their corresponding R^2 values for CO_2 adsorption using linear regression method. According to Perez et al. [30], the Langmuir constant k_L and Freundlich constant k_F are all related to the adsorption affinity. Its value decreases with increases in temperature, which signify physisorption behavior. This CO_2 adsorption capacity reduction can be interpreted by Le Chatelier's principle, where for an exothermic process (physisorption), low temperature is preferred during adsorption. The exothermic behavior of the CO_2 adsorption is aligned with the q_m value that is likely to decline with the rise in the adsorption temperature. The Langmuir adsorption model explains that adsorption does not occur after monolayer adsorbate formation on the adsorbent surface. This model assumes constant adsorption energies onto the surface and no adsorbate movement on the surface planes [29–31].

Table 4. Langmuir, Freundlich, and Temkin isotherm models via linearized technique.

Sample	Langmuir			Freundlich			Temkin			
	q_m	k_L	R^2	k_F	n	R^2	k_T	B	b_T	R^2
A6	57.47126	2.202532	0.9901	42.07266	1.67364	0.9902	28.81605	11.365	218.1099	0.9708
A5	57.80347	1.478632	0.9772	36.19095	1.515611	0.9962	21.37372	10.73	231.0176	0.9487

Freundlich adsorption model is based on an empirical relationship, which explains adsorption isothermal variation with pressure. This model is often used to explain heterogeneous surface adsorption characteristics. The Temkin adsorption model accounts for the adsorbate–adsorbent interaction. This model assumes adsorption heat of all adsorbate molecules decreases linearly [29]. As for Freundlich constant, n, and sometimes known as heterogeneity factor, its value signifies the type of adsorption, where $n > 1$ is for physical adsorption while $n < 1$ corresponds to chemical adsorption. Both samples A5 and A6 have a R^2 value closer to unity for the Freundlich model compared to Langmuir and Temkin. This shows that the adsorption process occurs in heterogeneous surfaces and is not restricted to monolayer adsorption as recommended by Langmuir. The summary for Table 4 suggests that the Freundlich model gives the best fitting correlation to the experimental data, owing its R^2 value approaching unity as it permits the CO_2 adsorbate molecules to form a successive layer onto the surface of AC.

3.6. Kinetic Analysis

Figure 4 shows the straight line plot of $\log(q_e - q_t)$ versus t for RSS AC of sample A6 at 25 °C using Equation (4). The kinetic data are summarized in Table 5. The pseudo-first order kinetic model is established on the assumption that the rate of adsorption is proportional to the number of vacant sites available on the adsorbent surface and is used regularly in liquid–solid phase [29]. Due to its low R^2 value of 0.8392, this kinetic model does not fit well with the CO_2 adsorption experimental data. Sadaf et al. [32] verified that the pseudo-first order model was unsuitable in the adsorption process as it can be only applied during the beginning stage and not for the entire period.

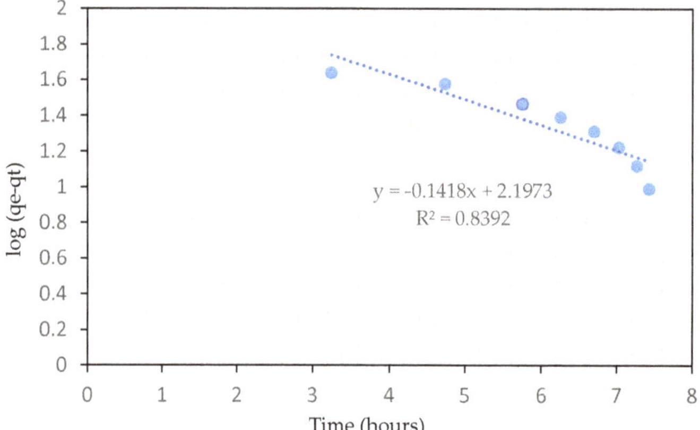

Figure 4. Pseudo-first order kinetic model of sample A6 RSS AC.

Table 5. Summary of kinetic models.

Kinetic Model	Parameter	Temperature (25 °C)
Pseudo-first order model	q_e (mg/g)	54.31
	k_1 (1/min)	0.221
	R^2	0.839
Pseudo-second order model	q_e (mg/g)	61.15
	k_2 (g/mg min)	7.82×10^{-4}
	h (mg/g min)	2.306
	R^2	0.939
Elovich model	β (g/mg)	0.022
	α (mg/g min)	2.92
	R^2	0.8188

For the pseudo-second order kinetics model, the rate of adsorption is assumed to be linearly related to the square of the number of vacant sites available on the adsorbent surface. This model has been used by many researchers for the modeling of experimental data of CO_2 adsorption kinetics [33,34]. A plot of t/q_t against time t will generate a straight line with $1/h$ and $1/q_e$ as y-interception and slope, respectively, if the model is applicable in the adsorption process. By taking the R^2 value shown in Figure 5, the pseudo-second order model fits the CO_2 adsorption profile with the regression coefficient value of 0.9388 compared to the first-order model of 0.8392. In addition, the magnitude of h that represents the rate of adsorption has a value of 2.306 mg/g min and its value is expected to decline with respect to the operating temperature. According to Simon et al. [29], the CO_2 molecules will gain an adequate amount of energies at elevated temperatures and be able to overcome the weak van der Waals bonding and finally will be moved back to the bulk gas phase. A similar trend was observed by Chao Ge et al. [25] when investigating the adsorption equilibrium capacity of CO_2 at temperatures between −47 and 28 °C using both pseudo-model kinetics. Results predicted by the pseudo-second order model were much closer and coincided with the experimental values (R^2 value greater than 0.996). Contrarily, larger deviation between the actual and calculated value resulted by using the pseudo-first order model and, thus, producing lower R^2 values.

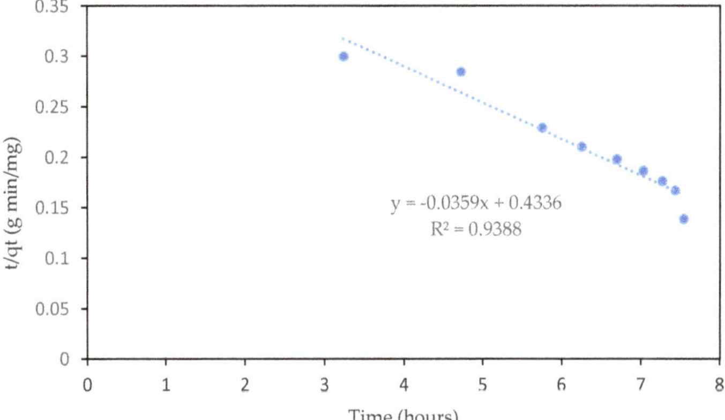

Figure 5. Pseudo-second order kinetic model of sample A6 RSS AC.

Figure 6 displays the Elovich plot of CO_2 adsorption using the same data. If it fits with the model, a straight line with gradient of $1/\beta$ and y-interception of $1/\beta \ln(\alpha\beta)$ will be produced. The Elovich kinetic model assumes that the rate of adsorption exponentially decreases with the increase in the amount of CO_2 adsorbed on the adsorbent surface without any interaction among the adsorbed species [35,36]. By applying boundary conditions and integrate in Equation (3), it results in a new equation as shown below:

$$q_t = \frac{1}{\beta}\ln(\alpha\beta) + \frac{1}{\beta}lnt \tag{4}$$

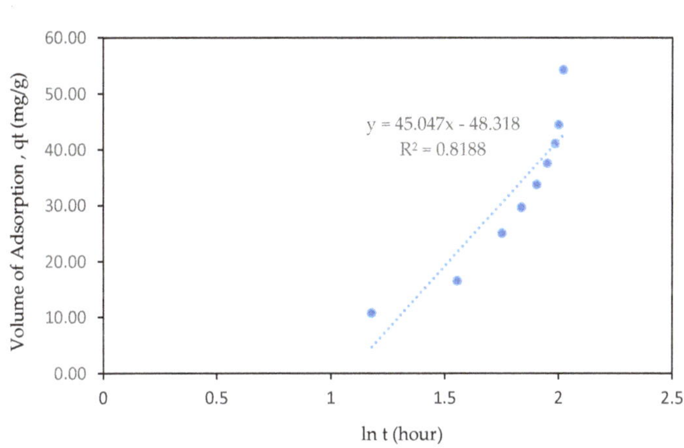

Figure 6. Elovich model of sample A6 RSS AC.

As demonstrated in the figure, the Elovich kinetic model gave a poor regression coefficient of 0.8188 compared to the other two pseudo-models.

Based on the summary of kinetic models shown in Table 5, it was noticed that the pseudo-first order and Elovich model are not well suited to fit the CO_2 adsorption kinetics data owing to their low R^2 values. Significant deviations are observed when determining equilibrium adsorption capacity using the two kinetic models. The pseudo-second order kinetics model fits the adsorption kinetics data, with a R^2 value of 0.939. It can be concluded that the three kinetics models are found to be suitable for fitting the present adsorption kinetics data in the following subsequent order: Pseudo-second order > pseudo-first order > Elovich.

3.7. Comparison Study with Other Biomass Activated Carbon Materials

Figure 7 shows the CO_2 adsorption comparison capacities on selected types of AC from agricultural waste ranging from coconut shell, banana peel, rice husk, palm shell, and coconut fiber [17,34]. From this relative study, it can be seen that the RSS AC of sample A6 has a noticeable higher adsorption capacity compared to banana peel and rice husk based AC. The highest CO_2 adsorption capacity is coconut shell at 78.77 mg/g, followed by coconut fiber at 60.2 mg/g and palm shell at 58.52 mg/g. RSS is ranked fourth with an adsorption capacity of 54.31 mg/g of CO_2. The usage of RSS waste biomass from rubber plantation for the development into AC is practical as it can overcomes the shortage of the non-renewable precursors, such as zeolites, metal-organic frameworks (MOF), mesoporous oxides, polymers, etc. To guarantee a long-term sustainability of this industry, Khalili et al. [37,38] acknowledged that the biomass-based AC may be synthesized from the renewable feedstock. In addition, the preparation of wastes from sustainable biomass precursors provides an environmentally friendly and sustainable passage for the advancement of CO_2 sorbent materials.

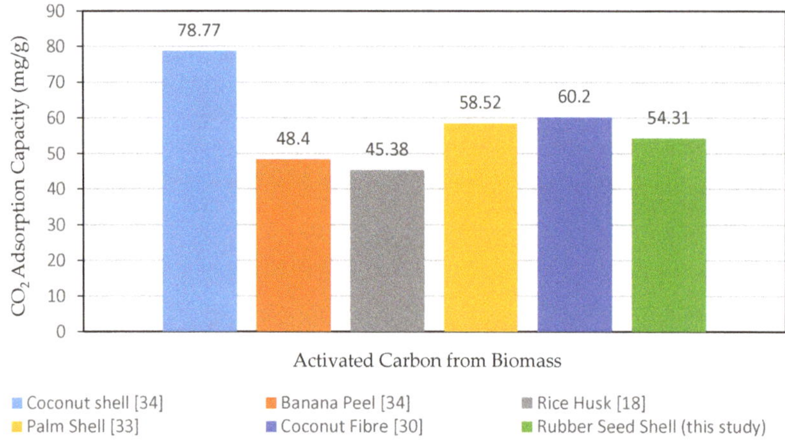

Figure 7. CO_2 adsorption capacity by different agricultural waste AC [18,30,33,34].

4. Conclusions

This study indicates the promising potential in producing a low-cost AC from RSS using KOH as activating agent. The resulting AC yield a high surface area, pore volume, and average diameter at an impregnation ratio of 1:2, activation temperature of 700 °C, and activation time of 120 min. According to the characterization analysis, the range of pore diameter (2–50 nm) is in the mesopores classification. The AC with a high surface area was also verified to have higher CO_2 adsorption capacity using a static volumetric instrument. Moreover, the adsorption capacity test proved that adsorption using RSS based AC has a high prospective in reducing CO_2 and is on par with some already existing conventional biomass AC. Based on isotherm model analysis, the Freundlich isotherm model fit best, and the kinetic analysis demonstrated that the CO_2 adsorption onto the AC obeys the pseudo-second order model due to its closest proximity R^2 value toward unity.

Author Contributions: Data curation, A.B.; formal analysis, A.B. and J.W.L.; funding acquisition, A.B.; methodology, A.B. and S.Y.; supervision, S.Y.; validation, A.B. and S.Y.; writing—original draft, A.B.; writing—review and editing, J.W.L. and P.L.S.

Funding: The authors extend their appreciation to the Ministry of Higher Education (MOHE), Government of Malaysia, under the Fundamental Research Grant Scheme, [FRGS No: FRGS/1/2018/TK10/UTP/02/9] for funding this research.

Acknowledgments: The authors gratefully thank the Universiti Teknologi PETRONAS and the Center for Biofuel and Biochemical Research (CBBR), UTP for providing financial assistance and support.

Conflicts of Interest: All authors declare no conflict of interest.

References

1. Amira, K.; Mohammad, A. Exploring the Impact of Renewable Energy on Climate Change in the GCC Countries. *Int. J. Energy Econ. Policy* **2019**, *9*, 124–130.
2. Chao, X.; Maria, S. Sustainable Porous Carbon Materials Derived from Wood-Based Biopolymers for CO_2 Capture. *Nanomaterials* **2019**, *9*, 103.
3. Olivier, J.G.J.; Janssens-Maenhout, G.; Muntean, M.; Peters, J.A.H.W. *Trends in Global CO_2 Emissions: 2016 Report*; PBL Netherlands Environmental Assessment Agency: The Hague, The Netherlands, 2016; pp. 1–86.
4. Shahid, S.; Minhans, A.; Che Puan, O. Assessment of greenhouse gas emission reduction measures in transportation sector in Malaysia. *J. Teknol.* **2014**, *70*, 1–8. [CrossRef]
5. Mahdi, F.; Olga, E.; Christian, B. Techno-economic assessment of CO_2 direct air capture plants. *J. Clean. Prod.* **2019**, *224*, 957–980.
6. Yuan, W.; Li, Z.; Alexander, O.; Martin, R.D.S. A Review of Post-combustion CO_2 Capture Technologies from Coal-fired Power Plants. *Energy Procedia* **2017**, *114*, 650–665.
7. Bryce, D.; Maohong, F.; Armistead, G.R. Amine-Based CO_2 Capture Technology Development from the Beginning of 2013-A Review. *ACS Appl. Mater. Interfaces* **2015**, *7*, 2137–2148.
8. Singh, V.K.; Kumar, E.A. Measurement and analysis of adsorption isotherms of CO_2 on activated carbon. *Appl. Therm. Eng.* **2016**, *97*, 77–86. [CrossRef]
9. Srinivas, B.N.; Kishore, P.; Rao, K.S.; Kumar, T.A. Preparation of surface modified activated carbons from rice husk and CO_2 adsorption studies. *IOSR J. Appl. Chem.* **2017**, *10*, 54–60.
10. Boonpoke, A.; Chiarakon, S.; Laosiripojana, N.; Towprayoon, S.; Chidthaisong, A. Synthesis of activated carbon and MCM-41 from bagasse and rice husk and their carbon dioxide adsorption capacity. *J. Sustain. Energy Environ.* **2011**, *2*, 77–81.
11. Ademiluyi, F.T.; David-West, E.O. Effect of chemical activation on adsorption of heavy metals using activated carbons from waste materials. *Int. Sch. Res. Netw.* **2012**. [CrossRef]
12. Borhan, A.; Hoong, P.K.; Taha, M.F. Biosorption of heavy metal ions, oil and grease from industrial waste water by banana peel. *Appl. Mech. Mater.* **2014**, *625*, 749–752. [CrossRef]
13. Jahagindar, A.A.; Ahmed, M.N.Z.; Devi, D.V. Adsorption of chromium on activated carbon prepared from coconut shell. *Int. J. Eng. Res. Appl.* **2012**, *2*, 364–370.
14. El-Sadaawy, M.; Abdelwahab, O. Adsorptive removal of nickel from aqueous solutions by activated carbons from doum seed coat. *Alex. Eng. J.* **2014**, *53*, 399–408. [CrossRef]
15. Zabihi, M.; Haghighi, A.A.; Ahmadpour, A. Studies on adsorption of mercury from aqueous solution on activated carbons prepared from walnut shell. *J. Hazard. Mater.* **2010**, *174*, 251–256. [CrossRef]
16. Yahya, M.; Al-Qodah, Z.; Ngah, C. Agriculture bio-waste materials as potential sustainable precursors used for activated carbon production: A review. *Renew. Sustain. Energy Rev.* **2015**, *46*, 218–235. [CrossRef]
17. Borhan, A.; Taha, M.F.; Hamzah, A.A. Characterization of activated carbon from wood sawdust via chemical activation using potassium hydroxide. *Adv. Mater. Res.* **2014**, *832*, 132–137. [CrossRef]
18. Rashidi, N.A.; Yusup, S.; Borhan, A. Isotherm and Thermodynamic Analysis of Carbon Dioxide on Activated. *Procedia Eng.* **2016**, *148*, 630–637. [CrossRef]
19. Al-Marri, M.J.; Al-Saad, M.K.; Saad, M.A.; Cortes, D.J.; Khader, M.M. Thermodynamics of CO_2 adsorption on polyethyleneimine mesoporous silica and activated carbon. *J. Phys. Chem. Biophys.* **2017**, *1*, 1–5.
20. Jhonatan, R.G.R.; Juan Carlos, M.P.; Liliana, G.G. Kinetic and Equilibrium Study of the Adsorption of CO_2 in Ultramicropores of Resorcinol-Formaldehyde Aerogels Obtained in Acidic and Basic Medium. *J. Carbon Res.* **2018**, *4*, 1–19.
21. Eka, H.D.; Tajul, A.Y.; Wan, N.W.A. Potential use of Malaysian rubber (*Hevea brasiliensis*) seed as food, feed and biofuel. *Int. Food Res. J.* **2010**, *17*, 527–534.
22. Borhan, A.; Hamidi, M.N.R. Modification of Rubber-Seed Shell Activated Carbon using Chitosan for Removal of Cu^{2+} and Pb^{2+} from Aqueous Solution. In Proceedings of the International Symposium on Green and Sustainable Technology (ISGST 2019), Perak, Malaysia, 23–26 April 2019; Volume 2157, p. 020024.

23. Cao, Q.; Xie, K.C.; Liv, Y.K.; Bao, W.R. Process effects of activated carbon with large specific area from corncob. *Bioresour. Technol.* **2012**, *97*, 110–115. [CrossRef] [PubMed]
24. Ogungbenro, A.E.; Quang, D.V.; Al-Ali, K.; Abu-Zahra, M.R.M. Activated carbon from date seeds for CO_2 capture applications. *Energy Procedia* **2017**, *114*, 2313–2321. [CrossRef]
25. Chao, G.; Dandan, L.; Shaopeng, C.; Jie, G.; Jianjun, L. Highly Selective CO_2 Capture on Waste Polyurethane Foam-Based Activated Carbon. *Processes* **2019**, *7*, 1–15.
26. Mays, T.J. A new classification of pore sizes. *Stud. Surf. Sci. Catal.* **2007**, *160*, 57–62.
27. Esteves, I.A.A.C.; Lopes, M.S.S.; Nunes, P.M.C.; Mota, J.B.P. Adsorption of natural gas and biogas components on activated carbon. *Sep. Purif. Technol.* **2008**, *62*, 281–296. [CrossRef]
28. Hauchhum, L.; Mahanta, P. Kinetic, thermodynamics and regeneration studies for CO_2 adsorption onto activated carbon. *Int. J. Adv. Mech. Eng.* **2014**, *4*, 27–32.
29. Simon, J.C.; Bushra, A.D.; Nannan, S.; Cheng-gong, S.; Colin, E.S.; Kaixi, L.S.; Joseph, W. Carbon Dioxide Separation from Nitrogen/Hydrogen Mixtures over Activated Carbon Beads: Adsorption Isotherms and Breakthrough Studies. *Energy Fuels* **2015**, *29*, 3796–3807.
30. Perez, N.; Sanchez, M.; Rincon, G.; Delgado, L. Study of the behaviour of metal adsorption in acid solutions on lignin using a comparison of different adsorption isotherms. *Lat. Am. Appl. Res.* **2015**, *37*, 157–162.
31. Li, J.; Hitch, M. Carbon dioxide sorption isotherm study on pristine and acid-treated olive and its application in the vacuum swing adsorption process. *Miner* **2015**, *5*, 259–275. [CrossRef]
32. Sadaf, S.; Bhatti, H.N. Evaluation of peanut husk as a novel, low cost biosorbent for the removal of Indosol Orange RSN dye from aqueous solutions: Batch and fixed bed studies. *Clean Technol. Environ. Policy* **2014**, *16*, 527–544. [CrossRef]
33. Jia, Z.; Li, Z.; Ni, T.; Li, S. Adsorption of low-cost absorption materials based on biomass (Cortaderia selloana flower spikes) for dye removal: Kinetics, isotherms and thermodynamic studies. *J. Mol. Liq.* **2017**, *229*, 285–292. [CrossRef]
34. Rashidi, N.A.; Yusup, S.; Hameed, B.H. Kinetic studies on CO_2 capture using lignocellulosic based activated carbon. *Energy* **2013**, *61*, 440–446. [CrossRef]
35. Balsamo, M.; Budinova, T.; Erto, A.; Lancia, A.; Petrova, B.; Petrov, N.; Tsyntsarski, B. CO_2 adsorption onto synthetic activated carbon: Kinetic, thermodynamic and regeneration studies. *Sep. Purif. Technol.* **2013**, *116*, 214–221. [CrossRef]
36. Shieu, A.; Hu, S.C.; Chang, S.M.; Ko, T.Y.; Hsieh, A.; Chan, A. Adsorption kinetics and breakthrough of carbon dioxide for the chemical modified activated carbon filter used in the building. *Sustanability* **2017**, *9*, 1533. [CrossRef]
37. Khalili, S.; Khoshandam, B.; Jahanshahi, M. Optimization of production conditions for synthesis of chemically activated carbon produced from pine cone using response surface methadology for CO_2 adsorption. *RSC Adv.* **2015**, *5*, 94115–94129. [CrossRef]
38. Ismat, H.A.; Mohammed, K.A.; Mohammad, I.K.; Mohd, D.; Majed, M.A. Exploring Adsorption Process of Lead (II) and Chromium (VI) Ions from Aqueous Solutions on Acid Activated Carbon Prepared from Juniperus procera Leaves. *Processes* **2019**, *7*, 1–14.

Article

Thermophysical Properties and CO_2 Absorption of Ammonium-Based Protic Ionic Liquids Containing Acetate and Butyrate Anions

Normawati M. Yunus [1,*], Nur Hamizah Halim [2], Cecilia Devi Wilfred [1], Thanabalan Murugesan [3], Jun Wei Lim [2] and Pau Loke Show [4]

1 Department of Fundamental and Applied Sciences, Center of Research in Ionic Liquids (CORIL), Institute of Contaminant Management for Oil and Gas, Universiti Teknologi PETRONAS, Seri Iskandar 32610, Perak, Malaysia
2 Department of Fundamental and Applied Sciences, Universiti Teknologi PETRONAS, Seri Iskandar 32610, Perak, Malaysia
3 Department of Chemical Engineering, Universiti Teknologi PETRONAS, Seri Iskandar 32610, Perak, Malaysia
4 Department of Chemical and Environmental Engineering, Faculty of Science and Engineering, University of Nottingham Malaysia, Jalan Broga Semenyih 43500, Selangor, Malaysia
* Correspondence: normaw@utp.edu.my; Tel.: +605-368-7689; Fax: +605-365-5905

Received: 17 October 2019; Accepted: 4 November 2019; Published: 5 November 2019

Abstract: Ionic liquids, which are classified as new solvents, have been identified to be potential solvents in the application of CO_2 capture. In this work, six ammonium-based protic ionic liquids, containing ethanolammonium [EtOHA], tributylammonium [TBA], bis(2-ethylhexyl)ammonium [BEHA] cations, and acetate [AC] and butyrate [BA] anions, were synthesized and characterized. The thermophysical properties of the ammonium-based protic ionic liquids were measured. Density, ρ, and dynamic viscosity, η, were determined at temperatures between 293.15 K and 363.15 K. The density and viscosity values were correlated using empirical correlations and the thermal coefficient expansion, α_p, and molecular volume, V_m, were estimated using density values. The thermal stability of the ammonium-based protic ionic liquids was investigated using thermogravimetric analyzer (TGA) at a heating rate of 10 °C·min^{-1}. The CO_2 absorption of the ammonium-based ionic liquids were measured up to 20 bar at 298.15 K. From the experimental results, [BEHA][BA] had the highest affinity towards CO_2 with the mol fraction of CO_2 absorbed approaching 0.5 at 20 bar. Generally, ionic liquids with butyrate anions have better CO_2 absorption than that of acetate anions while [BEHA] ionic liquids have higher affinity towards CO_2 followed by [TBA] and [EtOHA] ionic liquids.

Keywords: ammonium-based protic ionic liquids; density; thermal expansion coefficient; viscosity; thermal stability; CO_2 absorption

1. Introduction

Natural gas consists mainly of methane as well as other higher alkanes in varied amounts. It is mainly used as a fuel and as a raw material in petrochemical industries [1]. While natural gas is principally a mixture of combustible hydrocarbons, many natural gases also contain impurities, such as carbon dioxide, CO_2, hydrogen sulfide, H_2S, and water. Refining processes are required to remove all of these unwanted impurities from natural gas. Besides water and higher-molecular-weight hydrocarbons, one of the most crucial parts of gas processing is the elimination of CO_2 and this process is normally done by means of chemical absorption techniques using alkanolamine solutions. Despite the successful practice of using alkanolamines for CO_2 removal, several disadvantages have been identified, such as solvent loss and degradation as well as corrosion issues [2]. In view of these issues, there is a need to develop new alternative, yet effective solvents for the same purposes. Ionic liquids have emerged as new solvents that have potential to be used for CO_2 removal due to their special features, namely non-detectable vapor pressure, wide liquid range, and remarkable thermal stability. Ionic liquids are low melting salts with typical melting points of below 100 °C. Furthermore, there are countless possible combinations of cations and anions that can yield ionic liquids and this flexibility is utilized to design ionic liquids based on the application. Ionic liquids are used in many applications, such as in chemical reactions and separation processes [3–6]. In the field of CO_2 capture using ionic liquids, initial work on the CO_2-ionic liquid system was done in 2001 [7] and, following this discovery, extensive works have been done to explore the absorption of CO_2 in ionic liquids under various operating conditions [8–15]. Imidazolium-based ionic liquids were used in most of the study of CO_2 absorption. They are highly unsymmetrical and therefore have low melting points. Recently, protic ionic liquids have been used in the study of CO_2 capture [16,17]. Protic ionic liquids are formed by a proton transfer between an equimolar amount of a Brønsted acid and a Brønsted base. The modern type of protic ionic liquids have been described by Ohno and co-workers [18] and an extensive review on the properties and applications of protic ionic liquids has been provided by Greaves and Drummond [19] who also noted the ability of protic ionic liquids to support amphiphile self-assembly [20,21]. The interest in the ionicity of protic ionic liquids has come to light due to the unlikeliness of complete proton transfer between the acid and base contributing to the presence of a neutral acid and base mixture [19,22,23]. MacFarlane and Seddon [24] proposed a limit of a 1% neutral species presence in an ionic liquid to be called 'pure ionic liquid'. The term pseudo-protic ionic liquid was suggested by Doi et al. [25] after they discovered that a mixture of *N*-methylimidazole and acetic acid exhibited electrical conductivity behavior even though the mixture was mostly dominated by neutral species rather than ions when inspected using Raman spectroscopy.

Nevertheless, the simple synthesis pathway of protic ionic liquids, i.e., a one-step neutralization reaction, and their proven ability to absorb CO_2 motivated us to explore this type of ionic liquids in the field of CO_2 capture. Nonetheless, physical properties, such as density, viscosity, and thermal stability of ionic liquids, are very important prior to using these new solvents in any applications. Precise understanding on the thermophysical properties is important as it is required to evaluate the suitability of ionic liquids to be used at an industrial scale [26]. For instance, viscosity is an important property for the design of industrial processes involving heat and mass transfer and dissolution of compounds in fluids [27]. Therefore, the aim of our work was to synthesize several new ammonium-based protic ionic liquids using a 1-step neutralization reaction, measure their thermophysical properties, and, lastly, test their ability to capture CO_2. In this work, six ammonium-based protic ionic liquids, containing acetate and butyrate anions, were synthesized using solvent-free, 1-step neutralization reaction. The density, dynamic viscosity, and thermal stability of these ionic liquids were determined. The density values enable the estimation of thermal expansion coefficient and the molecular volume of the ionic liquids. To assess the capability of these ionic liquids towards CO_2, absorption of CO_2 was done using a solubility cell and the screening was done in the CO_2 pressure range up to 20 bar and at 298.15 K. Results showed that the ammonium-based protic ionic liquids synthesized in this study have the potential to absorb CO_2.

2. Materials and Methods

2.1. Chemicals

Three amines and two organic acids from Merck Sdn. Bhd. were used in the production of the ammonium-based protic ionic liquids. All chemicals were of analytical grade. The amines and acids CAS numbers, abbreviations, and grades are as follows: ethanolamine (141–43–5, 99%), bis(2–ethylhexyl)amine (106–20–7, 99%), tributylamine (102–82–9, 99%), acetic acid (64–19–7, 99.8%), and butyric acid (107–92–6, 99%).

2.2. Synthesis

For the synthesis of each of the ammonium-based protic ionic liquids, an equimolar amount of the acid was added dropwise to the amine at ambient conditions and the mixture was consistently stirred for 24 h to facilitate mixing. The resulting solution was dried under vacuum at 65 °C for 6 h to remove remaining reactants. The final product was kept in a seal container until further use. The combinations of two acids and three amines produce six ammonium-based protic ionic liquids. Table 1 shows the structures and the abbreviation used for the ionic liquids. All ionic liquids exist as liquids except [BEHA][AC], which exists as a solid compound at room temperature.

Table 1. Structures of cations and anions, names and abbreviations.

Structure	Name and Abbreviation
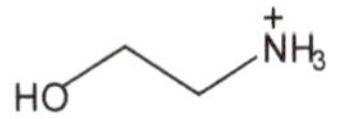 $HO-CH_2-CH_2-NH_3^+$	Ethanolammonium, [EtOHA]
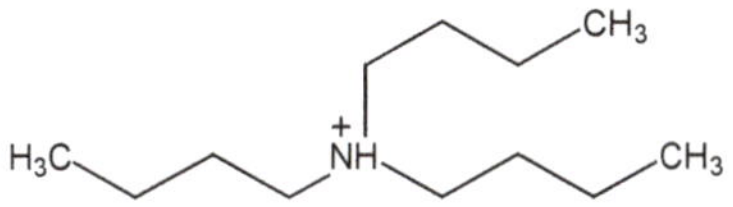 $(C_4H_9)_3NH^+$	Tributylammonium, [TBA]
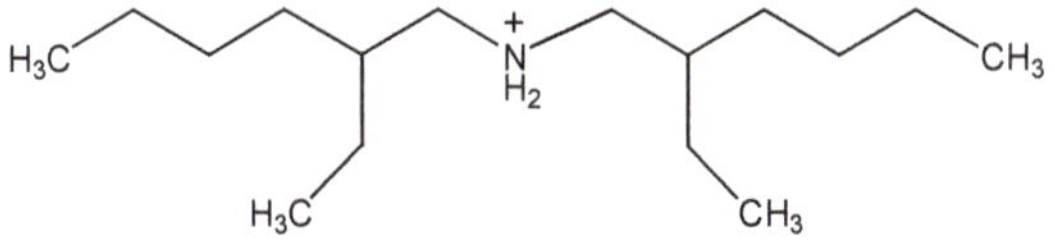 $[CH_3(CH_2)_3CH(C_2H_5)CH_2]_2NH_2^+$	Bis(2-ethylhexyl)ammonium, [BEHA]
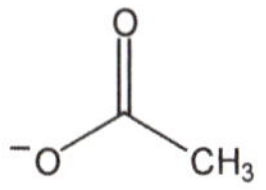 CH_3COO^-	Acetate, [AC]
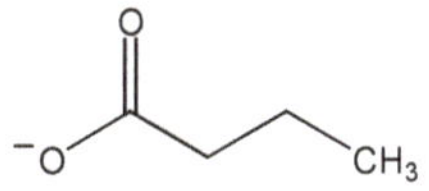 $CH_3CH_2CH_2COO^-$	Butyrate, [BA]

2.3. NMR and Water Content

The structure of the ammonium-based protic ionic liquids was analyzed and confirmed via nuclear magnetic resonance (NMR) spectroscopy. About 5 mg sample of ionic liquid was dissolved in 6 mL deuterated solvent and the sample's purity was determined using 500 MHz Bruker NMR Oxford Instrument. Coulometric Karl Fischer autotitrator DL39 from Mettler was used to determine the water content of the ionic liquids.

2.4. Thermophysical Characterization

The viscosity and density of the ammonium-based protic ionic liquids were determined simultaneously using Anton Parr Stabinger Viscometer SVM3000 in the temperature range of 293.15 K to 363.15 K. The temperature measurement's accuracy was within 0.02 K while the reproducibility of the viscosity and density measurements were 0.35% and $\pm 5.10^{-4}$ $g \cdot cm^{-3}$, respectively [28]. The decomposition temperatures of the ionic liquids were examined by means of thermogravimetric analyzer, TGA Perkin Elmer STA 6000. About 10 mg of sample was loaded into a platinum pan and the sample was heated at a heating rate of 10 $°C \cdot min^{-1}$ under nitrogen flow.

2.5. CO_2 Absorption Measurement

The ability of the ammonium-based protic ionic liquids to absorb CO_2 was investigated based on a pressure drop technique using a solubility cell as described in our previous publication [29]. The solubility cell consists of an equilibrium cell and a gas vessel immersed in a thermostatic bath. In a pressure drop method, the gas with a known pressure at constant volume is allowed to be in contact with the ionic liquid in the equilibrium cell and the pressure drop is monitored as the gas absorbs into the ionic liquid until equilibrium is attained. In a typical experiment, the equilibrium cell was loaded with a pre-weighed amount of the ionic liquid and the equilibrium cell was evacuated to remove any gases. In the gas vessel, CO_2 was allowed to stabilize before being quickly charged into the equilibrium cell. The CO_2-ionic liquid system was assumed to achieve equilibrium when the pressure attained a constant value. The system was maintained in that conditions for an additional two hours to ensure equilibration. Equation (1) was used to calculate the amount of CO_2 absorbed in the ionic liquid, n_2 [30]:

$$n_2 = \frac{P_{ini} V_{total}}{Z_2(P_{ini}, T_{ini}) R T_{ini}} - \frac{P_{eq}\left[V_{total} - V_{liq}\right]}{Z_2(P_{eq}, T_{eq}) R T_{eq}}, \tag{1}$$

where P_{ini} and T_{ini} are the initial pressure and temperature of the system, P_{eq} and T_{eq} are the pressure and temperature of the system at equilibrium, V_{total} is the volume of the equilibrium cell, V_{liq} is the volume of ionic liquid, R is the gas constant, and Z_2 represents the compressibility factor of the gas. Z_2 can be calculated using Soave–Redlich–Kwong equation of state [31]. The mole fraction of CO_2 absorbed in the ionic liquid (x_2) was calculated using Equation (2):

$$x_2 = \frac{{n_2}^{liq}}{\left({n_2}^{liq} + {n_1}^{liq}\right)}, \tag{2}$$

where ${n_2}^{liq}$ represents the mole of dissolved CO_2 and ${n_1}^{liq}$ is the mole of the ionic liquid.

3. Results and Discussion

In this work, six ammonium-based protic ionic liquids—ethanolammonium acetate [EtOHA][AC], ethanolammonium butyrate [EtOHA][BA], tributylammonium acetate [TBA][AC], tributylammonium butyrate [TBA][BA], bis(2-ethylhexyl)ammonium acetate [BEHA][AC], and bis(2-ethylhexyl)ammonium butyrate [BEHA][BA]—were synthesized and characterized. All the ammonium-based protic ionic liquids exist as liquids at room temperature except [BEHA][AC], which is a solid. The NMR and water content of each ionic liquids are presented as follows:

[EtOHA][AC]: ^{1}H NMR (500 MHz, D_2O): δ 3.613 [t, 2H, H_2(-OH)], 2.922 [t, 2H, CH_2(-NH_2)], 1.726 [s, 3H, CH_3]. Water content: 2.93%

[EtOHA][BA]: ^{1}H NMR (500 MHz, D_2O): δ 3.669 [t, 2H, CH_2(-OH)], 2.988 [t, 2H, CH_2(-NH_2)], 2.010 [t, 2H, CH_2(-COOH)], 1.375–1.448 [m, 2H, CH_2], 0.747 [t, 3H, CH_3]. Water content: 2.06%

[BEHA][AC]: ^{1}H NMR (500 MHz, D_2O): δ 2.861 [d, 4H, CH_2(-NH)], 1.805 [s, 3H, CH_3], 1.613–1.677 [m, 2H, CH], 1.181–1.309 [m, 16H, CH_2], 0.786 [t, 12H, CH_3]. Solid

[BEHA][BA]: ^{1}H NMR (500 MHz, D_2O): δ 2.856 [d, 4H, CH_2(-NH)], 2.062 [t, 2H, CH_2(-COOH)], 1.582–1.660 [m, 2H, CH], 1.412–1.486 [m, 2H, CH_2(AC)], 1.236–1.339 [m, 16H, CH_2], 1.214-1.229 [t,t 15H, CH_3]. Water content: 0.15%

[TBA][AC]: ^{1}H NMR (500 MHz, D_2O): δ 3.002 [t, 4H, CH_2(-NH)], 1.788 [s, 3H, CH_3], 1.508–10571 [m, 6H, CH_2], 1.207–1.282 [m, 6H, CH_2], 0.805 [t, 9H, CH_3]. Water content: 0.47%

[TBA][BA]: ^{1}H NMR (500 MHz, D_2O): δ 3.003 [t, 6H, CH_2(-NH)], 2.031 [t, 2H, CH_2(-COOH)], 1.510–1.573 [m, 6H, CH_2], 1.394–1.468 [m, 2H, CH_2(AC)], 1.211-1.285 [m, 6H, CH_2], 0.809 [t, 9H, CH_3], 0.768 [t, 3H, CH_3(AC)]. Water content: 0.23%

3.1. Thermophysical Properties

The experimental density and dynamic viscosity values for all liquid samples of synthesized ammonium-based protic ionic liquids are presented in Tables 2 and 3. The experimental densities of [EtOHA][AC], [EtOHA][BA], [BEHA][BA], [TBA][AC], and [TBA][BA] as a function of temperature are shown in Figures 1–3. [BEHA][AC] is not included as it exists as solid. As can be seen from Figure 1, the density of all five ammonium-based protic ionic liquids decreased gradually and linearly with increasing temperature over the range of temperature studied. An increase in temperature caused higher mobility of the ions which, in turn, weakens the intermolecular forces between the constituent ions and correspondingly increases the unit volume for these ions [32]. The density of these ammonium-based protic ionic liquids was slightly affected by the length of the alkyl chain of the anion in which the density of ionic liquids with the [AC] anion was higher than of ionic liquids with the [BA] anion for a fixed cation, as shown in Figure 2a,b. This observation is consistent with the literature in which it has been shown that the density value drops as the alkyl chain gets longer [28,32–37]. Our experimental density value of [EtOHA][AC] is in good agreement with Kurnia et al. [35] and Hosseini et al. [38] with the value differences of less than 0.2% and 0.8%, respectively. Generally, effective arrangement of ions in a liquid can increase the density of the liquid due to a greater number of ions available in a unit volume [39]. Based on our experimental results, as shown in Figure 3, ionic liquids with the [EtOHA] cation have higher density values compared to the rest, suggesting a better packing of the ions due to the small size of the cation. [TBA][BA] had the lowest density values at all temperatures due to the combined effects of branching of the cation, which increases the asymmetricity and steric hindrance of the ionic liquid [32], along with the large size of the alkyl chain of the [BA] anion.

Table 2. Density (ρ) values of ionic liquids from 293.15 K to 363.15 K.

T/K	ρ ($g \cdot cm^{-3}$)				
	[EtOHA][AC]	[EtOHA][BA]	[BEHA][BA]	[TBA][AC]	[TBA] [BA]
293.15	1.1468	1.0772	0.8695	0.9118	0.8591
303.15	1.1410	1.0725	0.8613	0.9035	0.8531
313.15	1.1359	1.0657	0.8532	0.8952	0.8445
323.15	1.1305	1.0579	0.8451	0.8869	0.8356
333.15	1.1250	1.0516	0.8368	0.8785	0.8268
343.15	1.1194	1.0452	0.8285	0.8700	0.8185
353.15	1.1137	1.0387	0.8203	0.8614	0.8102
363.15	1.1079	1.0324	0.8124	0.8527	0.8020

Table 3. Dynamic viscosity (η) values of ionic liquids from 293.15 K to 363.15 K.

T/K	η (mPa·s)				
	[EtOHA][AC]	[EtOHA][BA]	[BEHA][BA]	[TBA][AC]	[TBA] [BA]
293.15	385.760	1814.000	26.061	1.8482	8.368
303.15	197.860	883.310	15.364	1.5090	5.832
313.15	110.780	464.320	9.780	1.1485	4.225
323.15	66.947	261.220	6.628	1.0573	3.185
333.15	43.099	155.690	4.706	0.9041	2.457
343.15	29.214	97.348	3.478	0.7886	1.949
353.15	20.677	63.345	2.652	0.6928	1.579
363.15	15.157	42.651	1.874	0.6149	1.287

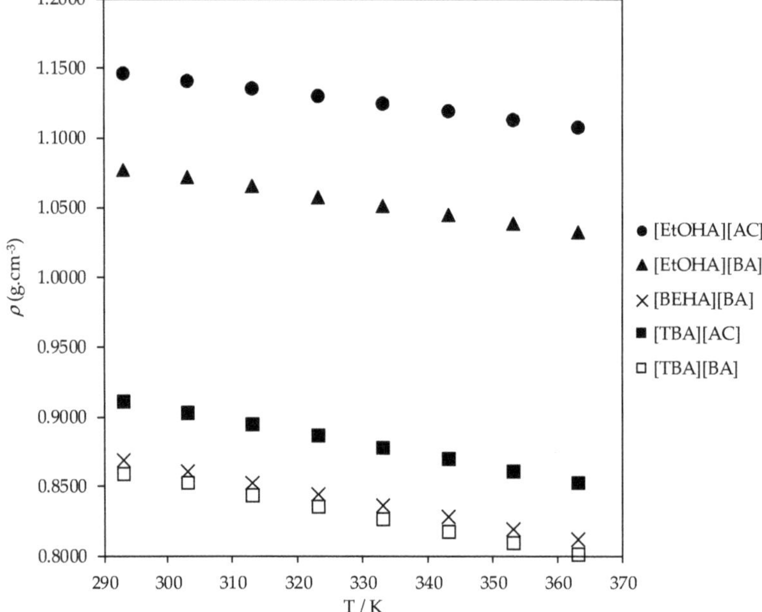

Figure 1. Plot of experimental density (ρ) values of ammonium-based protic ionic liquids.

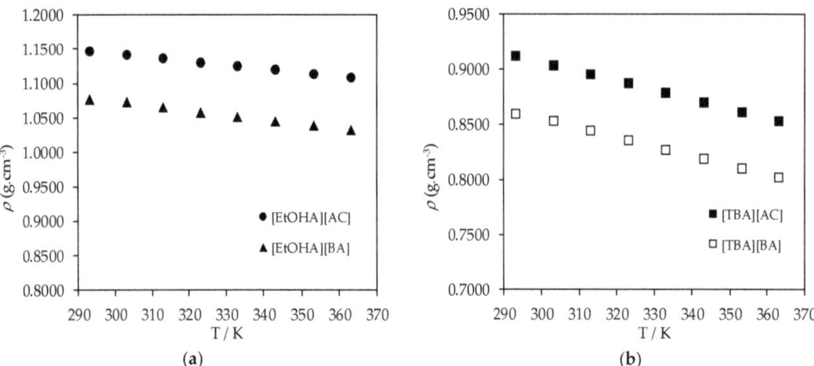

Figure 2. (**a**) Plot of experimental density (ρ) values of [EtOHA][AC] and [EtOHA][BA] and (**b**) plot of experimental density (ρ) values of [TBA][AC] and [TBA][BA] as a function of temperature.

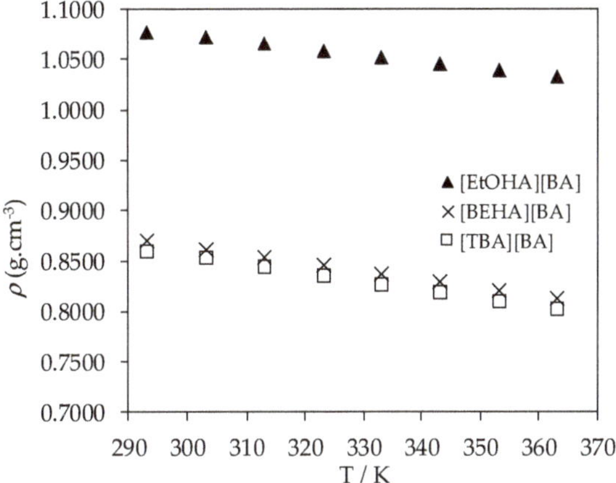

Figure 3. Plot of experimental density (ρ) values of [EtOHA][BA], [BEHA][BA], and [TBA][BA] as a function of temperature.

The dynamic viscosity of the ammonium-based protic ionic liquids, presented in Figure 4, dropped significantly as the temperature increased and the viscosity of ionic liquids with a [BA] anion was higher than that of ionic liquids with an [AC] anion for each type of cation studied in this work, as shown in Figure 5. The longer the alkyl chain in the ionic liquid structure, the higher the viscosity of the ionic liquids due to the increase in van der Waals attraction between the aliphatic alkyl chains [35]. On the other hand, [EtOHA] ionic liquids have remarkable high viscosity values compared to [BEHA] and [TBA] ionic liquids due to the presence of the hydroxyl group which enables a hydrogen bonding interaction between the structures of the ions.

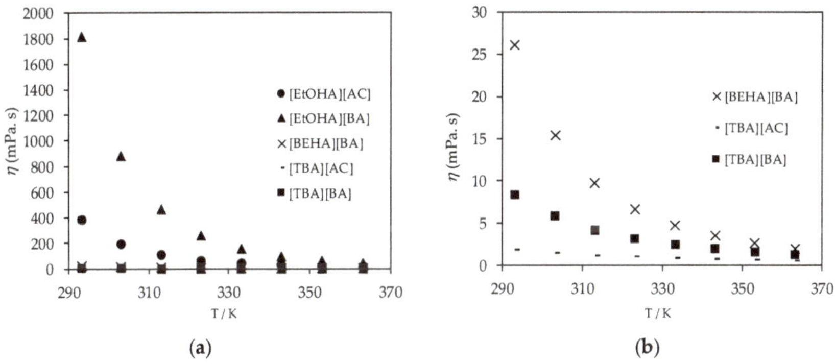

Figure 4. (**a**) Plot of experimental viscosity (η) values of ammonium based ionic liquids and (**b**) plot of experimental viscosity (η) values of [BEHA][BA], [TBA][AC], and [TBA][BA].

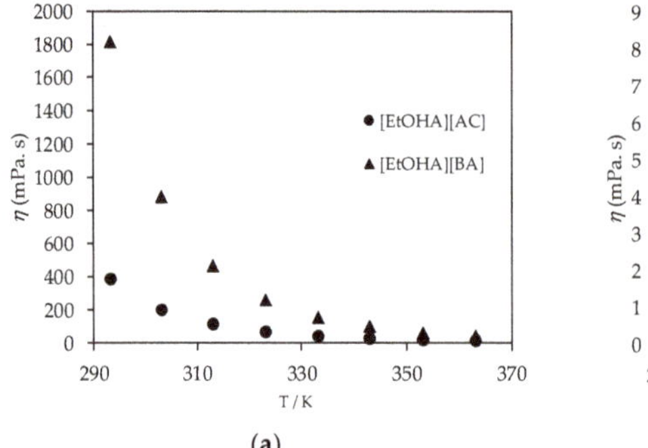

(a)

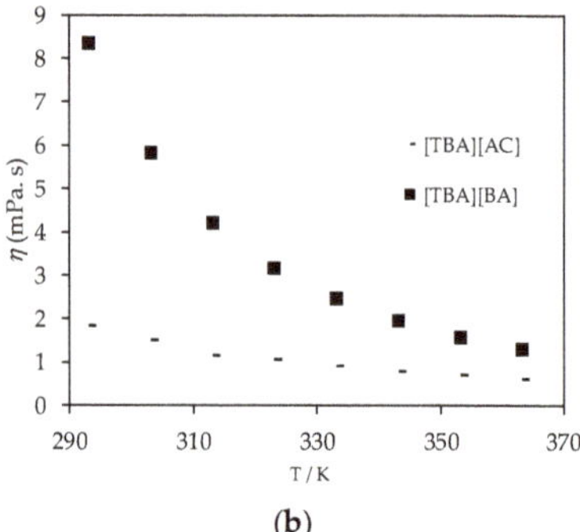

(b)

Figure 5. (**a**) plot of experimental viscosity (η) values of [EtOHA][AC] and [EtOHA][BA] ionic liquids and (**b**) plot of experimental viscosity (η) values of [TBA][AC] and [TBA][BA].

The values of density, ρ and dynamic viscosity, η were fitted using Equations (3) and (4) [28]:

$$\rho = A_0 + A_1 T, \quad (3)$$

$$\lg \eta = A_2 + A_3/T, \quad (4)$$

where ρ is the density, η is the dynamic viscosity of the ionic liquids, T is temperature in K, and A_0, A_1, A_2, and A_3 are correlation coefficients determined using the method of least squares. The calculated correlation coefficients together with standard deviations, SD are presented in Tables 4 and 5. The standard deviations, SD, were calculated using Equation (5) in which Z_{expt} and Z_{calc} are experimental and calculated values, respectively, while n_{DAT} is the number of experimental points:

$$SD = \sqrt{\frac{\sum_{i}^{n_{DAT}} (Z_{\exp t} - Z_{calc})^2}{n_{DAT}}}. \quad (5)$$

Table 4. Fitting parameters of Equation (3) for density (ρ) correlation and standard deviation (SD) from Equation (5).

Ionic Liquid	A_0	A_1	SD
[EtOHA][AC]	1.3087	−0.0006	0.016
[EtOHA][BA]	1.2702	−0.0007	0.015
[BEHA][BA]	1.1093	−0.0008	0.006
[TBA][AC]	1.1592	−0.0008	0.014
[TBA][BA]	1.1051	−0.0008	0.011

Table 5. Fitting parameters of Equation (4) for dynamic viscosity (η) correlation and standard deviation (SD) from Equation (5).

Ionic Liquid	A_2	A_3	SD
[EtOHA][AC]	−4.7222	2127.7	0.031
[EtOHA][BA]	−5.1998	2469.4	0.021
[BEHA][BA]	−4.4045	1696.4	0.020
[TBA][AC]	−2.1829	713.31	0.014
[TBA][BA]	−3.2888	1229.3	0.011

The thermal expansion coefficient, α_p, for the ammonium-based protic ionic liquids can be calculated using Equation (6) [28] while the molecular volume, V_m can be estimated from Equation (7) in which M is the molar mass of the ionic liquid and N_A represents the Avogadro's number [32,36,37]:

$$\alpha_p = -1/\rho \cdot (\partial\rho/\partial T)_p = -(A_1)/(A_0 + A_1 T) \text{ and} \tag{6}$$

$$V_m = M/N_A \cdot \rho. \tag{7}$$

The calculated thermal expansion coefficients and molecular volume values of the ammonium-based protic ionic liquids are presented in Tables 6 and 7. The calculated values lie in the range of (5.3 to 9.8)·10^{-4} K^{-1} for all five ionic liquids. The thermal expansion coefficients were found to be quite consistent over the temperature range studied and therefore are considered to be temperature independent. The pattern of the results is consistent with other types of ionic liquids [27,28,36,40,41]. The molecular volume, V_m, of the [BA] ionic liquid was greater than that of [AC] for a fixed cation and this may be attributed to the presence of additional CH_2 groups [36,37]. In this work, the V_m decreased in the sequence of [BEHA] > [TBA] > [EtOHA] for a fixed anion, i.e., [BA].

Table 6. Thermal expansion coefficients (α_p) of the ionic liquids calculated using Equation (6).

T/K	$\alpha_p \times 10^{-4}/(K^{-1})$				
	[EtOHA][AC]	[EtOHA][BA]	[BEHA][BA]	[TBA][AC]	[TBA] [BA]
293.15	5.3	6.6	9.1	8.7	9.2
303.15	5.3	6.6	9.2	8.7	9.3
313.15	5.4	6.7	9.3	8.8	9.4
323.15	5.4	6.7	9.4	8.9	9.4
333.15	5.4	6.8	9.5	9.0	9.5
343.15	5.4	6.8	9.6	9.0	9.6
353.15	5.5	6.8	9.7	9.1	9.7
363.15	5.5	6.9	9.8	9.2	9.8

Table 7. Molecular volume (V_m) of the ionic liquids calculated using Equation (7).

T/K	nm^3				
	[EtOHA][AC]	[EtOHA][BA]	[BEHA][BA]	[TBA][AC]	[TBA] [BA]
293.15	0.1754	0.2300	0.6295	0.4470	0.5287
303.15	0.1763	0.2310	0.6355	0.4511	0.5324
313.15	0.1771	0.2325	0.6416	0.4553	0.5378
323.15	0.1780	0.2342	0.6477	0.4596	0.5436
333.15	0.1788	0.2356	0.6542	0.4640	0.5493
343.15	0.1797	0.2371	0.6607	0.4685	0.5549
353.15	0.1807	0.2386	0.6673	0.4732	0.5606
363.15	0.1816	0.2400	0.6738	0.4780	0.5663

The thermal decomposition (T_d) of the ammonium-based protic ionic liquids were measured at a heating rate of 10 °C·min^{-1}. The T_d was approximately determined by the intersection of the baseline weight from the beginning of the measurement and the tangent of the weight against the temperature curve as the decomposition process occurs. The T_d of the ionic liquids are presented in Table 8 and the thermal decomposition curves for [EtOHA][BA] and [BEHA][BA] are given in Figure 6. The thermal stability of the ammonium-based protic ionic liquids in this study varied with the ion combination. The T_d of [BA] ionic liquid was higher than that of [AC] ionic liquid for every type of cation studied while [EtOHA] ionic liquids displayed the highest T_d followed by [BEHA] and [TBA]. However, ammonium-based protic ionic liquids in this work and from the literature [36] tend to possess lower thermal stability compared to other ionic liquids, such as imidazolium and pyridinium ionic liquids [28,41,42]. However, generalization must not be made as the thermal stability depends largely on the combination of the cation and anion of the ionic liquids.

Table 8. Thermal decomposition (T_d) temperature of the ionic liquids.

	[EtOHA][AC]	[EtOHA][BA]	[BEHA][AC]	[BEHA][BA]	[TBA][AC]	[TBA] [BA]
T/°C	147.23	237.47	124.89	146.06	111.46	120.58

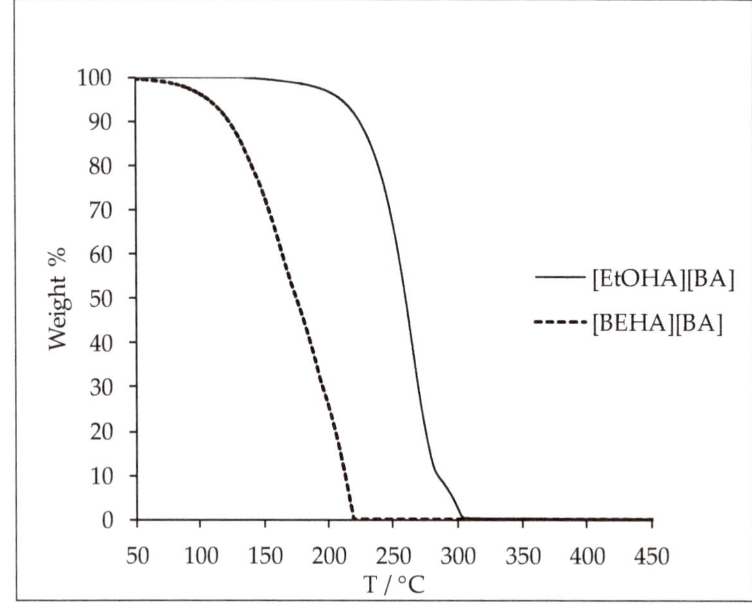

Figure 6. Plot of thermal decomposition of [EtOHA][BA] and [BEHA][BA].

3.2. CO_2 Absorption

The experimental results of CO_2 absorption in the ammonium-based protic ionic liquids are shown in Figure 7. Generally, the CO_2 absorption in these ammonium-based protic ionic liquids increased with pressure following Henry's law; the solubility of a gas in a liquid is proportional to the partial pressure of the gas above the surface of the liquid. The mol fraction of CO_2 absorbed in the ammonium-based protic ionic liquids was in the range of about 0.02 to 0.48 and up to 20 bar at 298.15 K. The effects of cation structure on the CO_2 absorption in the ionic liquids are shown in Figure 8. For a fixed anion, the solubility of CO_2 increased in the sequence of [EtOHA] < [TBA] < [BEHA] where the mol fraction of CO_2 absorbed in [BEHA][BA] was 0.486 in comparison to 0.328 and 0.307 in [TBA][BA] and [EtOHA][BA], respectively. Meanwhile, Figure 9 indicates a slight increase in the CO_2 solubility when the anion of a common cation was changed from [AC] to [BA]. Based on our experimental results, there is a relationship between absorption of CO_2 with the density and the molecular volume of the ionic liquids. As the density decreases and molecular volume increases, the fractional free volume increases and, thus, the solubility of CO_2 increases [43,44]. By using a common anion, the CO_2 absorption in [BEHA][AC] was compared with 1-ethyl-3-methylimidazolium acetate, [EMIM][AC] [45] and 1-butyl-3-methylimidazolium acetate, [BMIM][AC] [11]. At about 20 bar and 298.15 K, there was only a marginal difference between the CO_2 solubility in [BEHA][AC] compared to that of [EMIM][AC] and [BMIM][AC]. This result shows a positive indication that our newly synthesized [BEHA][AC] and [BEHA][BA] ionic liquids have comparable performance towards CO_2 capture when compared to more established type of ionic liquids. However, more experimental investigation and data are needed to further evaluate the potential ability of our ammonium-based protic ionic liquids in the application of CO_2 capture.

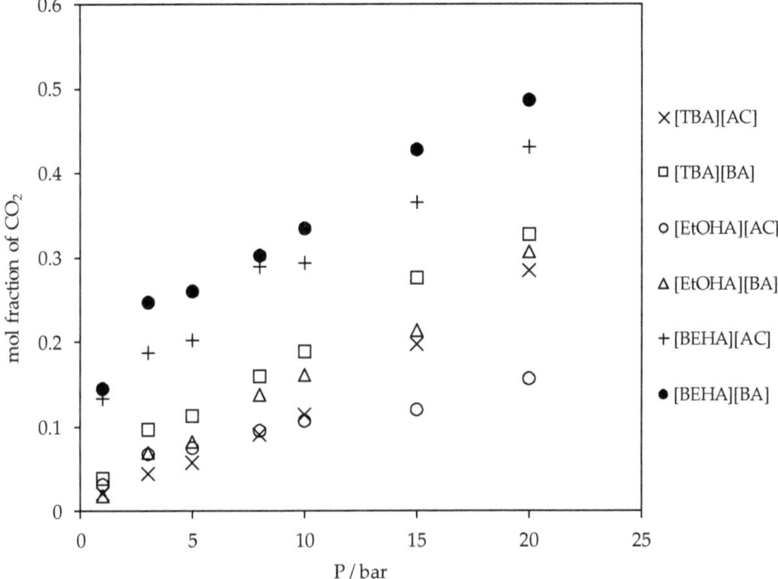

Figure 7. Plot of CO_2 absorption in ammonium-based protic ionic liquids at 298.15 K.

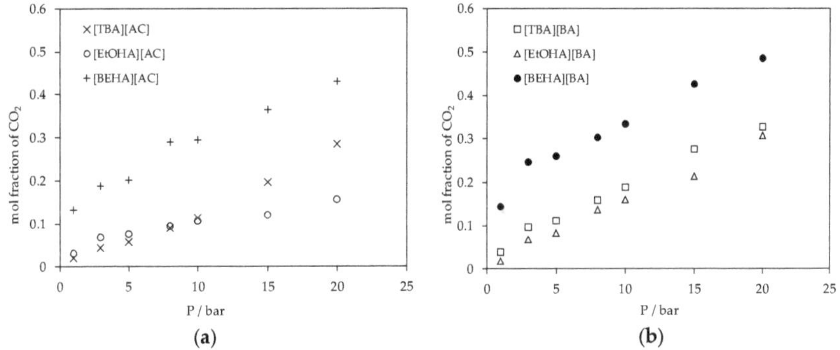

Figure 8. (**a**) Plot of CO_2 absorption in ammonium-based protic ionic liquids with [AC] anion and (**b**) plot of CO_2 absorption in ammonium-based protic ionic liquids with [BA] anion at 298.15 K.

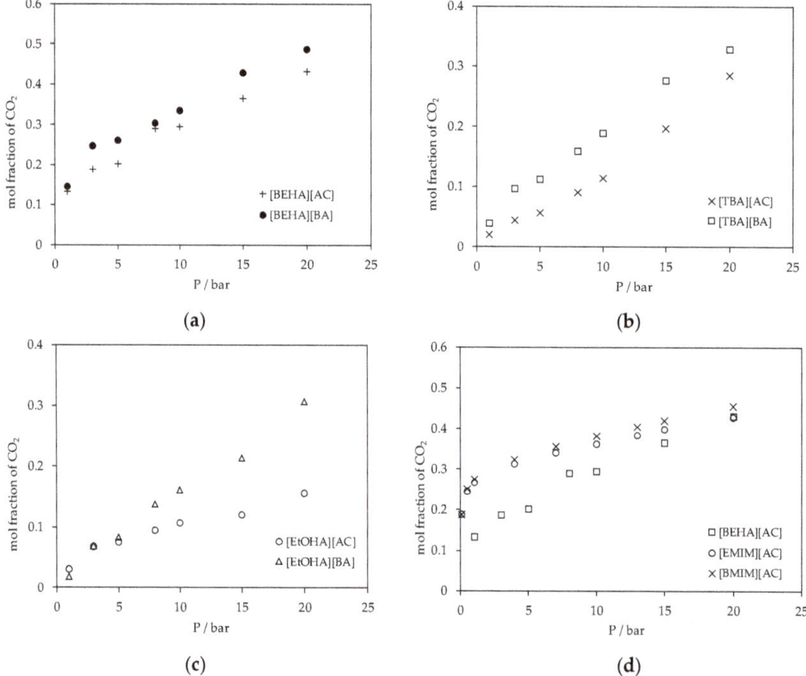

Figure 9. (**a**) Plot of CO_2 absorption in ammonium-based protic ionic liquids with the [BEHA] cation; (**b**) plot of CO_2 absorption in ammonium-based protic ionic liquids with the [TBA] cation; (**c**) plot of CO_2 absorption in ammonium-based protic ionic liquids with the [EtOHA] cation and; and (**d**) plot of CO_2 absorption in [BEHA][AC], [EMIM][AC] [45] and [BMIM][AC] [11] at 298.15 K.

4. Conclusions

Six ammonium-based protic ionic liquids were successfully synthesized via solvent-free 1-step neutralization reaction. The density, viscosity, and decomposition temperature were measured. The thermal expansion coefficient and the molecular volume were calculated using the density values. The density and viscosity values were inversely proportional with temperature in the range of temperature studied at atmospheric pressure. The density decreased when the alkyl chain of the anion increased, while the viscosity increased with the alkyl chain of the anion. The decomposition temperature of the ammonium-based protic ionic liquids was affected by the combination of cation and anion and [EtOHA] ionic liquids had the highest thermal stability when compared to the other ionic liquids. The absorption of CO_2 in the six ammonium-based protic ionic liquids was measured at 298.15 K and up to a pressure of 20 bar. The CO_2 absorption values in the ammonium-based protic ionic liquids increased with pressure and both the cation and anion affected the solubility of CO_2 in the ionic liquids. The amount of CO_2 absorbed was affected by the length of the alkyl chain of the anion while [BEHA] ionic liquids displayed higher CO_2 absorption capacity compared to [TBA] and [EtOHA] ionic liquids. Results indicate the potential of the ammonium-based protic ionic liquids to be used as solvents for CO_2 capture.

Author Contributions: Conceptualization, N.M.Y. and N.H.H.; methodology, N.H.H. and N.M.Y.; validation, N.H.H.; formal analysis, T.M.; resources, C.D.W. and T.M.; data curation, N.M.Y. and T.M.; writing—original draft preparation, N.M.Y. and N.H.H.; writing—review and editing, J.W.L. and P.L.S.; supervision, N.M.Y.; project administration, N.M.Y. and C.D.W.; funding acquisition, N.M.Y.

Funding: This research was funded by the Ministry of Higher Education (MOHE), Government of Malaysia, under the Fundamental Research Grant Scheme, [FRGS No: FRGS/1/2016/STG01/UTP/03/1] entitled *The impact of cations and anions structures of ionic liquids on the CO_2 absorption from natural gas* and also YUTP grant (cost centre 015LC0-054).

Acknowledgments: Financial assistance and support from Universiti Teknologi PETRONAS and Center of Research in Ionic Liquids (CORIL), UTP are greatly acknowledged.

Conflicts of Interest: All authors declare no conflict of interest.

References

1. Mokhatab, S.; Poe, W.A.; Speight, J.G. *Handbook of Natural Gas Transmission and Processing*; Gulf Processing Publishing, Elsevier: Amsterdam, The Netherlands, 2006.
2. Davis, J.; Rochelle, G. Thermal degradation of monoethanolamine at stripper conditions. *Energy Procedia* **2009**, *1*, 327–333. [CrossRef]
3. Jin, X.; Feng, J.; Li, S.; Song, H.; Yu, C.; Zhao, K.; Kong, F. A novel homogeneous catalysis-liquid/solid separation system for highly effective recycling of homogeneous catalyst based on a phosphine-functionalized polyether guanidium ionic liquid. *Mol. Catal.* **2019**, *475*, 110503. [CrossRef]
4. Zhang, P.; Jiang, Z.; Cui, Y.; Xie, G.; Jin, Y.; Guo, L.; Xu, Y.; Zhang, Q.; Li, X. Catalytic performance of ionic liquid for dehydrochlorination reaction: Excellent activity and unparallel stability. *Appl. Catal. B* **2019**, *255*, 117757. [CrossRef]
5. Abejón, R.; Rabadán, J.; Lanza, S.; Abejón, A.; Garea, A.; Irabien, A. Supported ionic liquid membranes for separation of lignin aqueous solutions. *Processes* **2018**, *6*, 143. [CrossRef]
6. Baicha, Z.; Salar-García, M.J.; Ortiz-Martínez, V.M.; Hernández-Fernández, F.J.; de los Ríos, A.P.; Maqueda Marín, D.P.; Collado, J.A.; Tomás-Alonso, F.; El Mahi, M. On the selective transport of nutrients through polymer inclusion membranes based on ionic liquids. *Processes* **2019**, *7*, 544. [CrossRef]
7. Blanchard, L.A.; Gu, Z.; Brennecke, J.F. High-pressure phase behavior of ionic liquid/CO_2 systems. *J. Phys. Chem. B* **2001**, *105*, 2437–2444. [CrossRef]
8. Anthony, J.L.; Maginn, E.J.; Brennecke, J.F. Solubilities and thermodynamic properties of gases in the ionic liquid 1-n-butyl-3-methylimidazolium hexafluorophosphate. *J. Phys. Chem. B* **2002**, *106*, 7315–7320. [CrossRef]
9. Kim, Y.S.; Choi, W.Y.; Jang, J.H.; Yoo, K.-P.; Lee, C.S. Solubility measurement and prediction of carbon dioxide in ionic liquids. *Fluid Phase Equilib.* **2005**, *228–229*, 439–445. [CrossRef]
10. Muldoon, M.J.; Aki, S.N.V.K.; Anderson, J.L.; Dixon, J.K.; Brennecke, J.F. Improving carbon dioxide solubility in ionic liquids. *J. Phys. Chem. B* **2007**, *111*, 9001–9009. [CrossRef]
11. Shiflett, M.B.; Kasprzak, D.J.; Junk, C.P.; Yokozeki, A. Phase behavior of {carbon dioxide + [bmim][Ac]} mixtures. *J. Chem. Thermodyn.* **2008**, *40*, 25–31. [CrossRef]
12. Gurkan, B.E.; de la Fuente, J.C.; Mindrup, E.M.; Ficke, L.E.; Goodrich, B.F.; Price, E.A.; Schneider, W.F.; Brennecke, J.F. Equimolar CO_2 absorption by anion-functionalized ionic liquids. *J. Am. Chem. Soc.* **2010**, *132*, 2116–2117. [CrossRef] [PubMed]
13. Sharma, P.; Park, S.D.; Baek, I.H.; Park, K.T.; Yoon, Y.I.; Jeong, S.K. Effects of anions on absorption capacity of carbon dioxide in acid functionalized ionic liquids. *Fuel Process. Technol.* **2012**, *100*, 55–62. [CrossRef]
14. Sistla, Y.S.; Khanna, A. Carbon dioxide absorption studies using amine-functionalized ionic liquids. *J. Ind. Eng. Chem.* **2014**, *20*, 2497–2509. [CrossRef]
15. Zoubeik, M.; Mohamedali, M.; Henni, A. Experimental solubility and thermodynamic modeling of CO_2 in four new imidazolium and pyridinium-based ionic liquids. *Fluid Phase Equilib.* **2016**, *419*, 67–74. [CrossRef]
16. Zhu, X.; Song, M.; Xu, Y. DBU-Based protic ionic liquids for CO_2 capture. *ACS Sustian. Chem. Eng.* **2017**, *5*, 8192–8198. [CrossRef]
17. Vijayaraghavan, R.; Oncsik, T.; Mitschke, B.; MacFarlane, D.R. Base-rich diamino protic ionic liquid mixtures for enhanced CO_2 capture. *Sep. Purif. Technol.* **2018**, *196*, 27–31. [CrossRef]
18. Hirao, M.; Sugimoto, H.; Ohno, H. Preparation of novel room-temperature molten salts by neutralization of amines. *J. Electrochem. Soc.* **2000**, *147*, 4168–4172. [CrossRef]
19. Greaves, T.L.; Drummond, C.J. Protic ionic liquids: Properties and applications. *Chem. Rev.* **2008**, *108*, 206–237. [CrossRef]

20. Greaves, T.L.; Drummond, C.J. Ionic liquids as amphiphile self-assembly media. *Chem. Soc. Rev.* **2008**, *37*, 1706–1726. [CrossRef]
21. Greaves, T.L.; Drummond, C.J. Solvent nanostructure, the solvophobic effect and amphiphile self-assembly in ionic liquids. *Chem. Soc. Rev.* **2013**, *42*, 833–1412. [CrossRef]
22. MacFarlane, D.R.; Forsyth, M.; Izgorodina, E.I.; Abbott, A.P.; Annat, G.; Fraser, K. On the concept of ionicity in ionic liquids. *Phys. Chem. Chem. Phys.* **2009**, *11*, 4962–4967. [CrossRef] [PubMed]
23. Shen, M.; Zhang, Y.; Chen, K.; Che, S.; Yao, J.; Li, H. Ionicity if protic ionic liquid: Quantitative measurement by spectroscopic methods. *J. Phys. Chem. B* **2017**, *121*, 1372–1376. [CrossRef] [PubMed]
24. MacFarlane, D.R.; Seddon, K.R. Ionic liquids–Progress on the fundamental issues. *Aust. J. Chem.* **2007**, *60*, 3–5. [CrossRef]
25. Doi, H.; Song, X.; Minofar, B.; Kanzaki, R.; Takamuku, T.; Umebayashi, Y. A new proton conductive liquid with no ions: Pseudo-protic ionic liquids. *Chem. Eur. J.* **2013**, *19*, 11522–11526. [CrossRef] [PubMed]
26. Aparicio, S.; Atilhan, M.; Karadas, F. Thermophysical properties of pure ionic liquids: Review of present situation. *Ind. Eng. Chem. Res.* **2010**, *49*, 9580–9595. [CrossRef]
27. Bhattacharjee, A.; Lopes-da-Silva, J.A.; Freire, M.G.; Coutinho, J.A.P.; Carvalho, P.J. Thermophysical properties of phosphonium-based ionic liquids. *Fluid Phase Equilib.* **2015**, *400*, 103–113. [CrossRef] [PubMed]
28. Yunus, N.M.; Abdul Mutalib, M.I.; Man, Z.; Bustam, M.A.; Murugesan, T. Thermophysical properties of 1-alkylpyridinium bis(trifluoromethylsulfonyl)imide ionic liquids. *J. Chem. Thermodyn.* **2010**, *42*, 491–495. [CrossRef]
29. Yunus, N.M.; Abdul Ghani, M.A.; Nik Mohamad Kamil, R. Synthesis, characterization and CO_2 solubility of [him][Tf_2N] and [hmim][Ac] ionic liquids. *AIP Conf. Proc.* **2014**, *1621*, 284–289.
30. Jacquemin, J.; Husson, P.; Majer, V.; Costa Gomes, M.F. Low-pressure solubilities and thermodynamics of solvation of eight gases in 1-butyl-3-methylimidazolium hexafluorophosphate. *Fluid Phase Equilib.* **2006**, *240*, 87–95. [CrossRef]
31. Soave, G. Equilibrium constants from a modified Redlich-Kwong equation of state. *Chem. Eng. Sci.* **1972**, *2*, 1197–1203. [CrossRef]
32. Gusain, R.; Panda, S.; Bakshi, P.S.; Gardas, R.L.; Khatri, O.P. Thermophysical properties of trioctylalkylammonium bis(salicylato)borate ionic liquids: Effect of alkyl chain length. *J. Mol. Liq.* **2018**, *269*, 540–546. [CrossRef]
33. Pinkert, A.; Ang, K.L.; Marsh, K.N.; Pang, S. Density, viscosity and electrical conductivity of protic alkanolammonium ionic liquids. *Phys. Chem. Chem. Phys.* **2011**, *13*, 5136–5143. [CrossRef] [PubMed]
34. Kurnia, K.A.; Wilfred, C.D.; Murugesan, T. Thermophysical properties of hydroxyl ammonium ionic liquids. *J. Chem. Thermodyn.* **2009**, *41*, 517–521. [CrossRef]
35. Machanová, K.; Boisset, A.; Sedláková, Z.; Anouti, M.; Bendová, M.; Jacquemin, J. Thermophysical properties of ammonium-based bis{(trifluoromethyl)sulfonyl}imide ionic liquids: Volumetric and transport properties. *J. Chem. Eng. Data* **2012**, *57*, 2227–2235. [CrossRef]
36. Chhotaray, P.K.; Gardas, R.L. Thermophysical properties of ammonium and hydroxylammonium protic ionic liquids. *J. Chem. Thermodyn.* **2014**, *72*, 117–124. [CrossRef]
37. Montalbán, M.G.; Bolívar, C.L.; Guillermo Díaz Baños, F.; Víllora, G. Effect of temperature, anion, and alkyl chain length on the density and refractive index of 1-alkyl-3-methylimidazolium-based ionic liquids. *J. Chem. Eng. Data* **2015**, *60*, 1986–1996. [CrossRef]
38. Hosseini, S.M.; Hosseinian, A.; Aparicio, S. An experimental and theoretical study on 2-hydroxyethylammonium acetate ionic liquid. *J. Mol. Liq.* **2019**, *284*, 271–281. [CrossRef]
39. Rios-Vera, R.M.; Sirieix-Plénet, J.; Gaillon, L.; Rizzi, C.; Ávila-Rodríguez, M.; Cote, G.; Chagnes, A. Physicochemical properties of novel cholinium ionic liquids for the recovery of silver from nitrate media. *RSC Adv.* **2015**, *5*, 78268–78277. [CrossRef]
40. Sarkar, A.; Sharma, G.; Singh, D.; Gardas, R.L. Effect of anion on thermophysical properties of *N*,*N*-diethanolammonium based protic ionic liquids. *J. Mol. Liq.* **2017**, *242*, 249–254. [CrossRef]
41. Vieira, N.S.M.; Luís, A.; Reis, P.M.; Carvalho, P.J.; Lopes-da-Silva, J.A.; Esperança, J.M.S.S.; Araújo, J.M.M.; Rebelo, L.P.N.; Freire, M.G.; Pereiro, A.B. Fluorination effects on the thermodynamic, thermophysical and surface properties of ionic liquids. *J. Chem. Thermodyn.* **2016**, *97*, 354–361. [CrossRef]

42. Shanchez-Ramirez, N.; Martins, V.L.; Ando, R.A.; Camilo, F.F.; Urahata, S.M.; Ribeiro, M.C.C.; Torresi, R.M. Physicochemical properties of three ionic liquids containing a tetracyanoborate anion and their lithium salt mixtures. *J. Phys. Chem. B* **2014**, *118*, 8772–8781. [CrossRef] [PubMed]
43. Gupta, K.M. Tetracyanoborate based ionic liquids for CO_2 capture: From ab initio calculations to molecular simulations. *Fluid Phase Equilib.* **2016**, *415*, 34–41. [CrossRef]
44. Shaikh, A.R.; Karkhanechi, H.; Kamio, E.; Yoshioka, T.; Matsuyama, H. Quantum mechanical and molecular dynamics simulations of dual-amino-acid ionic liquids for CO_2 capture. *J. Phys. Chem. C* **2016**, *120*, 27734–27745. [CrossRef]
45. Shiflett, M.B.; Yokozeki, A. Phase behavior of carbon dioxide in ionic liquids: [emim][acetate], [emim][trifluoroacetate], and [emim][acetate] + [emim][trifluoroacetate] mixtures. *J. Chem. Eng. Data* **2009**, *54*, 108–114. [CrossRef]

Article

Physical and Thermal Studies of Carbon-Enriched Silicon Oxycarbide Synthesized from Floating Plants

Guan-Ting Pan [1], Siewhui Chong [2], Yi Jing Chan [2], Timm Joyce Tiong [2], Jun Wei Lim [3], Chao-Ming Huang [4], Pradeep Shukla [5,*] and Thomas Chung-Kuang Yang [1,*]

1 Department of Chemical Engineering and Biotechnology, National Taipei University of Technology, Taipei 106, Taiwan; gtpan@ntut.edu.tw
2 Department of Chemical and Environmental Engineering, University of Nottingham Malaysia, Broga Road, Semenyih 43500, Malaysia; Faye.Chong@nottingham.edu.my (S.C.); Yi-Jing.Chan@nottingham.edu.my (Y.J.C.); Joyce.Tiong@nottingham.edu.my (T.J.T.)
3 Department of Fundamental and Applied Sciences, Centre for Biofuel and Biochemical Research, Institute of Self-Sustainable Building, Universiti Teknologi PETRONAS, Seri Iskandar 32610, Malaysia; Junwei.lim@utp.edu.my
4 Green Energy Technology Research Centre and Department of Materials Engineering, Kun Shan University, Tainan 710, Taiwan; charming@mail.ksu.edu.tw
5 Queensland Alliance for Environmental Health Sciences, Faculty of Health and Behavioural Sciences, The University of Queensland, St Lucia, QLD 4072, Australia
* Correspondence: Pradeep.shula@uq.edu.au (P.S.); ckyang@mail.ntut.edu.tw (T.C.-K.Y.); Tel.: +61-7-336-56195 (P.S.); +886-2-2771-2171 (T.C.-K.Y.)

Received: 6 September 2019; Accepted: 28 October 2019; Published: 2 November 2019

Abstract: In the present study, amorphous mesoporous silicon oxycarbide materials (SiOC) were successfully synthesized via a low-cost facile method by using potassium hydroxide activation, high temperature carbonization, and acid treatment. The precursors were obtained from floating plants (floating moss, water cabbage, and water caltrops). X-ray diffraction (XRD) results confirmed the amorphous Si–O–C structure and Raman spectra revealed the graphitized carbon phase. Floating moss sample resulted in a rather rough surface with irregular patches and water caltrops sample resulted in a highly porous network structure. The rough surface of the floating moss sample with greater particle size is caused by the high carbon/oxygen ratio (1: 0.29) and low amount of hydroxyl group compared to the other two samples. The pore volumes of these floating moss, water cabbage, and water caltrops samples were 0.4, 0.49, and 0.63 cm^3 g^{-1}, respectively, resulting in thermal conductivities of 6.55, 2.46, and 1.14 Wm^{-1} K^{-1}, respectively. Floating plants, or more specifically, floating moss, are thus a potential material for SiOC production.

Keywords: silicon oxycarbide; thermal conductivity; floating plants; SiOC; silica

---

1. Introduction

Silicon oxycarbide (SiOC) is an important material that finds application in semiconductor devices [1,2], electrode materials of lithium–ion batteries [3,4], micro electromechanical systems devices [5], high temperature sensors [6], conductive protective coatings [7], and super-capacitor [8] due to its superior mechanical properties. SiOC exhibits high strength [9], high chemical durability [10], excellent oxidation resistance under high temperature, good antioxidant crystallization property [11], and exceptional thermal expansion properties [12]. However, amongst these, thermal properties of silicon oxycarbide were rarely studied. Qiu et al. [13] reported a three-dimensional reticular macro-porous SiOC ceramic structure prepared by sol–gel process. The sample's porosity and specific surface area were reduced due to particle agglomeration, resulting in lower thermal conductivity that ranged from 0.041 to 0.062 Wm^{-1} K^{-1}. Recently, random porous structure has attracted an increasing interest due to its radiation tolerance and very high crystallization temperature, which in turn resulted in good thermal conductivity [14,15]. Gurlo et al. [16] reported that the thermal conductivity of SiOC material containing zirconium and hafnium was 1.3 $Wm^{-1} K^{-1}$ which was similar to that of silica. Mazo et al. [17] observed that silicon oxycarbide glasses that were synthesized by using spark plasma sintering had higher thermal conductivity values ($\approx$1.38 $Wm^{-1} K^{-1}$). Eom, et al. [18] prepared barium-added silicon oxycarbide (SiOC–Ba) via pyrolysis method and showed that the addition of Ba resulted in an increase in thermal conductivity values from 1.8 to 5.6 $Wm^{-1} K^{-1}$.

The traditional approach to producing amorphous SiOC microstructure has been via physical and chemical processing methods, such as powder metallurgy process [19], magnetron sputtering [20], sol–gel method [21,22], and chemical vapor deposition [23]. The raw materials used in synthesizing these SiOC materials were mainly chemical feedstocks such as tetraethoxysilane (TEOS)/hydroxyl-terminated polydimethylsiloxane (PDMS) organic–inorganic hybrid materials [17], polyxiloxane [18], methyltrimethoxysilane and dimethyldimethoxysilane [22], and dimethyl dimethoxy silane [21]. There are also limited studies on the thermal conductivity of SiOC materials. It could be expected that the material's structure affects its heat transfer properties as the type of particles and different pore sizes in the SiOC structure are the main reason to the different thermal conductivity abilities [12]. Low-cost sacrificial template has been widely used to generate replicated pore structures within such ceramic materials. Utilizing organic bio-mass, such as wood, or chemically modified biomass as a template for synthesizing SiOC with varying degree of porosity has been reported [24,25]. The use of selected biomass can also provide an added advantage of providing a source of silica for synthesis of SiOC.

These observations motivated our research for synthesizing SiOC using plant based biomass and to evaluate its physical properties and thermal conductivity. The main advantages of this method, as compared to other technologies, are its benign environmental nature and production of high-purity products via a low-cost process. In this study, silicon oxycarbide materials were synthesized by using a facile and inexpensive route from floating plants (floating moss, water cabbage, and water caltrops). The characterizations of these silicon oxycarbide materials were studied with regards to their structural, morphology, functional group, textural properties, and thermal conductivity.

2. Experimental

2.1. Preparation of Materials

All chemicals of analytical grade were purchased from Merck Chemicals. Three floating plants were used to synthesize carbon-enriched silicon oxycarbide (SiOC), namely, floating moss (Salvinia natans), water cabbage (Pistia stratiotes), and water caltrops (Trapa natans). The plants were collected from a pond at Kun Shan University, Tainan, Taiwan. The plants were washed and bone-dried in an oven at 100 °C before being grinded using a blender. For each sample synthesis, 0.071 mole potassium hydroxide (KOH) was dissolved in 40 mL of deionized water into which 1 gram of the dried and grounded biomass was added. The mixture was stirred for 2 h at 85 °C for activation before being annealed for 2 h in a tube furnace purged with nitrogen gas at 800 °C for carbonization. After that, the samples were added into 1.8 M hydrochloric acid (HCl) under magnetic stirring for 1 h. The resultant solid powder from the mixture solutions were washed by vacuum filtration until the pH value reached 7, and finally dried in vacuum at 110 °C overnight. The KOH activation enlarged the pores in these samples, thereby destroying their internal structure. These carbon-rich materials were then converted into pure carbon via high-temperature carbonization. Finally, acidic treatment (HCl) was used to alter the surface functional groups, surface morphology, and textural properties of the samples.

2.2. Characterization of Carbon-Enriched Silicon Oxycarbide

The crystalline structures of the carbon-enriched silicon oxycarbide samples were studied using an X-ray diffractometer (PANalytical X'Pert PRO, Almelo, the Netherlands) with copper K–alpha radiation (λ = 0.15418 nm) scanning from 10° to 70°. The structures of all samples were analyzed using Raman Spectroscopy (DONGWOO DM500i, Gyeonggi-do, Korea). The morphology and compositions of the SiOC samples were analyzed via a field emission scanning electron microscopy (FESEM, JEOL JSM-6700F, JEOL, Peabody, MA, USA) with energy-dispersive X-ray spectroscopy (EDX). The morphology imaging was carried out at 5 k magnification with an accelerating voltage of 12 kV and a working distance of 12.1 mm. TEM observations of precipitates were performed on selected samples, using a transmission electron microscopy (TEM, JEM2100F, Akishima, Japan) at 200 kV. Fourier transform infrared (FTIR) spectroscopy (Perkin Elmer Spectrum GX, Shelton, CT, USA) was carried out to study the sample's structure using a diffuse reflectance infrared Fourier transform accessory (DRIFT) equipped with a heating cartridge. The pore textural characterizations of the samples were measured by a volumetric sorption analyzer (Micromeritics ASAP 2020, Micromeritics Instrument Corporation, Norcross, GA, USA). An adsorption/desorption isotherm of nitrogen gas was recorded at −196 °C and the pore size distribution was investigated by using the Barrett–Joyner–Halenda (BJH) model for the specific surface area and pore volume of the samples. The thermal conductivity absolute value was directly determined by a thermal conductivity probe (Mathis Instruments Ltd., Fredericton, Canada).

3. Results and Discussion

3.1. Structural Analysis

XRD patterns of the SiOC samples are shown in Figure 1. The patterns correspond to an amorphous Si–O–C material feature with disordered carbon and with broad (002) and (100) at 2θ angles of 22.5° and 44°, respectively [26,27]. The peak at 44° is involved with small graphene sheets, and the peak at 22.5° means small graphene sheets stacked with random rotations or translations [27]. Raman spectra are also provided in Figure 2 for comparison. The Raman spectra of the prepared samples show the presence of two peaks at 1350 and 1582 cm^{-1}, corresponding to D- and G-bands. The relative intensity/area ratio of D- and G-bands can be used to analyze the carbon material's structure. The G-band is a reflection of the presence of sp2 carbon–carbon bond hybridization within a graphitic ring structure, whereas the D band reflect a disordered carbon structure. It can be seen that the G band of these samples is broader than the D band, indicating the high graphitized carbon structure in these samples [28,29].

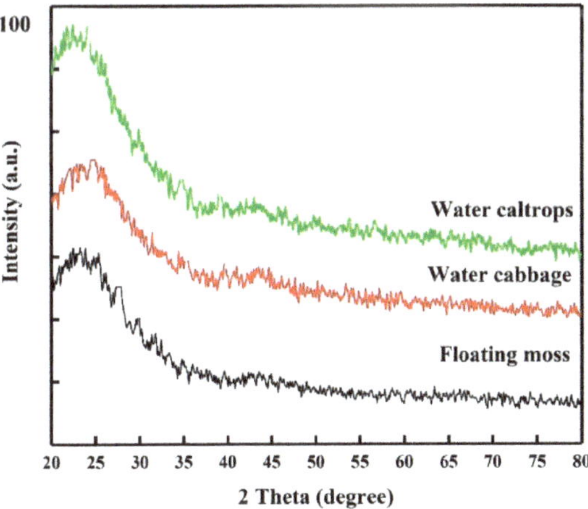

Figure 1. X-ray diffraction patterns of the silicon oxycarbide (SiOC) samples.

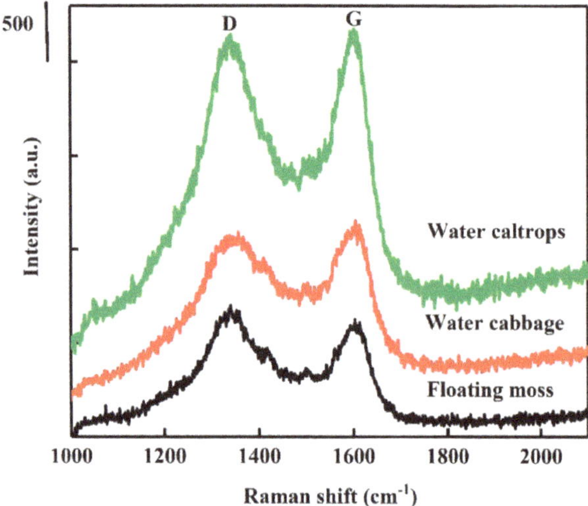

Figure 2. Raman spectra of the silicon oxycarbide (SiOC) samples.

3.2. Morphology and Composition Analysis

Figure 3 shows the field emission scanning electron microscopy (FESEM), TEM images, and EDX spectra of the prepared samples. Figure 3a shows that the floating mass sample contains a rough surface with irregular patches. Figure 3b, on the other hand, shows that the water cabbage sample contains a randomly distributed and overlapped structure. Contradictorily, Figure 3c shows an interconnected three-dimensional layered material for the water caltrops sample. Apart from the overall view by FESEM, all TEM images showed that these samples contained ordered hexagonal pore arrays with the majority of pores less than 1 nm.

The quantitative analysis was carried out with using energy dispersion analysis to determine the atomic ratios of C, O, and Si in these samples. As shown in Figure 3, the samples contain no contaminations other than C, O, and Si elements. Table 1 shows that the atomic ratios of C, O, and Si of the samples are 1:0.29:0.06, 1:1.13:0.48, and 1:2.24:1.07, respectively for the floating moss, water cabbage, and water caltrops samples. It is apparent that the element oxygen plays an important role to form structural layering. The surface roughness of the samples reduces with a decrease in the ratio of carbon/oxygen. The respective layers of water caltrops sample display a parallel arrangement due to the low ratio of carbon/oxygen. Shen et al. [30] reported that the surface structure becomes rougher when there is a deprivation of epoxide and hydroxyl groups, as well as a high C/O ratio.

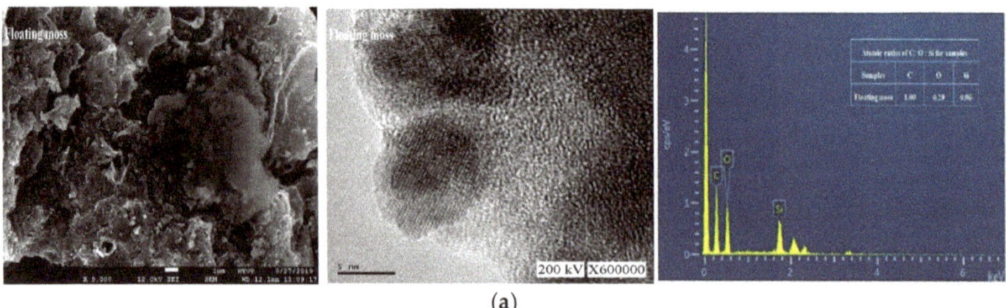

(a)

Figure 3. *Cont.*

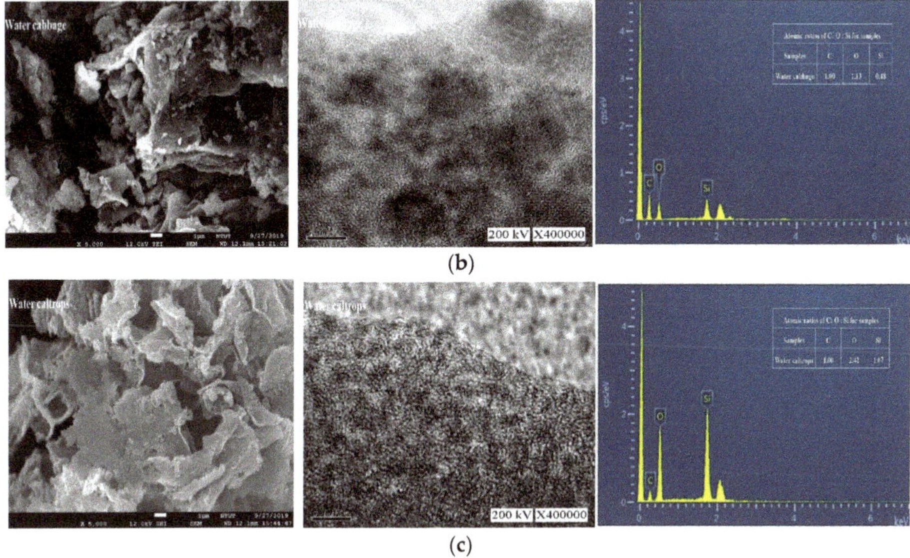

Figure 3. Field emission scanning electron microscopy (FESEM), transmission electron microscopy (TEM) and energy-dispersive X-ray spectroscopy (EDX) images (from left to right) of (**a**) Floating moss sample; (**b**) Water cabbage sample; and (**c**) Water caltrops sample.

Table 1. The C, O, and Si contents of the samples obtained by energy-dispersive X-ray spectroscopy (EDX) analysis.

Atomic Ratios			
Sample	**C**	**O**	**Si**
Floating moss	1	0.29	0.06
Water cabbage	1	1.13	0.48
Water caltrops	1	2.42	1.07

3.3. Fourier Transform Infrared Spectroscopy (FTIR) Spectra

Figure 4 shows the FTIR spectra of the SiOC samples. Before the IR spectra analysis, the adsorbed water was removed by heating to 250 °C to advance the understanding of the functional group properties. The stretching vibration of siloxane bond (Si–O–Si) and the Si–O stretching vibration of the silanol group appear weak at 800 cm^{-1} and 972 cm^{-1}, individually [31,32]. The asymmetric vibration of the Si–O–Si obviously appears at 1100 cm^{-1} [33] for water cabbage and water caltrops samples. However, in the floating moss sample, the Si–O–Si asymmetric vibration sharply reduces, due to the low amount of O and Si present in the structure. In addition, the floating moss sample has the lowest amount of hydroxyl groups, indicated by the low peak at 3450 cm^{-1}. The water caltrops sample has the highest amount of hydroxyl groups. The amount of hydroxyl groups corresponds to the particle size from FESEM morphology in which the floating moss sample with the lowest amount of hydroxyl groups has a larger particle size compared to the water cabbage and water caltrops samples. It is speculated that the particle size of the floating moss sample is larger as the hydroxyl groups of the sample may generate some surface charge, thus resulting in agglomeration [34].

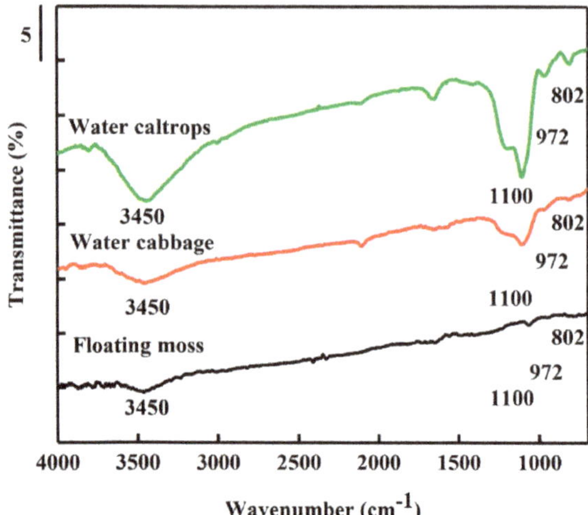

Figure 4. Fourier transform infrared (FTIR) spectra of the silicon oxycarbide (SiOC) samples.

3.4. Porosity and Surface Area Study

Figure 5a shows the nitrogen adsorption–desorption isotherms of the prepared samples. There is a very slow increase in N_2 adsorption up to 0.80 of the relative pressure (P/P_0) before the sharp rise in the adsorbed volume when capillary condensation occurred. A H_2-type hysteresis loop consisting of a triangular shape and a steep desorption branch of the isotherms can be found, indicating the existence of highly meso-pores with narrow mouths and wider bodies (ink-bottle pores) [35]. Figure 5b shows the corresponding pore size distributions of the prepared samples. There is a narrow pore size distribution centered at around 2.75 nm with a small meso-pore part in the floating moss sample and water cabbage sample. In addition, the water caltrops sample has the most porous structures with pore diameters ranged from 4 to 32 nm, indicating that it belongs to hierarchical porous structure. Table 2 shows the specific surface area, pore volume, and average pore diameter calculated using the Barrett–Joyner–Halenda (BJH) equation.

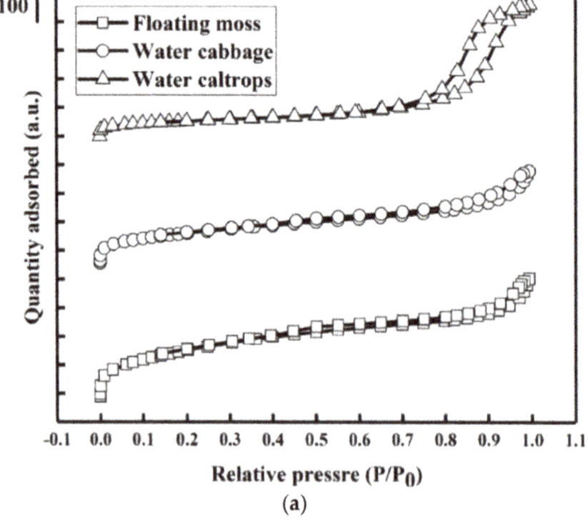

(a)

Figure 5. *Cont.*

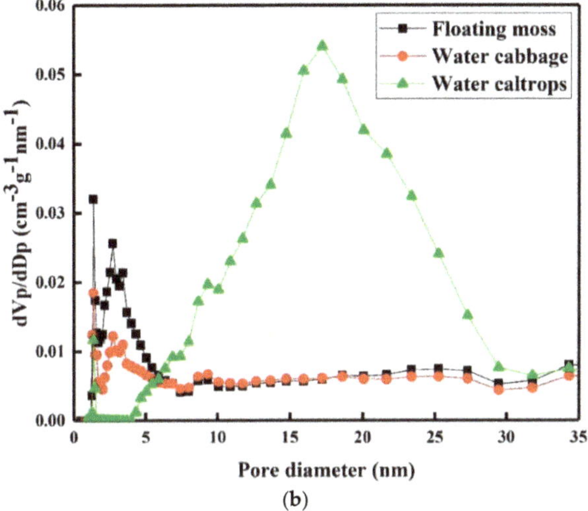

(b)

Figure 5. (**a**) Nitrogen adsorption/desorption isotherms and (**b**) pore size distributions of the silicon oxycarbide (SiOC) samples.

Table 2. Textural characterization for the silicon oxycarbide (SiOC) samples.

Sample	V_{pore} ($cm^3\ g^{-1}$)	S_{BET} ($m^2\ g^{-1}$)	D_p (nm)	Thermal Conductivity ($Wm^{-1}\ K^{-1}$)
Floating moss	0.40	894.81	7.00	6.55
Water cabbage	0.49	565.21	6.04	2.46
Water caltrops	0.63	213.00	4.75	1.14

The pore volumes (V_{pore}) of the floating moss, water cabbage, and water caltrops samples are respectively 0.4, 0.49, and 0.63 $cm^3\ g^{-1}$. The Brunauer–Emmett–Teller (BET) specific surface areas (S_{BET}) of the floating moss sample is greater than the water cabbage and water caltrops samples. The BJH adsorption-average pore diameters (D_p) are 7.00, 6.04, and 4.75 nm, respectively, for the floating moss, water cabbage, and water caltrops samples. The texture properties of the prepared samples agree with the surface morphology results in which the floating moss sample having the largest grain size has the lowest pore volume and largest pore diameter.

3.5. Thermal Conductivity

The thermal conductivities of these samples were studied using Yang et al.'s method [36]. The thermal conductivities of these samples are shown in Table 2 and compared with other literature in Table 3. As we can see, the thermal conductivity of the floating moss sample is the highest among these samples due to lowest pore volume and larger grain size [12]. The water caltrops sample has the lowest thermal conductivity among these samples, indicating that the surface hydroxyl groups on silica reduced the thermal conductivity of the sample due to the increased amount of voids [37]. Eom et al. [38] also reported that the equivalent porosity is the reason for low thermal conductivity. Thermal conductivity of these SiOC samples reduces in the order of floating moss (6.55 $Wm^{-1}\ K^{-1}$), water cabbage (2.46 $Wm^{-1}\ K^{-1}$), and water caltrops (1.14 $Wm^{-1}\ K^{-1}$). However, when compared with other literature (Table 3), the thermal conductivity of our floating-plant derived SiOC samples is considerably higher and comparable with those made from chemical feedstock. It is also a key finding that the surface hydroxyl groups on silica decreases the thermal conductivity. Nevertheless, among the three floating plants: floating moss, water cabbage, and water caltrops, the floating mass plant is the candidate with the most potential to replace the use of chemical precursors in making SiOC.

Table 3. Comparison of thermal conductivity with other literature.

No	Precursors	Method	Thermal Conductivity (Wm^{-1} K^{-1})	References
1	Silicon oxide carbide powders	Pyrolysis	1.8~2.7	[12]
2	Alkoxy silane	Sol–gel	0.041~0.078	[13]
3	Polysiloxane pyrolysis residue, and barium isopropoxide	Pyrolysis	1.8~5.6	[18]
4	Commercial poly(methylsilsesquioxane) starting polymer and zirconium acetylacetonate	Pyrolysis	2	[16]
5	Silicon oxycarbide (SiOC) powders	Spark plasma sintering	1.38	[17]
6	Floating plants	Pyrolysis	1.14~6.55	This paper

4. Conclusions

In this study, silicon oxycarbide materials were synthesized via a simple and low-cost method from floating plants. Three floating plants: floating moss, water cabbage, and water caltrops were used as the precursors of the SiOC powders. XRD shows the amorphous Si–O–C material structure as the major phase. The morphologies and textural properties of the SiOC samples were porous micro-structures. The maximum thermal conductivity of the prepared samples in this study was found to be 6.55 Wm^{-1} K^{-1} from floating moss. The surface hydroxyl groups of the SiOC samples may have possibly reacted with the silica atom, resulting in interconnecting bonds and thus larger grain size which enhances the thermal conductivity. Overall, the results demonstrated the feasibility of using floating plants as the precursor in making SiOC powder with high thermal conductivity via a facile and low-cost procedure, making it a suitable process for industrial applications.

Author Contributions: Conceptualization and writing, G.-T.P.; Review, S.C.; Validation, Y.J.C.; Sample analysis, T.J.T.; Funding acquisition, J.W.L.; Resources, C.-M.H.; Editing and finalization, P.S.; Supervision, T.C.-K.Y.

Funding: The financial support Universiti Teknologi PETRONAS via YUTP-FRG with the cost center 0153AA-E48 is gratefully acknowledged. Funding from Ministry of Education Malaysia through HICoE awarded to the Centre for Biofuel and Biochemical Research, Universiti Teknologi PETRONAS is as well duly acknowledged.

Acknowledgments: The authors would like to thank Kuan Ching Lee, Ho-Shin Shiu, Ping-Chun Lin, and Yi-Hsuan Lai for their assistance in conducting the experiments.

Conflicts of Interest: The authors declare no conflicts of interest.

References

1. Miyajima, H.; Masuda, H.; Watanabe, K.; Ishikawa, K.; Sekine, M.; Hori, M. Chemical bonding structure in porous SiOC films (k < 2.4) with high plasma-induced damage resistance. *Micro Nano Eng.* **2019**, *3*, 1–6.
2. Colombo, P.; Mera, G.; Riedel, R.; Soraru, G.D. Polymer-derived ceramics: 40 years of research and innovation in advanced ceramics. *J. Am. Ceram. Soc.* **2010**, *93*, 1805–1837. [CrossRef]
3. Wu, Z.; Lv, W.; Cheng, X.; Gao, J.; Qian, Z.; Tian, D.; Li, J.; He, W.; Yang, C. A Nanostructured Si/SiOC Composite Anode with Volume-Change-Buffering Microstructure for Lithium-Ion Batteries. *Chem. A Eur. J.* **2019**, *25*, 2604–2609. [CrossRef]
4. Wu, Z.; Cheng, X.; Tian, D.; Gao, T.; He, W.; Yang, C. SiOC nanolayers directly-embedded in graphite as stable anode for high-rate lithium ion batteries. *Chem. Eng. J.* **2019**, *375*, 121997. [CrossRef]
5. Duan, L.; Ma, Q.; Mei, L.; Chen, Z. Fabrication and electrochemical performance of nanoporous carbon derived from silicon oxycarbide. *Microporous Mesoporous Mater.* **2015**, *202*, 97–105. [CrossRef]
6. Roth, F.; Schmerbauch, C.; Ionescu, E.; Nicoloso, N.; Guillon, O.; Riedel, R. High-temperature piezoresistive C/SiOC sensors. *J. Sens. Sens. Syst.* **2015**, *4*, 133–136. [CrossRef]

7. Bik, M.; Stygar, M.; Jeleń, P.; Dąbrowa, J.; Leśniak, M.; Brylewski, T.; Sitarz, M. Protective-conducting coatings based on black glasses (SiOC) for application in solid oxide fuel cells. *Int. J. Hydrogen Energy* **2017**, *42*, 27298–27307. [CrossRef]
8. Tolosa, A.; Krüner, B.; Jäckel, N.; Aslan, M.; Vakifahmetoglu, C.; Presser, V. Electrospinning and electrospraying of silicon oxycarbide-derived nanoporous carbon for supercapacitor electrodes. *J. Power Source* **2016**, *313*, 178–188. [CrossRef]
9. Wu, N.; Wang, B.; Wang, Y. Enhanced mechanical properties of amorphous Si OC nanofibrous membrane through in situ embedding nanoparticles. *J. Am. Ceram. Soc.* **2018**, *101*, 4763–4772. [CrossRef]
10. Sorarù, G.D.; Modena, S.; Guadagnino, E.; Colombo, P.; Egan, J.; Pantano, C. Chemical durability of silicon oxycarbide glasses. *J. Am. Ceram. Soc.* **2002**, *85*, 1529–1536. [CrossRef]
11. Saha, A.; Raj, R. Crystallization maps for SiCO amorphous ceramics. *J. Am. Ceram. Soc.* **2007**, *90*, 578–583. [CrossRef]
12. Stabler, C.; Reitz, A.; Stein, P.; Albert, B.; Riedel, R.; Ionescu, E. Thermal properties of SiOC glasses and glass ceramics at elevated temperatures. *Materials* **2018**, *11*, 279. [CrossRef] [PubMed]
13. Qiu, L.; Li, Y.; Zheng, X.; Zhu, J.; Tang, D.; Wu, J.; Xu, C. Thermal-conductivity studies of macro-porous polymer-derived SiOC ceramics. *Int. J. Thermophys.* **2014**, *35*, 76–89. [CrossRef]
14. Santana, J.A.C.; Mora, E.E.; Price, L.; Balerio, R.; Shao, L.; Nastasi, M. Synthesis, thermal stability and the effects of ion irradiation in amorphous Si–O–C alloys. *Nucl. Instrum. Methods Phys. Res. Sect. B Beam Interact. Mater. At.* **2015**, *350*, 6–13. [CrossRef]
15. Nastasi, M.; Su, Q.; Price, L.; Santana, J.A.C.; Chen, T.; Balerio, R.; Shao, L. Superior radiation tolerant materials: Amorphous silicon oxycarbide. *J. Nucl. Mater.* **2015**, *461*, 200–205. [CrossRef]
16. Gurlo, A.; Ionescu, E.; Riedel, R.; Clarke, D.R. The Thermal Conductivity of Polymer-Derived Amorphous Si–O–C Compounds and Nano-Composites. *J. Am. Ceram. Soc.* **2016**, *99*, 281–285. [CrossRef]
17. Mazo, M.A.; Palencia, C.; Nistal, A.; Rubio, F.; Rubio, J.; Oteo, J.L. Dense bulk silicon oxycarbide glasses obtained by spark plasma sintering. *J. Eur. Ceram. Soc.* **2012**, *32*, 3369–3378. [CrossRef]
18. Eom, J.-H.; Kim, Y.-W.; Kim, K.J.; Seo, W.-S. Improved electrical and thermal conductivities of polysiloxane-derived silicon oxycarbide ceramics by barium addition. *J. Eur. Ceram. Soc.* **2018**, *38*, 487–493. [CrossRef]
19. Yan, X.; Wang, F.; Hattar, K.; Nastasi, M.; Cui, B. Novel amorphous SiOC dispersion-strengthened austenitic steels. *Materialia* **2019**, *6*, 100345. [CrossRef]
20. Zare, A.; Su, Q.; Gigax, J.; Shojaee, S.; Harriman, T.; Nastasi, M.; Shao, L.; Materer, N.; Lucca, D. Effects of ion irradiation on chemical and mechanical properties of magnetron sputtered amorphous SiOC. *Nucl. Instrum. Methods Phys. Res. Sect. B Beam Interact. Mater. At.* **2019**, *446*, 10–14. [CrossRef]
21. Zare, M.; Niroumand, B.; Maleki, A.; Allafchian, A.R. Sol-gel synthesis of amorphous SiOC nanoparticles from BS290 silicone precursor. *Ceram. Int.* **2017**, *43*, 12898–12903. [CrossRef]
22. Liu, C.; Meng, X.; Zhang, X.; Hong, C.; Han, J.; Han, W.; Xu, B.; Dong, S.; Du, S. High temperature structure evolution of macroporous SiOC ceramics prepared by a sol–gel method. *Ceram. Int.* **2015**, *41*, 11091–11096. [CrossRef]
23. Yu, S.; Tu, R.; Goto, T. Preparation of SiOC nanocomposite films by laser chemical vapor deposition. *J. Eur. Ceram. Soc.* **2016**, *36*, 403–409. [CrossRef]
24. Zollfrank, C.; Kladny, R.; Sieber, H.; Greil, P. Biomorphous SiOC/C-ceramic composites from chemically modified wood templates. *J. Eur. Ceram. Soc.* **2004**, *24*, 479–487. [CrossRef]
25. Pan, J.; Pan, J.; Cheng, X.; Yan, X.; Lu, Q.; Zhang, C. Synthesis of hierarchical porous silicon oxycarbide ceramics from preceramic polymer and wood biomass composites. *J. Eur. Ceram. Soc.* **2014**, *34*, 249–256. [CrossRef]
26. Wang, M.; Xia, Y.; Wang, X.; Xiao, Y.; Liu, R.; Wu, Q.; Qiu, B.; Metwalli, E.; Xia, S.; Yao, Y. Silicon oxycarbide/carbon nanohybrids with tiny silicon oxycarbide particles embedded in free carbon matrix based on photoactive dental methacrylates. *ACS Appl. Mater. Interfaces* **2016**, *8*, 13982–13992. [CrossRef]
27. Xue, J.; Myrtle, K.; Dahn, J. An Epoxy-Silane Approach to Prepare Anode Materials for Rechargeable Lithium Ion Batteries. *J. Electrochem. Soc.* **1995**, *142*, 2927–2935. [CrossRef]
28. Dibandjo, P.; Graczyk-Zajac, M.; Riedel, R.; Pradeep, V.; Soraru, G. Lithium insertion into dense and porous carbon-rich polymer-derived SiOC ceramics. *J. Eur. Ceram. Soc.* **2012**, *32*, 2495–2503. [CrossRef]

29. Fukui, H.; Ohsuka, H.; Hino, T.; Kanamura, K. A Si–O–C composite anode: High capability and proposed mechanism of lithium storage associated with microstructural characteristics. *ACS Appl. Mater. Interfaces* **2010**, *2*, 998–1008. [CrossRef]
30. Shen, L.; Zhang, L.; Wang, K.; Miao, L.; Lan, Q.; Jiang, K.; Lu, H.; Li, M.; Li, Y.; Shen, B. Analysis of oxidation degree of graphite oxide and chemical structure of corresponding reduced graphite oxide by selecting different-sized original graphite. *RSC Adv.* **2018**, *8*, 17209–17217. [CrossRef]
31. Amutha, K.; Ravibaskar, R.; Sivakumar, G. Extraction, synthesis and characterization of nanosilica from rice husk ash. *Int. J. Nanotechnol. Appl.* **2010**, *4*, 61–66.
32. Witoon, T.; Chareonpanich, M.; Limtrakul, J. Synthesis of bimodal porous silica from rice husk ash via sol–gel process using chitosan as template. *Mater. Lett.* **2008**, *62*, 1476–1479. [CrossRef]
33. Rida, M.A.; Harb, F. Synthesis and characterization of amorphous silica nanoparitcles from aqueous silicates uisng cationic surfactants. *J. Met. Mater. Miner.* **2014**, *24*, 37–42.
34. Chapter 3 Surface chemistry of porous silica. *J. Chromatogr. Libr.* **1979**, *16*, 57–146. [CrossRef]
35. Kruk, M.; Jaroniec, M. Gas adsorption characterization of ordered organic–Inorganic nanocomposite materials. *Chem. Mater.* **2001**, *13*, 3169–3183. [CrossRef]
36. Yao, W.; Guangsheng, G.; Fei, W.; Jun, W. Fluidization and agglomerate structure of SiO_2 nanoparticles. *Powder Technol.* **2002**, *124*, 152–159. [CrossRef]
37. Pan, G.-T.; Chong, S.; Yang, T.C.-K.; Yang, Y.-L.; Arjun, N. Surface modification of amorphous SiO_2 nanoparticles by oxygen-plasma and nitrogen-plasma treatments. *Chem. Eng. Commun.* **2016**, *203*, 1666–1670. [CrossRef]
38. Eom, J.-H.; Kim, Y.-W.; Raju, S. Processing and properties of macroporous silicon carbide ceramics: A review. *J. Asian Ceram. Soc.* **2013**, *1*, 220–242. [CrossRef]

Article

Conversion Technologies: Evaluation of Economic Performance and Environmental Impact Analysis for Municipal Solid Waste in Malaysia

Rabiatul Adawiyah Ali [1], Nik Nor Liyana Nik Ibrahim [1,*] and Hon Loong Lam [2]

1 Department of Chemical and Environmental Engineering, Faculty of Engineering, Universiti Putra Malaysia, Serdang, Selangor 43400, Malaysia; adwyhali@gmail.com

2 Department of Chemical and Environmental Engineering, The University of Nottingham Malaysia Campus, Semenyih, Selangor 43500, Malaysia; HonLoong.Lam@nottingham.edu.my

* Correspondence: niknorliyana@upm.edu.my

Received: 22 August 2019; Accepted: 24 September 2019; Published: 16 October 2019

Abstract: The generation of municipal solid waste (MSW) is increasing globally every year, including in Malaysia. Approaching the year 2020, Malaysia still has MSW disposal issues since most waste goes to landfills rather than being utilized as energy. Process network synthesis (PNS) is a tool to optimize the conversion technologies of MSW. This study optimizes MSW conversion technologies using a PNS tool, the "process graph" (P-graph). The four highest compositions (i.e., food waste, agriculture waste, paper, and plastics) of MSW generated in Malaysia were optimized using a P-graph. Two types of conversion technologies were considered, biological conversion (anaerobic digestion) and thermal conversion (pyrolysis and incinerator), since limited data were available for use as optimization input. All these conversion technologies were compared with the standard method used: landfilling. One hundred feasible structure were generated using a P-graph. Two feasible structures were selected from nine, based on the maximum economic performance and minimal environmental impact. Feasible structure 9 was appointed as the design with the maximum economic performance (MYR 6.65 billion per annum) and feasible structure 7 as the design with the minimal environmental impact (89,600 m^3/year of greenhouse gas emission).

Keywords: optimization; P-graph; municipal solid waste conversion technology

1. Introduction

Municipal solid waste (MSW) is material arising from human activities. It is generated commonly from different areas such as residential, commercial, and institutional zones, as well as public parks [1]. The generation of MSW is drastically increasing globally every year, by a factor of 2.6 [2]. In 2016, the world's MSW generated was around 2.01 billion tons, and this figure is expected to increase to 3.40 billion tons by 2050 [3].

In Asia, MSW generation is expected to reach 1.8 million tons every day in 2025, as more than 1 million tons of MSW is currently being generated every day [4]. Based on a survey conducted by the Malaysian government, MSW generation in Malaysia has increased from 23,000 tons/day in 2008 to 33,000 tons/day in 2012 [5]. The increases of MSW generation in Malaysia are caused by three significant factors: (i) the rapid increase in population; (ii) accelerated urbanization; and (iii) increased industrialization processes [6]. The total population of Malaysia in 2017, as mentioned by the World Bank [7], was 31.62 million. As the population increases, the per capita generation rate also increases. For Malaysia, the MSW per capita generation rate range was of 0.6–0.8 kg per capita per day between 2001 and 2005 [8]. This number is expected to increase to double digits by the year 2020 [9]. Based on the World Bank's report, the waste generation per capita in Malaysia increased by up to 1.00–1.49 kg per capita per day by September 2018 [3]. At present, 54% of the world's population lives in urban areas, and this percentage will increase to 66% or more by 2050 [10].

Increasing MSW generation has become the most prominent environmental issue as MSW may contain dangerous substances that are harmful to our ecosystem and increase the potential risk to our health. MSW must be appropriately disposed of and managed efficiently. Many significant environmental issues may arise from this kind of waste, such as the generation of greenhouse gases (GHGs) released from MSW. Besides, the increasing number of landfills can increase the numbers of rodents and insects that may cause diseases to humans. In recent years, more landfill sites are needed to dispose of all the MSW generated [11]. The main issue we face this traditional disposal method is shortage of landfill sites inland [12]. An essential component of a healthy society and a sustainable environment is an efficient waste management system [13].

The main purpose to manage MSW efficiently include reducing (i) the amount of MSW generated, (ii) the impact on the environment with a lower cost of disposal of MSW, and (iii) the impact on human health [14]. In MSW management in developing countries, five typical problems can be identified: (i) inadequate service coverage, (ii) operational inefficiency of services, (iii) limited utilization of recycling activities, (iv) poor management of non-industrial hazardous waste, and (v) shortage of landfill disposal sites [15]. The present waste management method in Malaysia depends on landfill [11]. Only 5.5% of MSW is recycled and 1% is composted, while the remaining waste goes to landfill [16]. Currently, there are 174 landfills around Malaysia [14]. The recycling rate increased from 5.5% in 2009 to 10.5% in 2012 [17]. Malaysia's recycling rate was 17.5% in 2016, which is still far from the target of 22% by 2020 [18].

One method to solve problems related to landfills is to introduce sustainable and efficient waste management [6]. Integrated waste recycling and various conversion technologies could be effective waste management strategies [19]. There are several steps in sustainable and efficient waste management [20]. Possible waste generation and their conversion technologies are illustrated in Figure 1.

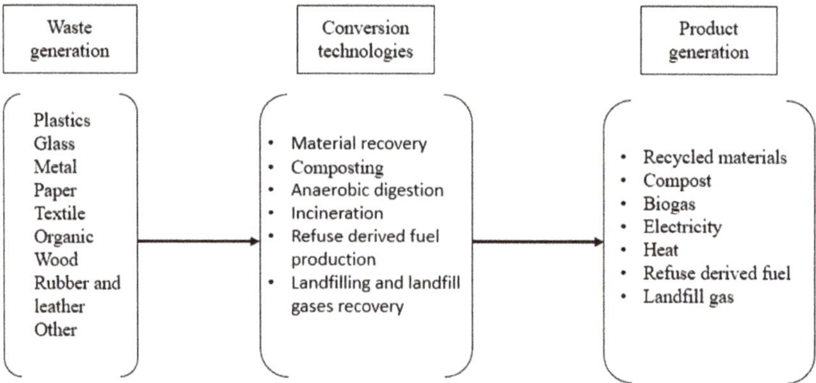

Figure 1. Possible waste generation relationship between their conversion technologies, reproduced with permission from [20]. Copyright Elsevier, 2017.

Based on Figure 1, there are many pathways for conversion technologies to manage MSW after it is collected from residences and processed before being sent to landfills. The three main steps to handle MSW are (i) the collection and transportation of MSW; (ii) the treatment and processing of MSW; and (iii) its final disposal [21]. Each stage has its own investment cost, operating cost, and energy recovery.

However, there are pros and cons to each conversion technology. It is useful to utilize analytical tools to synthesize a promising waste management strategy [22]. There are limited studies on the performance of conversion technology in the context of Malaysia. The optimization of MSW conversion technology will help decide the most favorable and useful method and pathway in managing MSW. Through optimization, we can introduce combined conversion technology to manage Malaysia's MSW.

A process network synthesis (PNS) problem is defined as specifying the raw materials, operating units, and desired products in chemical engineering problem, for example the conversion technologies problem. The PNS problem was developed as a mathematical model in which variables correspond to decisions, such as input and output flow rates, with a limitation corresponding to the mathematical description of the optimization criterion such as the material balance objective function [23]. The common problems in a PNS are (i) the reaction pathway; (ii) process design; (iii) the heat exchangers network; (iv) the water integration system; and (v) the separation unit [24]. The process graph, best known as the P-graph, is one method to solve the PNS problem [25]. The P-graph is a graphical optimization which is available in the software P-Graph Studio [26]. The P-graph is a bi-graph, meaning that its vertices are in disjunctive sets and there are no edges between vertices in the same set [27]. The vertices of the P-graph are denoted as the operating unit (O) and the material (M). This P-graph represents the material flow between the material and the operating unit. The P-graph methodology was originally developed for PNS problems in chemical engineering applications. The P-graph methodology is based on five axioms [28], as follows:

1. Every final product is represented in the graph.
2. A vertex of material/energy type (M-type) has no input if and only if it represents a raw material.
3. Every vertex of operating type (O-type) represents an operating unit defined in the synthesis problem.
4. Every vertex of O-type has at least one path leading to a vertex of the M-type representing a final product.
5. If a vertex of the M-type belongs to the graph, it must be input to or output from at least one vertex of the O-type in the graph.

To summarize, the P-graph methodology is composed of the following algorithms: (i) maximal structure generation (MSG); (ii) solution structure generation (SSG); and (iii) accelerated branch and bound (ABB) [29]. The MSG algorithm identifies a network structure, which is the union of all possible solution structures of the problem. It can be generated in polynomial time using the information specified in the five axioms. The SSG algorithm generates all combinatorically feasible solution structures or networks. Each solution is a subset of the maximal structure and represents a potential network configuration for the PNS problem. The ABB algorithm identifies the optimal structure based on the solution structures, in conjunction with additional problem-specific information.

The P-graph framework enables rigorous model building and the efficient generation of optimal solutions [29]. The PNS problem primarily utilizes unique information. The P-graph is known as a user-friendly decision-making tool for PNS. This helps in better design and better operations that lead to (i) lower capital and operating cost (CAPEX and OPEX); (ii) higher profitability through increased output and better quality of the product; (iii) reduced technology risk; and (iv) better health, safety, and environmental requirements. These factors may thus help in optimizing MSW conversion technology.

The main objective of optimization is maximizing the efficiency of production by minimizing the cost of production. Therefore, it is essential to optimize MSW conversion technologies using a process graph to evaluate the selected pathway. Table 1 shows different types of optimization models for solid waste management based on previous studies. Data are tabulated based on the optimization method used, the objective of the study, the focus of the study, and the optimization of economic performance (EP) and environmental impact (EI).

Table 1. Optimization models for solid waste management.

Method	References	Objectives	Focus		Optimization on	
			Energy System	Waste Management	Economy	Environment
Linear Programming	[30,31]	To maximize the economic utility of energy consumers	/		/	
Mixed Integer Linear Programming	[32]	To determine the optimal processing network waste-to-energy system	/		/	/
Non-Linear Programming	[33]	To maximize profit, while minimizing waste through source reduction		/	/	
Hybrid Model	[34]	Multi-objective programming and cost–benefit criteria on global warming impact in waste management		/	/	/
P-graph Model	[35]	To utilize organic and dry fractions of municipal waste	/	/	/	/

Although there are various models for optimizing MSW conversion technologies, we still cannot manage MSW efficiently in Malaysia, as there are no integrated conversion technologies for solid waste treatment. Therefore, this study aimed to simulate the feasibility of MSW conversion technologies and analyzed the EP and EI of MSW conversion technologies. The study framework was based on the following factors:

1. Type of resources;
2. Type of product;
3. Selection of conversion technologies;
4. Generation of GHGs;
5. Capital and operating expenses of conversion technologies.

The proposed processing network was designed using the P-graph model. There are a few types of MSW conversion technologies, including landfill, anaerobic digestion, incineration, and pyrolysis. The selected optimized conversion technology for MSW was then further assessed with respect to the impacts on feedstock and products on GHG emissions, demand, and prices. Two different scenarios were considered in this case study, which was designed for maximum EP and design for minimal EI.

2. Materials and Methods

Figure 2 shows the intracellular synthesis procedure for the process graph (P-graph). The procedure starts with the identification of materials and streams to yield the optimal MSW conversion technology network.

The intracellular synthesis procedure started with the identification of materials, streams, and operating units. After that, data input was required to generate the maximal superstructure and solution structure. The procedure ended with an optimal MSW conversion technology network.

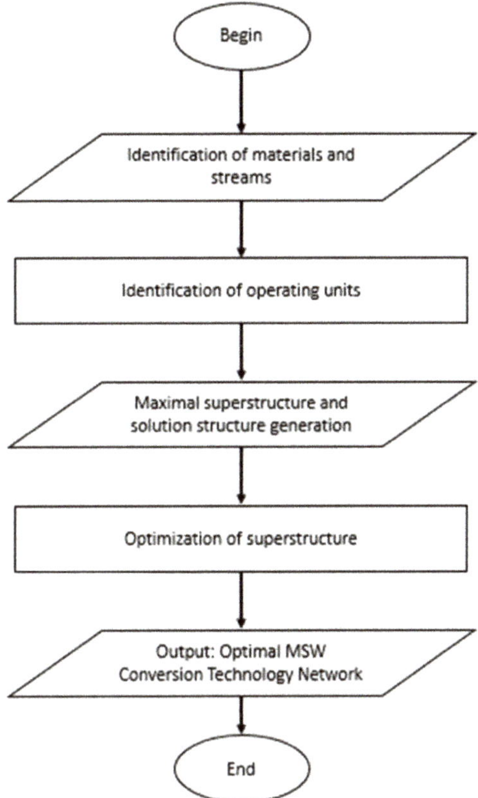

Figure 2. Intracluster synthesis procedure for the process graph (P-graph), reproduced with permission from [36]. Copyright Elsevier, 2010.

2.1. Identification of Materials and Streams

This step produced the details for the inputs and outputs of the system. In this study, there were four types of process feedstock. There are six types of outputs or products, along with their intermediate products, as illustrated in Figure 3.

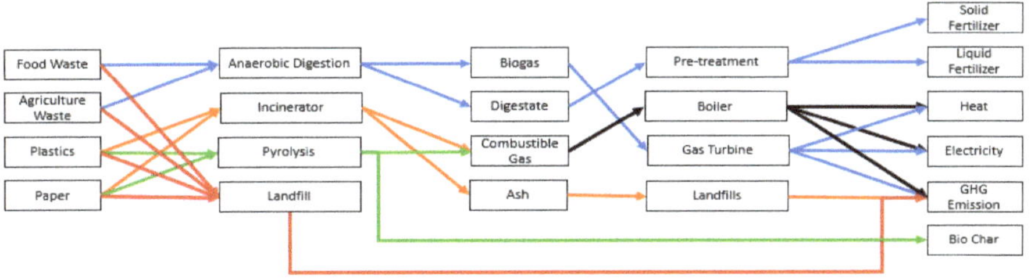

Figure 3. Superstructure for process flow managing MSW in this case study.

Figure 4 shows the circle nodes for the materials and the directed arrows represent as the streams. The number attached on the arrow signifies the consumption or production rate that represents the relationship between a material and an operating unit. Table 2 shows the list of raw materials, intermediate products, and products of the conversion technologies.

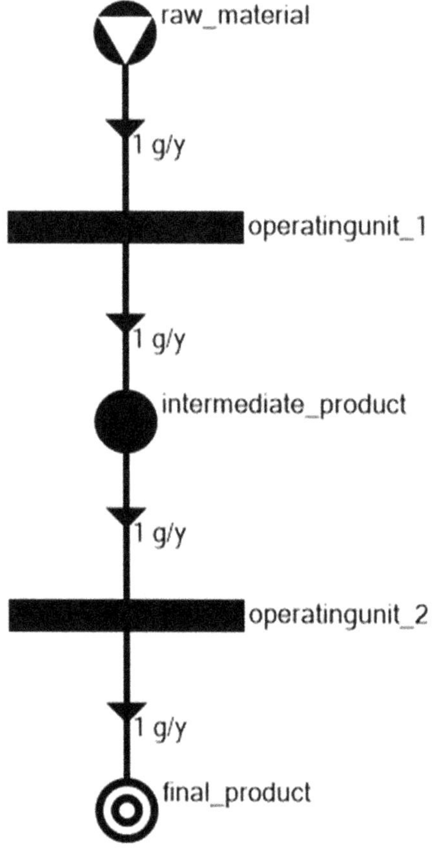

Figure 4. Graphical representation of materials in P-graph.

Table 2. Materials and products for the P-graph.

	Symbols	P-graph Classification	Description
1	Agricultural_Waste	Raw Material	Agricultural waste
2	Ash	Intermediate Product	Ash from plastic and paper in incinerator
3	Bio_Char	Output	Bio char
4	Biogas	Intermediate Product	Biogas production for anaerobic digestion of food and agriculture waste
5	Combustible_Gas	Intermediate Product	Combustible gas from plastic and paper in incinerator and pyrolysis
6	Digestate	Intermediate Product	Digestate of food and agriculture waste
7	Electricity	Output	Electricity generated
8	Food_Waste	Raw Material	Food waste
9	GHG_Emission	Output	Greenhouse gas emission
10	Heat	Output	Heat generated
11	Liquid_Fertilizer	Output	Liquid Fertilizer
12	Paper	Raw Material	Paper
13	Plastic	Raw Material	Plastic
14	Solid_Fertilizer	Output	Solid fertilizer

2.2. Identification of Operating Units

For this case study, 11 operating units were included in the flowsheet-generation problem shown to be solved algorithmically with P-graphs in Table 3. For this case study, anaerobic digestion, incineration, and pyrolysis were identified as the MSW conversion technologies in the model as an operating unit. The relationships between raw materials and operating units involved are shown in Figure 3.

Table 3. List of operating units for the P-graph.

	Symbols	P-graph Classification	Description
1	Boiler	Operating Unit	Boiler of combustible gas from incinerator and pyrolysis of plastic and paper
2	Digester_1	Operating Unit	Anaerobic digester for food waste
3	Digester_2	Operating Unit	Anaerobic digester for agricultural waste
4	Gas Turbine_1	Operating Unit	Gas turbine for anaerobic digestion of food and agriculture waste
5	Incinerator_1	Operating Unit	Incinerator for paper
6	Incinerator_2	Operating Unit	Incinerator for plastic
7	Landfill_1	Operating Unit	Landfill for MSW
8	Landfill_2	Operating Unit	Landfill for from incinerator of plastic and paper
9	Pre-Treatment	Operating Unit	Pre-treatment of digestate for anaerobic digestion of food and agriculture waste
10	Pyrolysis_1	Operating Unit	Pyrolysis of paper
11	Pyrolysis_2	Operating Unit	Pyrolysis of plastic

2.3. Input Data for Waste and Related Conversion Technologies

For this study, four types of MSW were chosen, based on the largest MSW composition generated in Malaysia: food waste, agricultural waste, plastics, and paper. The composition of MSW is shown in Table 4. Organic waste is the main component of MSW in Malaysia, representing up to 50% of the total waste.

Table 4. Composition generation of MSW in Malaysia [5].

	Type of Municipal Solid Waste	Composition of Municipal Solid Waste (%)
1	Food Waste	44.5
2	Plastics	13.2
3	Diapers	12.1
4	Paper	8.5
5	Agriculture Waste	5.8
6	Glass	3.3
7	Cloth	3.1
8	Steel/Metal	2.7
9	Rubber	1.8
10	Other	5.2

The study was conducted using data based on a literature review of MSW generation and MSW conversion technology. Data used were based on different studies and resources. The three types of conversion technologies considered in this case were pyrolysis, incineration, and anaerobic digestion, as limited data were available to be used for the optimization process. All these technologies were compared with the common method of MSW disposal in Malaysia, which are landfills. The allocation of waste to conversion technologies is illustrated in Table 5.

Both types of organic waste (i.e., food and agricultural) undergo an anaerobic digestion process. For inorganic waste, plastics and paper undergo two thermal treatments pyrolysis and incineration. After treatment, all waste is disposed of in landfills.

Table 5. Waste allocation to conversion technology.

	Food	Agriculture	Plastics	Paper
Anaerobic Digestion	/	/		
Incineration			/	/
Pyrolysis			/	/
Landfill	/	/	/	/

2.4. Optimization of Superstructure

The results of the solution structure generated from the previous step are then utilized in the selection of an optimal network using the solution structure generation and linear programming (SSG + LP) algorithm, allowing for the design of optimal process networks based on the solution structures in conjunction with additional problem-specific information, such as flow rates and costs. Consequently, the solution that provides the selected pathways with the best and near-optimum solutions is obtained, as shown in Figure 5. The selected optimized conversion technology for MSW was then further used to access the impacts on feedstock and products on GHG emissions, demand, and prices. Two different scenarios were considered: scenario 1: a design for maximum EP; and scenario 2: a design for minimal EI.

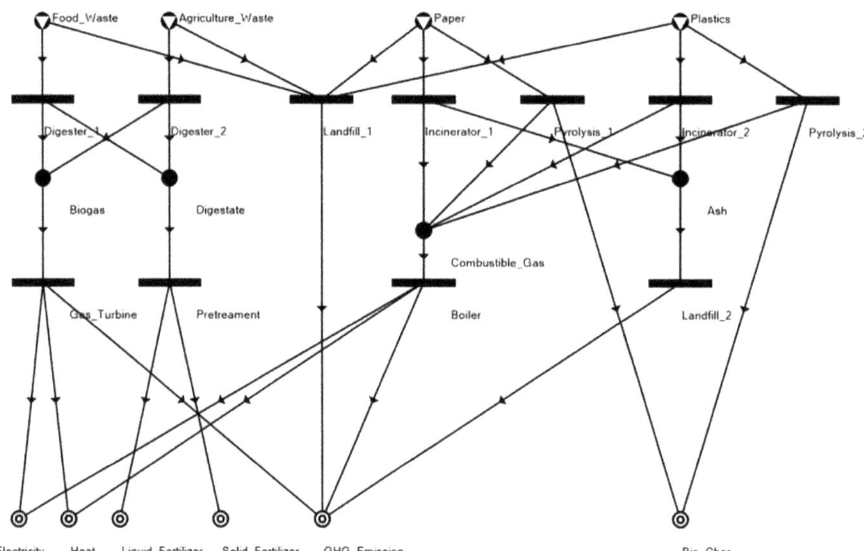

Figure 5. A P-graph representation of the municipal solid waste process network.

3. Results and Discussion

One hundred feasible structures were generated using the P-graph with the SSG + LP algorithm. The SSG generates all combinatorically, feasible solution structures or networks. Each solution is a subset of the maximal structure and represents a potential network configuration for the PNS problem. The LP is the process of finding the best solution under specific conditions.

Of the 100 feasible structures generated, nine were selected to identify and analyze their EP and EI. These nine feasible structures convert all types of waste into the final products. Four types of MSW (food waste, agriculture waste, plastics, and paper) were converted using three different types of conversion technologies (i.e., anaerobic digestion, incineration, and pyrolysis) to generate six main products, i.e., solid fertilizer, liquid fertilizer, heat, electricity, GHGs, and biochar.

3.1. Comparison of Different Feasible Structures

Figure 6 shows the profit generated for each feasible pathway. The profit generated was calculated by the total gain of the product minus the total cost of raw materials. The highest profit generated was feasible in structures 7 and 9. Both these feasible structures gave the same profit value, which was MYR 6.65 billion per annum. However, as shown in Table 6, feasible structure 7 did not generate two products: electricity and heat. The products generated by feasible structure 7 were solid fertilizer, liquid fertilizer, GHGs, and biochar; while feasible structure 9 generated all six products: solid and liquid fertilizers, heat, electricity, GHGs, and biochar.

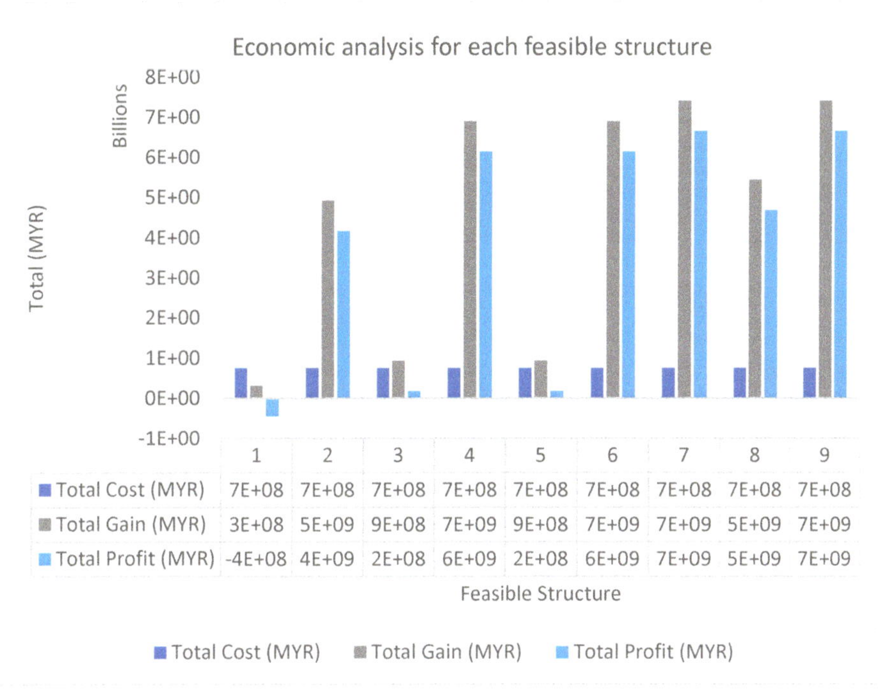

	1	2	3	4	5	6	7	8	9
Total Cost (MYR)	7E+08	7E+08	7E+08	7E+08	7E+08	7E+08	7E+08	7E+08	7E+08
Total Gain (MYR)	3E+08	5E+09	9E+08	7E+09	9E+08	7E+09	7E+09	5E+09	7E+09
Total Profit (MYR)	-4E+08	4E+09	2E+08	6E+09	2E+08	6E+09	7E+09	5E+09	7E+09

Figure 6. Economic analysis for each feasible structure.

Table 6. Products generated by each feasible structure.

	Product					
	Electricity	**Heat**	**Solid Fertilizer**	**Liquid Fertilizer**	**GHG Emission**	**Biochar**
Feasible Structure 1	/	/			/	/
Feasible Structure 2	/	/	/	/	/	/
Feasible Structure 3	/	/			/	/
Feasible Structure 4	/	/	/	/	/	/
Feasible Structure 5	/	/			/	/
Feasible Structure 6	/	/	/	/	/	/
Feasible Structure 7			/	/	/	/
Feasible Structure 8			/	/	/	/
Feasible Structure 9	/	/	/	/	/	/

The lowest profit was generated by feasible structure 1. The value gained from the product was much lower than the cost of raw materials. The cost of the raw materials was MYR 746 million, while the value gained from the product was MYR 301 million. There was around a 59.7% loss from this feasible structure because no fertilizer was generated from the digestate (a by-product of anaerobic digestion), as anaerobic digestion has a massive conversion of around 75% from raw materials to digestate before undergoing treatment to convert it into two types of fertilizer: solid and liquid.

For each feasible structure, at least five types of operating units were used. For a feasible structure that generates electricity and heat, at least one operating unit involved either a boiler or a gas turbine. For the generation of liquid and solid fertilizers, the digestate must undergo pre-treatment before it can be sold as products. Table 7 shows the operating units that affected the volume of GHGs generated. The GHG emissions were generated from landfills, gas turbines, and boilers. From landfills, 0.1605 m^3 GHGs per ton of MSW were released into the surroundings. The different types of technology involved in converting MSW gives a different ratio of GHG emissions produced.

Table 7. Operating units that affected the volume of GHG emissions.

	Landfill_1	Landfill_2	Boiler	Gas_Turbine
Feasible structure 1	/	/	/	/
Feasible structure 2	/	/	/	/
Feasible structure 3			/	/
Feasible structure 4			/	/
Feasible structure 5		/	/	/
Feasible structure 6		/	/	/
Feasible structure 7		/		
Feasible structure 8	/	/		
Feasible structure 9			/	

The lowest three structures that generated GHGs were feasible structures 7, 8, and 9 with values of 89,600, 140,000, and 16,500 m^3/year as shown in Figure 7. The GHGs generation was affected by only one piece of operating equipment for feasible structures 7 and 9. Since the conversion of GHG emissions from boilers was only 0.1 m^3 per combustible gas, the generation of GHGs for feasible structure 9 was lower than for feasible structure 7, which was affected by the generation of GHGs from landfills as mentioned earlier. Although feasible structure 8 had two pieces of equipment that affected the generation of GHGs, it was one of the top three feasible structures with a low generation of GHGs. Comparing feasible structures 3 and 4, both had a lower generation of GHGs because both used a gas turbine that affected their GHG generation. The gas turbine had a higher conversion of GHG emission, which was 0.505 m^3 per biogas. The highest GHG volumes were generated from feasible structures 1, 2, 5, and 6. All these feasible structures used three to four operating units, which affected the volume of GHGs generated.

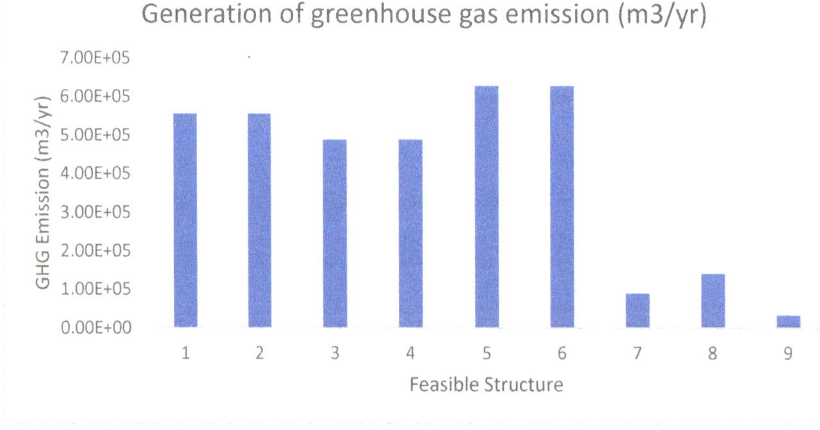

Figure 7. Generation of greenhouse gas for each feasible structure.

3.2. Scenario 1: Maximum EP

Feasible structure 9, as shown in Figure 8, was selected as the structure with the maximum EP. Based on Figure 6, the maximum EP of the selected pathways was estimated to be MYR 6.65 billion per annum or considering the total population of Malaysia in 2017, as mentioned by the World Bank [7] (31.62 million), it was estimated to be MYR 210 per person. For this feasible structure, both organic wastes (i.e., food and agriculture waste) underwent anaerobic digestion in operating units Digester_1 and Digester_2, which produced biogas and digestate. The digestate was separated into two types of fertilizer, liquid and solid, after pre-treatment. However, biogas did not undergo further treatment to convert it into electricity and heat. Plastics were burned in operating unit Pyrolysis_2 to be converted into biochar. Paper was burned in Incinerator_1 to produce both ash and combustible gas. The combustible gas was used in the boiler to convert it into electricity and heat. From this feasible structure, GHGs were produced from the boiler. The highest EP yield products, which were electricity, heat, solid fertilizer, liquid fertilizer, GHG emissions, and biochar, had flow rates of 82,700 kWh/year, 215,000 J/year, 1,030,000 tons/year, 4,110,000 tons/year, 33,100 m^3/year, and 795,000 tons/year, respectively.

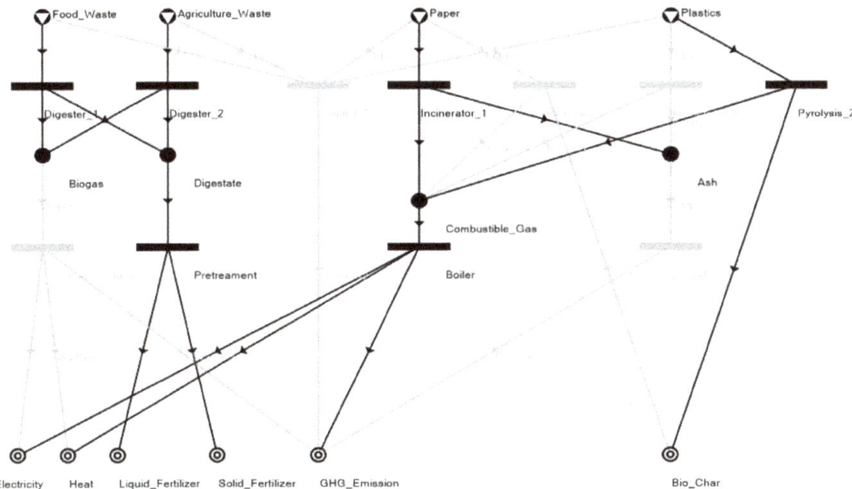

Figure 8. Feasible structure with the maximum economic performance.

The capital cost for a year was estimated to be MYR 123,587,000,000. This capital cost value is for the first year only. The profit margin was calculated as the net profits divided by the revenue. The profit margin for this feasible structure was 89.9%. The payback period of this pathway was 16.7 years. The payback period was calculated to identify the time required to earn back the investment money on the project.

3.3. Scenario 2: Minimal EI

Feasible structure 7, as shown in Figure 9, was selected as the structure with the minimal EI. Based on Figure 6, the minimal EI of the selected pathways was estimated to be MYR 6.65 billion per annum or considering the total population of Malaysia in 2017 as mentioned by the World Bank (31.62 million), it was estimated to be MYR 210 per person. This selected feasible structure was the same as the feasible structure of the maximum EP. However, feasible structure 7 produced less GHGs as GHGs are emitted only from landfills. For this feasible structure, both types of organic waste, food and agriculture waste also underwent anaerobic digestion in operating units Digester_1 and Digester_2, which produced biogas and digestate. The digestate was again separated into two types of fertilizer, liquid and solid, after pre-treatment. However, biogas did not undergo further treatment to convert it into electricity and heat. Plastics were also burned in operating unit Pyrolysis_2 to be converted into biochar. Paper was burned in Incinerator_1 to produce both ash and combustible gas Combustible gas was not converted into electricity and heat, and ash was removed to landfills. From this feasible structure, GHG emissions were produced from landfills only. The highest EP yield products were solid fertilizer, liquid fertilizer, GHG emissions, and biochar at flow rates of 1,030,000 tons/year, 4,110,000 tons/year, 89,600 m^3/year, and 795,000 tons/year, respectively.

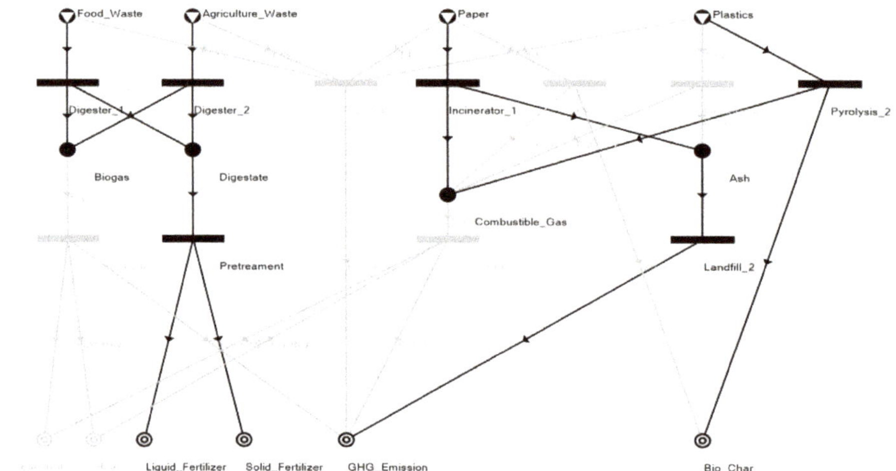

Figure 9. Feasible structure with the least environmental impact.

4. Conclusions

In this study, for the feasibility of MSW conversion technologies by PNS, "process graphs" were simulated. One hundred feasible structures were generated and nine of these were selected randomly for further analysis. Next, the EP and EI of the MSW conversion technologies were analyzed. Feasible structure 9 was chosen as the design with the maximum EP, with a total annual profit gain of MYR 6.65 billion, requiring up to 16 years as the payback period, with a constant flow of products. Feasible structure 7 was chosen as the design with the minimal EI, generating 89,600 m^3/year of GHGs for the whole of Malaysia.

Further studies should be conducted using a real case study. This method of study would be more convenient for optimizing a small case study, focusing primarily on a district rather than on the whole country. This study was a feasible-structure-based preliminary study to treat MSW in the whole country. For example, the payback period of conversion technologies could take up to 16 years because of the high cost of developing conversion technologies to manage the whole country's MSW.

Author Contributions: Conceptualization, R.A.A. and N.N.L.N.I.; methodology and software, R.A.A. and H.L.L.; writing-original draft preparation, R.A.A.; writing-review and editing, R.A.A. and N.N.L.N.I; supervision, N.N.L.N.I.

Funding: This research received no external funding.

Conflicts of Interest: The authors declare no conflict of interest.

References

1. Fodor, Z.; Klemeš, J.J. Waste as alternative fuel—Minimising emissions and effluents by advanced design. *Process Saf. Environ. Prot.* **2012**, *90*, 263–284. [CrossRef]
2. Tozlu, A.; Özahi, E.; Abuşoğlu, A. Waste to energy technologies for municipal solid waste management in Gaziantep. *Renew. Sustain. Energy Rev.* **2016**, *54*, 809–815. [CrossRef]

3. Silpa, K.; Lisa, Y.C.; Perinaz, B.T.; Frank, V.W. *What a Waste 2.0: A Global Snapshot of Solid Waste Management to 2050*; Urban Development; World Bank: Washington, DC, USA, 2018.
4. Chen, Y.-C. Effects of urbanization on municipal solid waste composition. *Waste Manag.* **2018**, *79*, 826–836. [CrossRef] [PubMed]
5. Solid Waste Management and Public Cleaning Corporation (SWCorp). *Strategic Plan SWCorp 2014–2020*; Ministry of Housing and Local Government Malaysia: Putrajaya, Malaysia, 2014.
6. Tan, S.T.; Lee, C.T.; Hashim, H.; Ho, W.S.; Lim, J.S. Optimal process network for municipal solid waste management in Iskandar Malaysia. *J. Clean. Prod.* **2014**, *71*, 48–58. [CrossRef]
7. World Bank. Total Population of Malaysia [Graph]. 2018. Available online: https://data.worldbank.org/indicator/SP.POP.TOTL?cid=GPD_1&locations=MY (accessed on 16 July 2019).
8. Economic Planning Unit (EPU). *Ninth Malaysian Plan 2006–2010*; Ministry of Finance Malaysia: Putrajaya, Malaysia, 2006.
9. Tarmudi, Z.; Abdullah, L.; Osman, A.; Tap, A.O.M. An overview of municipal solid wastes generation in Malaysia. *J. Teknol.* **2009**, *51*, 1–15. [CrossRef]
10. Erasu, D.; Feye, T.; Kiros, A.; Balew, A. Municipal solid waste generation and disposal in Robe town, Ethiopia. *J. Air Waste Manag. Assoc.* **2018**, *68*, 1391–1397. [CrossRef]
11. Performance Management and Delivery Unit (PEMANDU). *Solid Waste Management Final Lab Report 2015*; Ministry of Housing and Local Government Malaysia: Putrajaya, Malaysia, 2015.
12. Dong, C.; Jin, B.; Li, D. Predicting the heating value of MSW with a feed forward neural network. *Waste Manag.* **2003**, *23*, 103–106. [CrossRef]
13. Bello, H. Impact of changing lifestyle on municipal solid waste generation in residential areas: Case study of Qatar. *Int. J. Waste Resour.* **2018**, *8*, 2. [CrossRef]
14. Samsudin, M.D.; Don, M.M. Municipal solid waste management in Malaysia: Current practices, challenges and prospect. *J. Teknol.* **2013**, *62*, 95–101. [CrossRef]
15. Zurbrügg, C.; Schertenleib, R. *Main Problem and Issues of Municipal Solid Waste Management in Developing Countries with Emphasis on Problem Related to Disposal by Landfill*; Swiss Federal Institute for Environmental Science & Technology (EAWAG): Dübendorf, Sweden, 1998.
16. Perithamby, A.; Hamid, F.; Khidzir, K. Evolution of solid waste management in Malaysia: Impact and implication of solid waste. *Mater. Cycle Waste Manag.* **2009**, *11*, 96–103. [CrossRef]
17. National Solid Waste Management Department (JPSPN). *Survey on Solid Waste Composition, Characteristics & Existing Practice of Solid Waste Recycling in Malaysia*; Ministry of Housing and Local Government Malaysia: Putrajaya, Malaysia, 2013.
18. Solid Waste Management and Public Cleaning Corporation (SWCorp). *Waste Management in Malaysia: Towards a Holistic Approach*; Ministry of Housing and Local Government Malaysia: Putrajaya, Malaysia, 2017.
19. Muhammad, R.; Yousef, S.; Ali, A.; Ali, E. Optimal processing route for the utilization and conversion of municipal solid waste into energy and valuable products. *J. Clean. Prod.* **2017**, *174*, 857–867. [CrossRef]
20. Rodionov, M.; Nakata, A.T. Design of an optimal waste utilization system: A case study in St. Petersburg, Russia. *Sustainability* **2011**, *3*, 1486–1509. [CrossRef]
21. Joao, A.; Paulo, F. Assessing the costs of municipal solid waste treatment technologies in developing Asian countries. *Waste Manag.* **2017**, *69*, 592–608. [CrossRef]
22. Seadon, J.K. Sustainable waste management system. *J. Clean. Prod.* **2010**, *18*, 1639–1651. [CrossRef]
23. Lam, H.L. Extended P-graph applications in supply chain and Process Network Synthesis. *Curr. Opin. Chem. Eng.* **2013**, *2*, 475–486. [CrossRef]
24. Biegler, L.; Grossmann, I. Retrospective on optimization. *Comput. Chem. Eng.* **2004**, *28*, 1169–1192. [CrossRef]
25. Lam, H.L.; Klemeš, J.J.; Kravanja, Z.; Varbanov, P.S. Software tools overview: Process integration, modelling and optimisation for energy saving and pollution reduction. *Asia Pac. J. Chem. Eng.* **2011**, *6*, 696–712. [CrossRef]
26. Friedler, F.; Ng, K.M. Process systems engineering. *Curr. Opin. Chem. Eng.* **2012**, *1*, 418–420. [CrossRef]
27. Jozsef, T. P-Graph-based Workflow Modelling. *Acta Polytech. Hung.* **2007**, *4*, 75–88.
28. Friedler, F.; Tarján, K.; Huang, Y.W.; Fan, L.T. Graph-theoretic approach to process synthesis: Axioms and theorems. *Chem. Eng. Sci.* **1992**, *47*, 1973–1988. [CrossRef]

29. Lam, H.L.; Raymond, R.T.; Kathleen, B.A. Implementation of P-graph modules in undergraduate chemical engineering degree programs: Experiences in Malaysia and the Philippines. *J. Clean. Prod.* **2016**, *138*, 254–265. [CrossRef]
30. Munster, M.; Meibom, P. Long term affected energy production of waste to energy technologies identified by use of energy system analysis. *Waste Manag.* **2010**, *30*, 2510–2519. [CrossRef] [PubMed]
31. Munster, M.; Meibom, P. Optimization of use of waste in the future energy system. *Energy* **2011**, *36*, 1612–1622. [CrossRef]
32. Ng, W.; Varbanov, P.; Klemeš, J.; Hegyháti, M.; Bertok, B.; Heckl, I.; Lam, H. Waste to energy for small cities: Economic versus carbon footprint. *Chem. Eng. Trans.* **2013**, *35*, 889–894. [CrossRef]
33. Shadiya, O.O.; Satish, V.; High, K.A. Process enhancement through waste minimization and multi objective optimization. *J. Clean. Prod.* **2012**, *31*, 137–149. [CrossRef]
34. Chang, N.-B.; Qi, C.; Islam, K.; Hossain, F. Comparisons between global warming potential and cost–benefit criteria for optimal planning of a municipal solid waste management system. *J. Clean. Prod.* **2012**, *20*, 1–13. [CrossRef]
35. Walmsley, T.G.; Varbanov, P.S.; Klemeš, J.J. Networks for utilising the organic and dry fractions of municipal waste: P-graph approach. *Chem. Eng. Trans.* **2017**, *61*, 1357–1362. [CrossRef]
36. Lam, H.L.; Varbanov, P.S.; Klemeš, J.J. Optimization of regional energy supply chains utilising renewables: P-graph approach. *Comput. Chem. Eng.* **2010**, *34*, 782–792. [CrossRef]

Article

Gaussian Process Methodology for Multi-Frequency Marine Controlled-Source Electromagnetic Profile Estimation in Isotropic Medium

Muhammad Naeim Mohd Aris [1,*], Hanita Daud [1], Sarat Chandra Dass [2] and Khairul Arifin Mohd Noh [3]

1 Department of Fundamental and Applied Sciences, Universiti Teknologi PETRONAS, 32610 Seri Iskandar, Malaysia; hanita_daud@utp.edu.my

2 School of Mathematical and Computer Sciences, Heriot-Watt University Malaysia, 62200 Putrajaya, Malaysia; s.dass@hw.ac.uk

3 Department of Geosciences, Universiti Teknologi PETRONAS, 32610 Seri Iskandar, Malaysia; khairula.nmoh@utp.edu.my

* Correspondence: muhammad_naeim@yahoo.com; Tel.: +60-17204-4042

Received: 18 August 2019; Accepted: 19 September 2019; Published: 27 September 2019

Abstract: The marine controlled-source electromagnetic (CSEM) technique is an application of electromagnetic (EM) waves to image the electrical resistivity of the subsurface underneath the seabed. The modeling of marine CSEM is a crucial and time-consuming task due to the complexity of its mathematical equations. Hence, high computational cost is incurred to solve the linear systems, especially for high-dimensional models. Addressing these problems, we propose Gaussian process (GP) calibrated with computer experiment outputs to estimate multi-frequency marine CSEM profiles at various hydrocarbon depths. This methodology utilizes prior information to provide beneficial EM profiles with uncertainty quantification in terms of variance (95% confidence interval). In this paper, prior marine CSEM information was generated through Computer Simulation Technology (CST) software at various observed hydrocarbon depths (250–2750 m with an increment of 250 m each) and different transmission frequencies (0.125, 0.25, and 0.5 Hz). A two-dimensional (2D) forward GP model was developed for every frequency by utilizing the marine CSEM information. From the results, the uncertainty measurement showed that the estimates were close to the mean. For model validation, the calculated root mean square error (RMSE) and coefficient of variation (CV) proved in good agreement between the computer output and the estimated EM profile at unobserved hydrocarbon depths.

Keywords: multiple frequency marine controlled-source electromagnetic technique; Gaussian process; uncertainty quantification; computer experiment, electromagnetic profile estimation

---

1. Introduction

Nowadays, the controlled-source electromagnetic (CSEM) technique is a significant application to detect and discover hydrocarbon-filled reservoirs based on the principles of electromagnetic (EM) propagation. For decades, CSEM application has been widely exercised in onshore geophysical exploration (e.g., [1,2]). The efficiency of this application to characterize offshore hydrocarbon reservoirs has also been proven by many oil and gas companies around the world. Li and Key [3] stated that at early stage, marine CSEM application was employed to study the electrical conductivity of the upper mantle and oceanic crust (e.g., [4–8]). Studies related to the commercial application of the marine CSEM technique in offshore hydrocarbon exploration can be found in [9–17]. Previously, the seismic sounding survey, which employs acoustic waves, was solely utilized to map geological structures that have different acoustic properties [18]. This survey was very important to hydrocarbon exploration due to its capability of providing information of the subsurface. According to [19], seismic data interpretation provides good resolution of the subsurface structures underneath the seabed; however, it has deficiencies. It is said that seismic surveys are unable to distinguish the fluid content inside the reservoirs, whether brine (conductive seawater) or hydrocarbon. Zaid et al. [18] mentioned that seismic sounding is not compatible to the direct detection of the pore fluid reservoirs. Note that EM and seismic techniques are sensitive to two different properties of subsurface; thus, the marine CSEM technique was developed as a complementary interpretational tool to specifically characterize the target reservoirs.

The marine CSEM technique also is referred to as a seabed logging (SBL) application. This is thoroughly described by [10]. This application is particularly able to reduce ambiguities in data interpretation in hydrocarbon exploration. Andreis and MacGregor [16] stated that by studying the reflected EM signal, resistive mediums such as hydrocarbon, gas, and hydrate can be discovered to depths of several kilometers from the seabed. In addition, the resistivity of the subsurface in offshore environments is commonly identified by robust anisotropy because of the sedimentation factor [20]. Note that for a medium that is horizontally stratified, the subsurface is generally less resistive in the horizontal (parallel) direction than in the vertical (perpendicular) direction [20]. Offshore hydrocarbon reservoirs are normally embedded in a high conductive medium unlike the common case of onshore hydrocarbon reservoirs [21]. Hydrocarbon-filled reservoirs are known to have very high electrical resistivity compared to its surroundings, such as of saline water and sedimentary rocks. These structures are very conductive. From [22], hydrocarbon is known to have electrical resistivity between 30 and 500 Ohm meter, whereas the resistivity values of seawater and sediment are 0.5–2 Ohm meter and 1–2 Ohm meter, respectively. If a target reservoir is brine saturated, it is normally a few orders of magnitude less electrically resistive than a hydrocarbon-filled reservoir. From these characteristics, the resistivity of the subsurface can be resolved via data of electric (E-) and magnetic (H-) fields obtained from the marine CSEM survey. The measurement of amplitude and phase of E- and H- fields can be utilized to determine the geological subsurface. Li and Key [3] mentioned that the amplitude and phase of an EM field will vary depending on the resistivity of the structures beneath the seabed, the depth of seawater, and the source–receiver offset.

In the marine CSEM technique, data collected can be interpreted in two different groups depending on the domain—either in time-domain or frequency-domain. Analyzing data in time- or frequency-domains would theoretically give the same output/information [23]. Reyes et al. [24] mentioned that frequency-domain marine CSEM is normally used for the case of oil prospecting. In the frequency-domain application, an antenna/transmitter of towed EM dipole is generally used to generate a low-frequency EM field, and returned/reflected signals recorded by receivers placed on the seabed are utilized for resistivity distribution analysis. The choice of frequency is very crucial in marine CSEM application. In a standard configuration, marine CSEM surveys use a deep-towed horizontal electric dipole (HED) transmitter to emit a low-frequency EM wave which is usually between 0.1 Hz and 10 Hz to an array of seabed receivers, and normally, the transmitter is towed at 30–50 m above the seabed [25]. Practically, the low-frequency of EM waves is used in a deep water environment as the signal transmission due to the fact that low-frequency is able to yield farther penetration through seawater columns into sedimentary rocks. Next, for the receiver, there are two types of receiver configurations—inline and broadside. Inline configuration is when the separation distance is parallel to the direction of antenna, whereas broadside is when the source–receiver offset is normal to the antenna's direction [26]. Due to the characteristics of subsurface conductivity, EM signals spread with a higher rate through the seafloor than through the seawater. The EM energy, which is transmitted from the towed-source, spreads in all directions and is quickly attenuated in conductive medium such as sediment. The occurrence of all possible signal contributions is depicted in Figure 1.

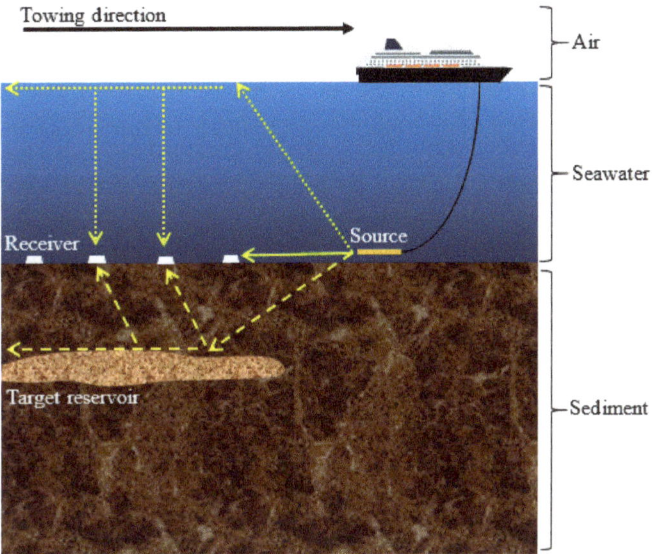

Figure 1. Basic layout of marine CSEM application in hydrocarbon exploration. The source is towed nearly to the seabed receivers. A target reservoir is expected to be embedded in the conductive sedimentary rocks. The EM signal spreads in all direction through the seawater, air–seawater interface, and sediment before being recorded by the EM receivers.

Based on Figure 1, the straight-line arrow denotes the direct-wave travelling from the source to the seabed receiver without any interaction with the geological subsurface beneath the seabed. Researchers in [27,28] mentioned that direct-wave dominates the data collection at short source–receiver separation offsets. Next, the dotted-line arrow represents the reflected and refracted wave from the source upwards to the air–seawater interface and vertically going back through the seawater to the receiver (i.e., air-wave). This wave travels with high rate of velocity (propagates with no attenuation) to the water surface since air is an infinitely electrical resistive medium. Seawater depth can influence the measured EM signal. Air-wave contribution increases as the depth of seawater decreases. Weiss [29] asserted that this contribution becomes significant in a seawater depth of roughly less than 300 m. Both these signals, direct- and air- waves, do not contain any information about hydrocarbon-filled reservoirs. Last signal contribution is denoted as dashed-line arrow. It is known as the reflected and refracted wave (i.e., guided-wave). This wave diffuses outward from the source through seawater column and then through the high resistive formations with less attenuation. The transmitted wave has to enter the formations at certain angle which is between 0° and 11° in order to set the guided mode [28]. This reflected and refracted wave strongly dominates the recordings at intermediate source–receiver offsets (~3 to ~8 km). The detection of the guided-wave is the basis of the marine CSEM survey.

In the context of geophysics forward modeling, electrical resistivity has an important role in oil and gas exploration. Numerical modeling is a crucial component that provides information of the electrical resistivity of the subsurface. There are various computational techniques exercised in EM applications such as the Finite Element (FE), the Finite Difference (FD) and the Method of Moment (MOM) [30], while researchers in [20] have said that FE, FD and integral equation (IE) methods are among the most famous numerical techniques for modeling EM data. According to [20], the FE method is more reliable for EM forward modeling in a complex geological structure when compared to FD and IE methods. Based on the literature, FE is a usual numerical technique exercised in CSEM modeling for hydrocarbon exploration. This computational method uses unstructured grids that can be easily conformed to irregular boundaries compared to the FD method. The MOM is less preferred as well in marine CSEM data interpretation since this method produces more complex derivations of governing equations than the FE method. The traditional FD method is easier to implement and maintain than the FE method, but the method is based on structured grids. This means that grid refinement is not possible, and hence it affects the overall computational processes [31]. Unstructured grids have long been exercised in various fields such as engineering and applied mathematics; however, this feature has only recently been used in the EM geophysical field, as exemplified by the use of FE method code in marine EM surveys (e.g., [3,32–34]). This feature can realistically replicate the complexities of geological structures [19].

Even though this numerical technique is very powerful, the ad hoc design of meshes in FE is time-consuming. Li and Key [3] mentioned that the most time-consuming task in their code (FE algorithms) is the solutions of the linear equation systems. The study could take a very lengthy computational time if they used all wavenumbers in the mesh refinement for a full solution. Bakr et al. [35] also stated that the most time-consuming tasks in FE are evaluating the integrals and solving the linear equations. For typical simulations, a few million elements are involved in the linear equation systems [31]. According to [24], the execution of real-field simulations in EM problems needs the use of high-performance computing (HPC). This is because typical actual executions require more than hundred thousand realizations which involve millions of degrees of freedom for each process. The computational and memory requirements to solve such solutions may become a serious challenge. It can be more complicated for higher-dimensional EM forward modeling and inversion. Besides forward modeling, inversion is also a powerful way to recover the electrical conductivity profile beneath the seabed given measurements of EM fields acquired from real-field surveys. Not to mention, nowadays, inverse modeling comes with robust inversion schemes and incorporation of more procedures and measurements. This makes it possible to compute the EM fields at the seabed receivers precisely and provide accurate geometry resolution. However, it is said that inversion algorithms tend to be computationally expensive due to the forward modeling schemes. Indeed, an inversion process needs multiple EM forward solutions [35]. Furthermore, in terms of application, the marine CSEM technique generates huge amounts of data (captured by seabed EM receivers with a moving HED source); therefore, processing those data has become a challenging task to many geophysicists [9,36]. Modeling the marine CSEM data is related to the need of accurate representation of very complex geo-electrical models, and the algorithms used should be powerful and fast enough to be applied to repeated use of hundreds of iterations and multiple source–receiver positions. In addition, understanding the noisy CSEM data to quantify the uncertainties involved in EM modeling also is very crucial. Constable and Srnka [14] stated that the economic challenges (e.g., related to drilling) increase as the hydrocarbon exploration moves to deeper offshore environments. Thus, any additional data that can be obtained or collected will be advantageous to the exploration if there is a potential to de-risk a given expectation [19].

In order to seek the most favorable balance between the computational cost involved in the interpretation of EM geophysical data and the accuracy of the modeling, our interest is focused on processing the one-dimensional (1D) frequency-domain marine CSEM data using Gaussian process (GP) algorithms. We propose GP as a methodology for two-dimensional (2D) forward modeling of the marine CSEM technique to provide information on EM profiles when hydrocarbon is present at various depths in isotropic mediums. This forward modeling provides the uncertainty measurement of the estimation in terms of variance. Although the existing CSEM models (the existing numerical modeling techniques) provide robust representation of real-field models, this work has significant contribution for hydrocarbon detection as well. This attempt is very useful and helpful when collected sets of data in CSEM surveys are insufficient for the interpretation of higher-dimensional modeling and inversion. This analysis also could reduce time in the CSEM workflow since forward GP modeling is able to provide uncertainty quantifications without integrating or combining any other numerical quantifiers. On top of that, there is its simplicity, as GP only involves simple equations which means faster computation which only needs basic memory space. Note that this analysis utilizes the simulation datasets generated through a commercial software, namely, Computer Simulation Technology (CST). Information of the CST software can be found in [37]. The details and literature of GP application are thoroughly elaborated in next section.

2. Statistical Background: Gaussian Process in Computer Experiments

Gaussian Process (GP) is random function which has a property that any finite number of evaluations of the (random) function has a multivariate Gaussian distribution. GP is fully specified by a mean function, $m(x)$, and a covariance function, $k(x, x')$. The Gaussian distribution has mean and covariance values in the forms of vector and matrix evaluations, respectively [38]. Here, x represents all potential independent variables that influences the outputs/responses. GP is a non-parametric and probabilistic method for fitting functional forms based on domain observations. It differs from most of other black-box identification approaches where it does not approximate the modelled system by fitting the parameters of basis function, but rather searches for relationships among the measured data. This non-parametric regression method does not need a fixed discretization. This technique is able to provide predictive mean values and uncertainty of the estimation measured in terms of variance. This variance reflects the quality of the output/information. It is an important numerical measure when it comes to distinguishing GP from the other computational intelligence methods. According to [38], GP is suitable for modeling uncertain processes or data which are unreliable, noisy or contain missing values. GP has been used in many different applications. Studies related to application of GP in various fields can be found in [39–45]. In general, prior belief of spatial smoothness is specified through a covariance defined by similarity characteristics. Training observations are then considered as the realizations from the updated multivariate Gaussian (i.e., posterior). Thus, the conditional realizations from the posterior are simply the testing output at all untried or unobserved points. The mathematics behind this concept are thoroughly explained in the methodology section.

Computer experiments are well-known and not new in science, technology, and engineering. This medium is getting very popular for solving scientific and engineering problems. Nowadays, scientists prefer to use computer simulators rather than doing case studies or conducting any related physical experiments. It can be implemented in any circumstances including experiments that are impossible to do physically, with shorter time taken than in the real situation. Computer experiments are run by means of a complex code and highly developed theories of physics, mathematics, and engineering fields. Sacks et al. [46] described that experimenters usually aim to estimate/predict the output at unobserved input points, optimize the function of the output points, and calibrate the computer code to physical data. To this end, [46] and the subsequent works modelled the output of a computer model, $Y(x)$, based on input x as a sum of regression terms, β_j, and stochastic component, $Z(x)$. $Y(x)$ is defined in Equation (1).

$$Y(x) = \sum_{j=1}^{k} \beta_j f_j(x) + Z(x), \quad (1)$$

where $f_j(x)$ is a known function with $j = 1, 2, \ldots, k$, and $Z(x)$ is a random process with a zero-mean and a covariance. The most famous choice of the stochastic component, $Z(x)$, is a GP where the distribution of the GP is assumed to be a normal (Gaussian) distribution with a mean and a covariance function. According to [47], GP is used as the surrogate model for any complex mathematical models which consume a lot of time to solve. GP is flexible in representing the computer output, $Y(x)$, and it is feasible to obtain analytical formulas of the predictive distribution and to design the equations.

From the reported literature, there are a few studies calibrating CST computer output of marine CSEM applications with GP (e.g., [48,49]). However, these studies present 1D GP modeling of SBL applications where they only considered univariate independent variables. Besides, the work only focused on predicting the presence of the hydrocarbon layer at a known depth, while we consider various depths of hydrocarbon at observed and unobserved depth levels. This is because the location of hydrocarbon reservoirs can be anywhere and is uncertain in real-field environments. Aris et al. [50] also described forward GP modeling of SBL applications, but the paper only focused on one transmission frequency and no error was considered in the presented GP modeling. Since CST computer output is assumed to generate very clean data, considering the error in modeling is very important to marine CSEM data processing. Thus, this attempt is novel in two ways; first, it proposes GP methodology to process marine CSEM data calibrated with CST computer output at multiple transmission frequencies at which hydrocarbon is present at all possible depth levels (250–2750 m); second, this is a data-dependent analysis where it utilizes an uncertainty quantification provided by the GP in marine CSEM data processing with error considerations before in-depth analysis. This may enhance the EM data interpretation where the EM profile is estimated with the measurement of variance at various possible depths of the hydrocarbon layer, which helps decision-making for hydrocarbon detection in marine CSEM applications.

3. Methodology

The methodological flow is based on a three-step procedure; (i) synthetic seabed logging (SBL) modeling using Computer Simulation Technology (CST) software, (ii) developing two-dimensional (2D) forward Gaussian Process (GP) models for multiple EM transmission frequencies, and (iii) model validation using the root mean square error (RMSE) and the coefficient of variation (CV).

3.1. *Synthetic SBL Modeling Using CST Software*

We designed synthetic models of typical marine CSEM application for hydrocarbon exploration which have various depths of hydrocarbon at multiple frequencies by using CST software. The transmissions were tested at frequencies of excitation current of 0.125, 0.25, and 0.5 Hz. Note that in the CST software, Maxwell's equations are discretized using the Finite Integration Method (FIM). FIM solves the Maxwell's equations in a finite calculation domain in grid cells to probe the resistivity contrast. For this study, the SBL model is a three-dimensional (3D) canonical structure, which consists of background layers such as air, seawater, and sediment. The model is designed with an air–seawater interface at z = 300 and the seawater thickness is fixed at 1000 m (deep offshore environment). A 200 m-thick horizontal resistive layer (hydrocarbon) is embedded in the sediment layer with various depths from the seabed. The thickness of sediment above the hydrocarbon layer, known as overburden thickness, is varied from 250 m to 2750 m with an increment of 250 m each. The thickness of the overburden layer indicates the depth of hydrocarbon reservoir. Thicker overburden layers mean a deeper location of the hydrocarbon. Both the background and hydrocarbon layers are considered as isotropic. Figure 2 shows the stratified illustration of the horizontal layers used in this study.

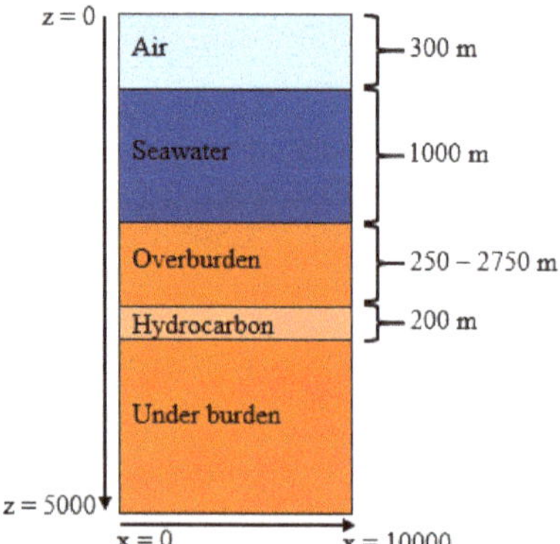

Figure 2. Illustration of SBL models used in this study. The thicknesses of air, seawater, and hydrocarbon are 300, 1000, and 200 m, respectively. The depth of the hydrocarbon layer (thickness of overburden layer) is varied from 250 m to 2750 m with an increment of 250 m. The total height and length of the models are 5000 and 10,000 m, respectively.

The replication of the 3D structures of the SBL model (length, x: 10,000 m; width; y: 10,000 m; height, z: 5000 m) can be referred to in [50]. The electrical conductivities of air, seawater, sediment, and hydrocarbon are tabulated in Table 1. Their properties are taken from [48].

Table 1. Electrical conductivity of every layer considered in the SBL models. Conductivity is the reciprocal of resistivity. The electrical conductivity of the hydrocarbon layer is lower than its surroundings, which are seawater and sediment. It means that the hydrocarbon layer is parameterized with reliable electrical resistivity.

Material	Electrical Conductivity (Sm^{-1})
Air	1.0×10^{-11}
Seawater	1.63
Sediment	1.00
Hydrocarbon	2.0×10^{-3}

The EM signal is transmitted by a HED source located with the orientation of x-direction in the seawater with coordinates (5000, 5000, 1270). This means that the inline transmitter pointing along x-axis is positioned at x, y = 5000, and a height of 30 m above the seabed. The HED source is held stationary at the center of the model. The values of the EM field are measured along an inline profile through the SBL model. In this study, the current strength of the HED source is fixed at 1250 A. The source–receiver separation distances (offsets) are varied along the replication model. An array of 1000 seabed receivers is placed along the seabed at x ranges of 0–10,000 m and y = 5000 m. This means, receivers are positioned along the seabed for every 10 m from 0–10,000 m of x-orientation. As a demonstration, Figure 3 is the mesh view of the replication of the marine CSEM model in isotropic medium, replicated by the CST software at a hydrocarbon depth of 250 m.

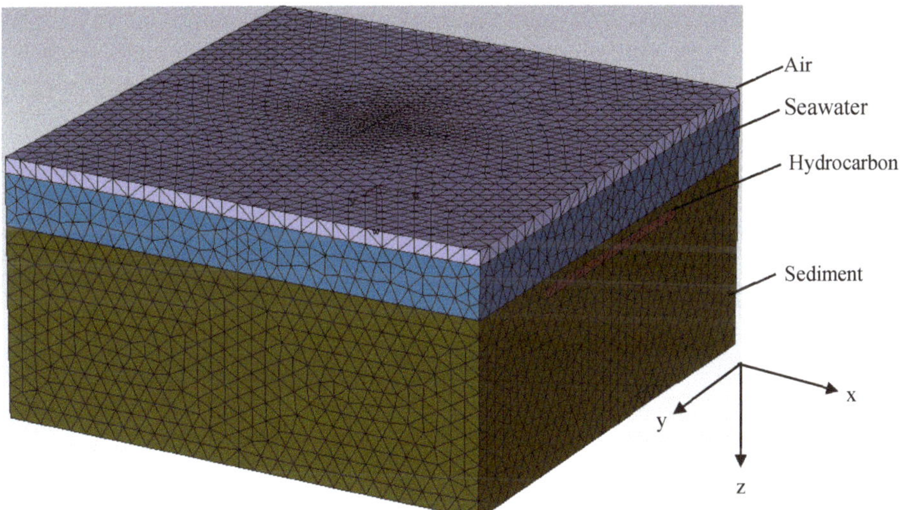

Figure 3. Mesh view of the 3D SBL model at hydrocarbon depth of 250 m replicated by CST software. Both background and hydrocarbon layers are set as isotropic. The hydrocarbon layer is designed with length (x) of 10,000 m, width (y) of 5000 m and height (z) of 200 m.

3.2. Developing 2D Forward GP Models at Multiple EM Transmission Frequencies

Let $Y(x_i)$ be the CST computer output at k different input specifications for every frequency used (0.125, 0.25, and 0.5 Hz), where $i = 1, 2, \ldots, k$. The input variable, x, can be univariate or multivariate, but in this study, we exercise a bivariate independent variable. Source–receiver separation distance (offset), s, and depth of hydrocarbon layer from the seabed, h, are considered as input variables where $x = (s, h)$. In this paper, we focus on processing non-normalized CST output, $Y(x)$, which is the magnitude of the E-field (amplitude) obtained from static source–receiver combinations where the transmitter is fixed at the center of the SBL model. For every frequency, we have source–receiver separation distances, $s_i = \{i = 1, 2, \ldots, 210\}$, and hydrocarbon depths, $h_i = \{i = 1, 2, \ldots, 11\}$, which are from 250 to 2750 m with and increment of 250 m each. Thus, in this study, we have $k = 210 \times 11 \times 3 = 6930$ different input specifications of CST computer output that are to be processed.

As mentioned earlier, GP is completely defined by a mean function, $m(x)$, and a covariance function, $k(x, x')$. The GP model on function f with a zero-mean function, $m(x) = 0$, and a covariance function, $k(x, x')$, can be written as

$$f(x) \sim G(0, k(x, x')). \tag{2}$$

An appropriate correlation function for our GP is selected. The choice of the correlation function is very crucial in computer experiments. It governs the smoothness of the sample path realizations of the GP and is dictated by CST computer output. We choose a popular covariance function which is the squared exponential (SE) function. This covariance function has been widely used in many applications of GP regression and it produces smooth functional estimates. The SE is defined in Equation (3).

$$k(x, x') = \sigma_f^2 . e^{\left(\frac{-|x-x'|^2}{2\ell^2}\right)}, \tag{3}$$

where σ_f and ℓ are signal variance and characteristic-lengths scale, respectively. These hyper-parameters need to be properly estimated, and this is usually by optimizing the marginal likelihood. By referring to Bayes' theorem, we assume that very little prior knowledge about these hyper-parameters are known, and this prior knowledge corresponds to the maximization of marginal log-likelihood. We have three different datasets (three different frequencies). For every dataset, there are 210 data points consisting of offset and hydrocarbon depths as the independent variables corresponding to the magnitude of the E-field as the dependent variable. According to [39], two-thirds of the total data should be considered as training data points and the remainder will be the testing points. Thus, in this paper, for every three data points, the first two data are set as the training data points. This procedure was implemented for every frequency used.

The GP regression model is assumed to generally have a relationship of the form $y_i = f(x_i) + \varepsilon$ where the prior joint distribution for the collection of random variables consist of training and testing points are defined as below

$$\begin{bmatrix} m \\ m_* \end{bmatrix} \sim G\left(0, \begin{bmatrix} K_\varepsilon & K_* \\ K_*^T & K_{**} \end{bmatrix}\right). \tag{4}$$

The vector $m \in \Re^{n_{train}}$ is observed at spatial locations $x \in \Re^{n_{train} \times n_d}$ where $x = (s, h)$. n_{train} denotes the number of training data points generated from CST software, while n_d is the spatial dimension exercised in this paper. Note that for every hydrocarbon depth, $n_{train} = 140$ and $n_d = 2$. Next, $m_* \in \Re^{n_{test}}$ is a vector that specifies the predicted values at particular spatial locations $x_* \in \Re^{n_{test} \times n_d}$ where $x_* = (s_*, h_*)$. n_{test} is the number of all desired observations (i.e., testing data points) where $n_{test} = 70$ for each depth. With 140 data per depth, a matrix $K \in \Re^{n_{train} \times n_{train}}$ is defined using Equation (3) for all pairs involved in the training points. Then, $K_\varepsilon \in \Re^{n_{train} \times n_{train}}$ is determined such that

$$K_\varepsilon = K + \sigma^2, \tag{5}$$

where σ^2 represents a diagonal covariance matrix of the specified additive noise at 5%. $K_* \in \Re^{n_{train} \times n_{test}}$, and $K_{**} \in \Re^{n_{test} \times n_{test}}$ are calculated using Equation (3) where these matrices define the correlation of training–testing data points and testing–testing data points, respectively. Hence, m_* is predicted at 70 locations of x_* per depth. Here, matrix K_{**} only has input from testing data points and it is derived from prior information. Thus, the posterior conditional GP as in Equation (1), given the information of x_*, x and m, is written as

$$p(m_*|x_*, x, m) = G(m_*|\mu_*, \Sigma_*) \tag{6}$$

Based on the theorem, for every frequency, the Gaussian probability for the random variable m_* with mean, μ_* (i.e., the estimated EM profile at observed and unobserved depths of hydrocarbon), and variance, Σ_* (uncertainty measurement in terms of ± two standard deviations), are defined in Equations (7) and (8), respectively. These equations are the main equations of GP regression.

$$\mu_* = K_*^T K_\varepsilon^{-1} m \tag{7}$$

$$\Sigma_* = K_{**} - K_*^T K_\varepsilon^{-1} K_* \tag{8}$$

3.3. Model Validation Using RMSE and CV

To validate our forward GP model, we calculated the root mean square error (RMSE) and coefficient of variation (CV) of the difference between data predicted by GP (estimates) and the data acquired from CST software. RMSE is able to calculate the difference between an estimate and the true value (observation) corresponding to the expected value of root squared loss. CV is calculated as well to evaluate the relative closeness between true values and estimates in percentage. Two random unobserved depths of hydrocarbon which are 900 m and 2200 m were selected for demonstration purposes. The SBL models with these depths of hydrocarbon were simulated separately. The CST computer output was considered as the true values, and the estimate values are the data predicted by GP at the same depths of hydrocarbon. The RMSE and CV between the true values, y_i, and the estimate values, y_i^*, are defined as below

$$RMSE = \sqrt{\frac{\Sigma(y_i - y_i^*)^2}{a}}, \tag{9}$$

$$CV = \frac{RMSE}{|\mu_{y_i^*}|} \times 100\%, \tag{10}$$

where a is the total number of testing data points, and $\mu_{y_i^*}$ is the absolute average of y_i^*. The estimates from the 2D forward GP model should match the simulation data acquired from the CST software at all observed and unobserved depths very well up until larger offset distances.

4. Results and Discussion

We implemented this analysis by using GP algorithms in a MATLAB code (built-in function) referred from [51]. We considered a typical synthetic frequency-domain marine CSEM study with three different transmission frequencies (0.125, 0.25, and 0.5 Hz), a HED source, and an array of receivers. Note that for every depth of hydrocarbon, the simulation was simultaneously run for the three transmission frequencies (three datasets per simulation). Every simulation process took ~15 min to generate the three different datasets. Hence, total computational time for the CST software to compute the EM fields for 11 hydrocarbon depths was approximately 165 min (~2 h and 45 min). In addition, in order to make the data interpretable, a logarithmic scale with base 10 was applied to the magnitude of the E-field for every frequency since it involves very small values. Figures 4–6 are the CST computer output at all input specifications for every frequency.

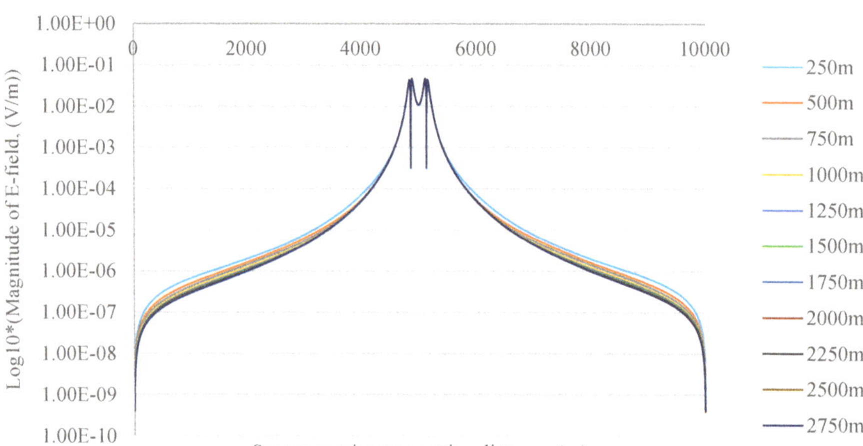

Figure 4. Log10 of magnitude of electric field at 0.125 Hz versus source–receiver separation distance (offset). Different hydrocarbon depths yield different EM responses. The offset is from 0 m (left of the SBL model) to 10,000 m (right of the SBL model).

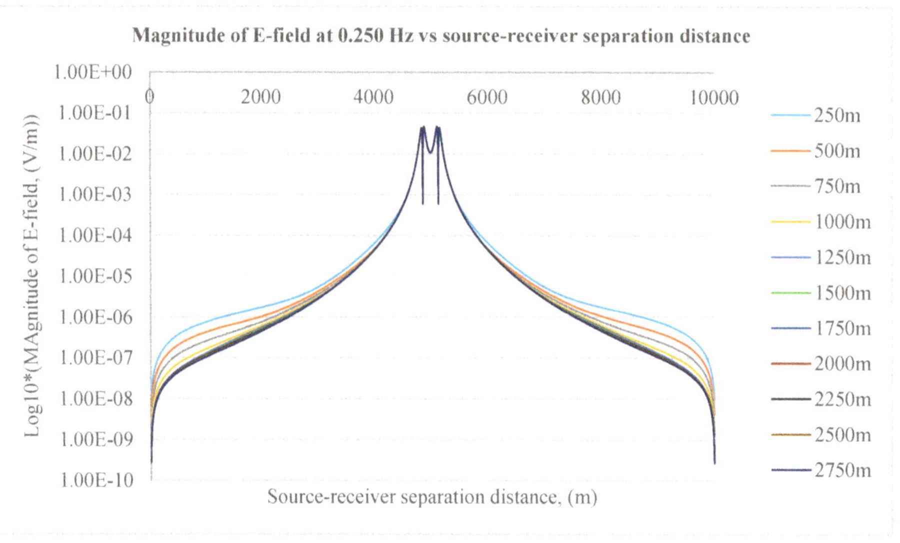

Figure 5. Log10 of magnitude of electric field at 0.25 Hz versus source–receiver separation distance (offset). Different hydrocarbon depths yield different EM responses. The offset is from 0 m (left of the SBL model) to 10,000 m (right of the SBL model).

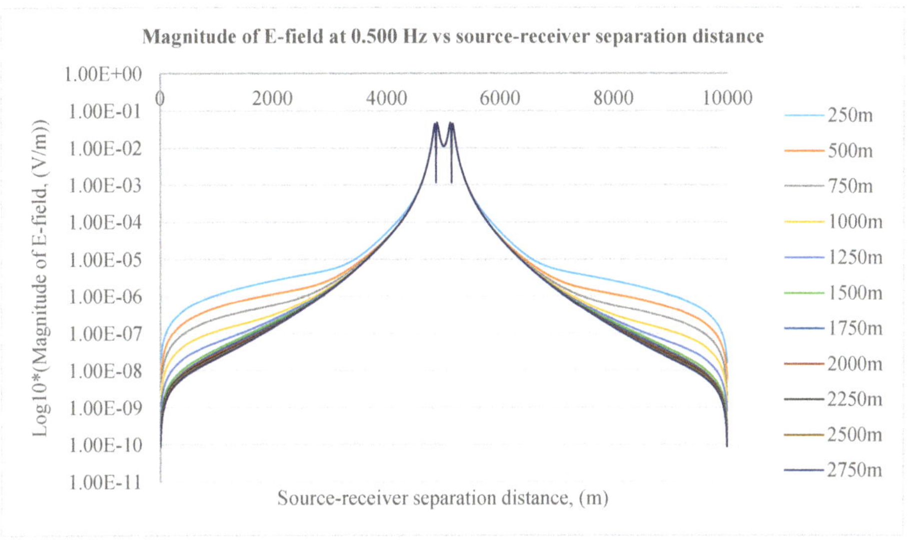

Figure 6. Log10 of magnitude of electric field at 0.5 Hz versus source–receiver separation distance (offset). Different hydrocarbon depths yield different EM responses. The offset is from 0 m (left of the SBL model) to 10,000 m (right of the SBL model).

From the results, the replicated SBL model is able to reflect the typical synthetic simulation model of EM application. The simulated datasets resulting from the CST software were in a good agreement with the behavior of real-field CSEM data on the effect of the source–receiver separation distance and variations of hydrocarbon depth to the magnitude of the E-field (amplitude). The strength of E-field is inversely proportional to the source–receiver offset and depth of hydrocarbon. If the source–receiver separation is placed further apart and the hydrocarbon layer is located deeper beneath the seabed, the E-field strength significantly decreases. The acquired responses vary in frequency as well. Here, based on figures above, the EM responses are symmetrical. The EM wave was transmitted from the source which was located at the center of the SBL model. The signal travelled equidistant from the source to the boundaries of the model (left and right of the SBL model). Due to this symmetrical setting, only data from 5000 to 10,000 m were considered for processing purposes. Next, from the figures as well, we can see that the magnitudes of the E-field for all hydrocarbon depths are indistinguishable (especially in Figure 6) at source–receiver offset smaller than ~7400 m. This happens because high transmission frequencies have high attenuation, thus the signal is not able to propagate farther than low-frequency EM wave. Thus, we generalized this analysis by utilizing data for the offset from ~7400 to ~9500 m.

From the CST computer output, we developed a 2D forward GP model for every frequency to provide EM profiles at the observed and unobserved depths of the hydrocarbon. Even though the offset distances considered in this paper are from ~7400 m to ~9500 m, the GP models were set to distances from ~2400 to ~4500 m in order to make it easy to interpret, since the EM signal was transmitted from $x = 5000$ (center of the SBL model). The 2D forward GP models for frequencies of 0.125, 0.25, and 0.5 Hz are depicted in Figures 7–9, respectively.

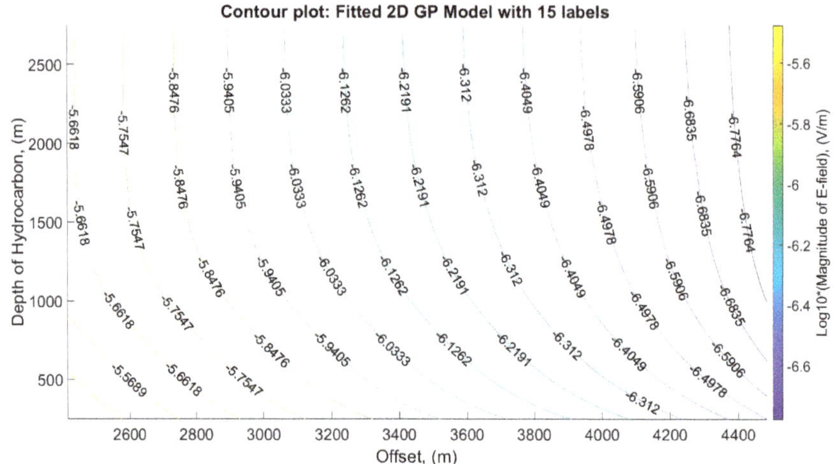

Figure 7. Contour plot of EM profiles (amplitude) for various offset distances (~2400 to ~4500 m) and hydrocarbon depths (250–2750 m) at a frequency of 0.125 Hz with 15 labels.

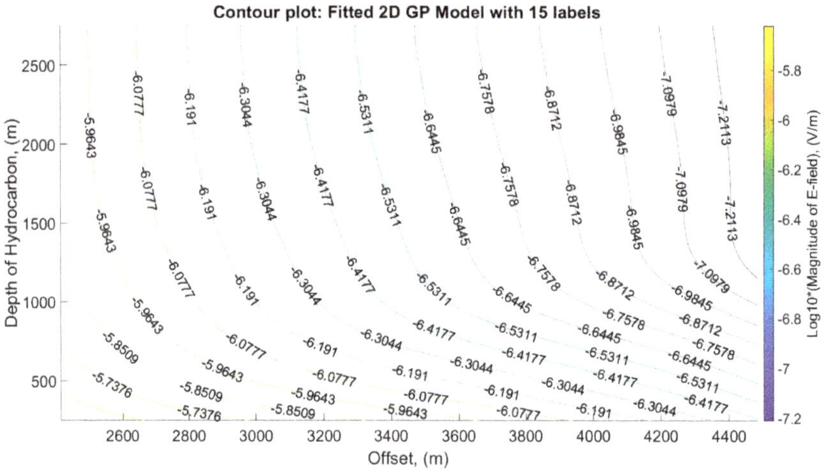

Figure 8. Contour plot of EM profiles (amplitude) for various offset distances (~2400 to ~4500 m) and hydrocarbon depths (250–2750 m) at a frequency of 0.25 Hz with 15 labels.

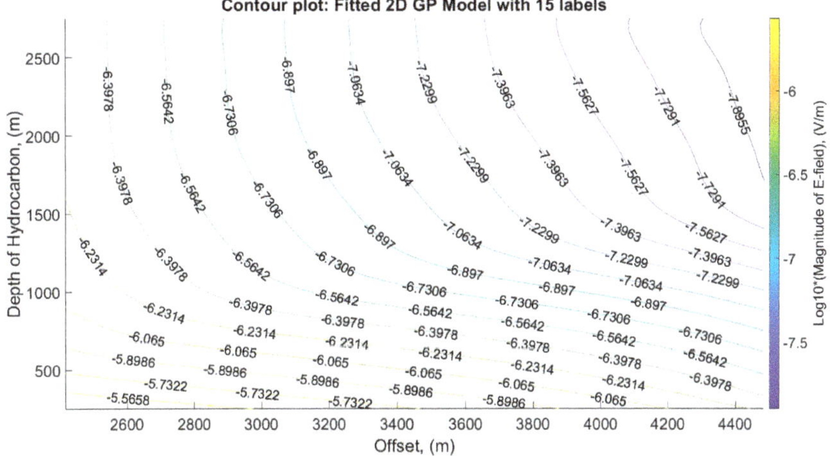

Figure 9. Contour plot of EM profiles (amplitude) for various offset distances (~2400 to ~4500 m) and hydrocarbon depths (250–2750 m) at a frequency of 0.5 Hz with 15 labels.

From the figures, we can see that the GP models are able to provide the information of EM profiles which are the magnitude of the E-field at all desired depths of hydrocarbon (observed and unobserved). Since variance is quantified in GP estimation, we tabulate the average of the variance of EM profiles for all observed depths of hydrocarbon in Table 2 to determine how far the data points are spread out from the mean value.

Table 2. The average of variance of EM responses (frequencies: 0.125, 0.25, and 0.5 Hz) at all tried depths of hydrocarbon (250–2750 m with an increment of 250 m each).

Depth (m)	Average of Variance		
	0.125 Hz	**0.25 Hz**	**0.5 Hz**
250	4.6402E–07	7.6034E–07	1.2504E–06
500	4.5931E–07	7.5443E–07	1.2418E–06
750	4.5602E–07	7.5090E–07	1.2372E–06
1000	4.5375E–07	7.4849E–07	1.2343E–06
1250	4.5283E–07	7.4699E–07	1.2326E–06
1500	4.5275E–07	7.4646E–07	1.2321E–06
1750	4.5283E–07	7.4699E–07	1.2326E–06
2000	4.5375E–07	7.4849E–07	1.2343E–06
2250	4.5602E–07	7.5090E–07	1.2372E–06
2500	4.5931E–07	7.5443E–07	1.2418E–06
2750	4.6402E–07	7.6034E–07	1.2504E–06

The confidence interval exercised by this paper is 95% of the data which lies within ± two standard deviations of the mean. Small values of variance indicate that the data points tend to be very close to the mean. Based on Table 2, all values of the average variances are very small. This implies that the 2D forward GP model is capable of fitting the marine CSEM data very well. Next, for better visualization, we depict a combination of 3D surface plots of the developed 2D forward GP models for every frequency in Figure 10.

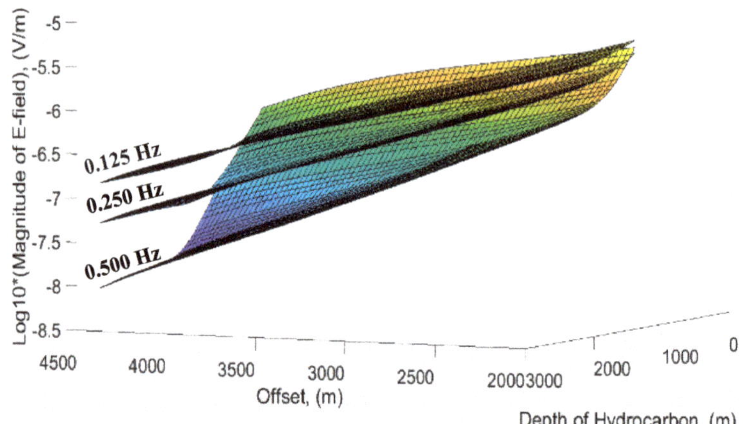

Figure 10. Combination of 3D surface plots of the 2D GP models for frequencies of 0.125, 0.25, and 0.5 Hz. The x-axis is the offset which is the source–receiver separation distance, the y-axis denotes the depth of hydrocarbon (250–2750 m), and the z-axis represents the log10 of magnitude of the electric field.

We determined the reliability of these 2D forward GP models in providing the information of EM profiles by calculating the RMSE and the CV between true (data generated through the CST software) and estimate values (data from the forward model) at unobserved/untried depths of hydrocarbon. In this section, random unobserved hydrocarbon depths (900 m and 2200 m) were selected. The EM profiles from these depths were compared with the CST computer output at the same depth levels. The RMSE and CV of the EM profiles at both depths are tabulated in Table 3.

Table 3. RMSE and CV analyses between EM profiles modelled by GP and EM profiles generated through CST software for all frequencies.

Depth (m)		0.125 Hz	0.25 Hz	0.5 Hz
900	RMSE	8.7419E–04	1.1284E–03	1.2946E–03
	CV (%)	1.4267E–02	1.7604E–02	1.9688E–02
2200	RMSE	5.8993E–04	7.4190E–04	1.3171E–03
	CV (%)	9.4615E–03	1.1241E–02	1.8402E–02

Based on Table 3, the RMSE values obtained are very small and all the CVs are generally less than 1%. This means that the modeling results of the 2D forward GP models are in good agreement with the responses acquired from the CST software even at the unobserved/untried depths of hydrocarbon.

5. Conclusions

We proposed a methodology of processing marine CSEM data using a statistical approach, Gaussian Process (GP). Based on the results, the EM responses estimated by GP are well fitted with the data generated from the CST software. The results (variance) proved that our proposed 2D forward GP model calibrated with computer simulation output is reliable for marine CSEM data-processing. In general, this 2D forward GP model, which contains EM profiles at various hydrocarbon depths, can be compared to surveyed data, and whichever estimate best matches the data measured from the survey will be the more likely case. The importance of this work lies in the application of GP methodology in multiple frequencies marine CSEM technique by developing a data-dependent model with uncertainty quantification to analyze the EM profiles and understand the geological structure underneath the seabed. It is too risky to directly make a decision in hydrocarbon exploration without any additional analysis. There are too many challenges involved especially when it comes to deeper offshore environments. Therefore, this methodology should be a data-processing tool that provides beneficial information to hydrocarbon exploration using marine CSEM techniques by utilizing the prior information obtained from real-field data before further analysis.

Author Contributions: Conceptualization, H.D. and S.C.D.; methodology, M.N.M.A.; software, H.D.; validation, S.C.D. and K.A.M.N.; writing—original draft preparation, M.N.M.A.; writing—review and editing, M.N.M.A. and K.A.M.N.; supervision, H.D., S.C.D. and K.A.M.N.; funding acquisition, H.D.

Funding: This research work was funded by International Grant (cost center: 015-ME0-012).

Acknowledgments: We would like to thank Universiti Teknologi PETRONAS for the Graduate Research Assistantship (GRA) Scheme. We are really grateful to have an open access of GPs algorithms, Gaussian Processes Machine Learning (GPML) Toolbox version 4.2, which are available at http://www.gaussianprocess.org/gpml/code/matlab/doc/manual.pdf. All data involved in the GP data processing are available at https://data.mendeley.com/datasets/bvwfy54j2d/1.

Conflicts of Interest: The authors declare no conflict of interest.

References

1. Ward, S.H.; Hohmann, G.W. *Electromagnetic Theory for Geophysical Applications*; SEG: Oklahoma City, Ok, USA, 1988.
2. Zhdanov, M.S.; Keller, G. *The Geoelectrical Methods in Geophysical Exploration*; Elsevier: Amsterdam, The Niederlande, 1994.
3. Li, Y.; Key, K. 2D marine controlled-source electromagnetic modeling: Part 1—An adaptive finite-element algorithm. *Geophysics* **2007**, *75*, WA51–WA62. [CrossRef]
4. Young, P.D.; Cox, C.S. Electromagnetic active source sounding near the East Pacific Rise. *Geophys. Res. Lett.* **1981**, *8*, 1043–1046. [CrossRef]
5. Evans, R.L.; Sinha, M.C.; Constable, S.C.; Unsworth, M.J. On the electrical nature of the axial melt zone at 13-degrees-N on the East Pacific Rise. *J. Geophys. Res.* **1994**, *99*, 577–588. [CrossRef]
6. Constable, S.; Cox, C.S. Marine controlled-source electromagnetic sounding—The PEGASUS experiment. *J. Geophys. Res.* **1996**, *101*, 5519–5530. [CrossRef]
7. MacGregor, L.M.; Constable, S.; Sinha, M.C. The RAMESSES experiment III: Controlled-source electromagnetic sounding of the Reykjanes Ridge at 57°45′N. *Geophys. J. Int.* **1998**, *135*, 773–789. [CrossRef]
8. MacGregor, L.M.; Sinha, M.; Constable, S. Electrical resistivity structure of the Value Fa Ridge, LauBasin, frommarine controlled-source electromagnetic sounding. *Geophys. J. Int.* **2001**, *146*, 217–236. [CrossRef]
9. Ellingsrud, S.; Eidesmo, T.; Johansen, S. Remote sensing of hydrocarbon layers by seabed logging: Results from a cruise offshore Angola. *Lead. Edge* **2002**, *21*, 972–982. [CrossRef]
10. Eidesmo, T.; Ellingsrud, S.; MacGregor, L.M.; Constable, S.; Sinha, M.C.; Johansen, S.; Kong, F.N.; Westerdahl, H. Sea Bed Logging (SBL), a new method for remote and direct identification of hydrocarbon filled layers in deep water areas. *First Break* **2002**, *20*, 144–152.
11. Hesthammer, J.; Boulaenko, M. The offshore EM challenge. *First Break* **2005**, *23*, 59–66.
12. Carazzone, J.J.; Burtz, O.M.; Green, K.E.; Pavlov, D.A. Three dimensional imaging of marine controlled source EM data. *SEG Expand. Abstr.* **2005**, *24*, 575.
13. Srnka, L.J.; Carazzone, J.J.; Ephron, M.S.; Eriksen, E.A. Remote reservoir resistivity mapping. *Lead Edge* **2006**, *25*, 972–975. [CrossRef]
14. Constable, S.; Srnka, L.J. An introduction to marine controlled-source electromagnetic methods for hydrocarbon exploration. *Geophysics* **2007**, *72*, WA3–WA12. [CrossRef]
15. Um, E.S.; Alumbaugh, D.L. On the physics of the marine controlled-source electromagnetic method. *Geophysics* **2007**, *72*, WA13–WA26. [CrossRef]
16. Andréis, D.; MacGregor, L. Controlled-source electromagnetic sounding in shallow water: Principles and applications. *Geophysics* **2008**, *73*, F21–F32. [CrossRef]
17. Zhdanov, M.S. Electromagnetic geophysics: Notes from the past and the road ahead. *Geophysics* **2010**, *75*, A49–A66. [CrossRef]

18. Zaid, H.M.; Yahya, N.B.; Akhtar, M.N.; Kashif, M.; Daud, H.; Brahim, S.; Shafie, A.; Hanif, N.H.H.M.; Zorkepli, A.A.B. 1D EM modeling for onshore hydrocarbon detection using MATLAB. *J. Appl. Sci.* **2011**, *11*, 1136–1142. [CrossRef]
19. Dunham, M.W.; Ansari, S.M.; Farquharson, C.G. *Application of 3D Marine CSEM Finite-Element Forward Modeling to Hydrocarbon Exploration in the Flemish Pass Basin Offshore Newfoundland, Canada*; SEG: Austin, TX, USA, 2016.
20. Cai, H.; Xiong, B.; Han, M.; Zhdanov, M. 3D controlled-source electromagnetic modeling in anisotropic medium using edge-based finite element method. *Comput. Geosci.* **2014**, *73*, 164–176. [CrossRef]
21. Persova, M.G.; Soloveichik, Y.G.; Domnikov, P.A.; Vagin, D.V.; Koshkina, Y.I. Electromagnetic field analysis in the marine CSEM detection of homogeneous and inhomogeneous hydrocarbon 3D reservoirs. *J. Appl. Geophys.* **2015**, *119*, 147–155. [CrossRef]
22. Gelius, L.J. Multi-component processing of sea bed logging data. *PIERS ONLINE* **2006**, *2*, 589–593. [CrossRef]
23. Gehrmann, R.A.S.; Schnabel, C.; Engels, M.; Schnabel, M.; Schwalenberg, K. Combined interpretation of marine controlled source electromagnetic and reflection seismic data in German North Sea: A case study. *Geophys. J. Int.* **2019**, *216*, 218–230. [CrossRef]
24. Reyes, O.C.; de la Puente, J.; Modesto, D.; Puzyrev, V.; Cela, J.M. A parallel tool for numerical approximation of 3D electromagnetic surveys in geophysics. *Comput. Sist.* **2016**, *20*, 29–39.
25. Guo, Z.; Liu, J.; Liao, J.; Xiao, J. Comparison of detection capability by the controlled source electromagnetic method for hydrocarbon exploration. *Energies* **2018**, *11*, 1839. [CrossRef]
26. Loseth, L.O.; Pedersen, H.M.; Schaug-Pettersen, T.; Ellingsrud, S.; Eidesmo, T. A scaled experiment for the verification of the SeaBed Logging method. *J. Appl. Geophys.* **2008**, *64*, 47–55. [CrossRef]
27. Daud, H.; Yahya, N.; Asirvadam, V. Development of EM simulator for sea bed logging applications using MATLAB. *Indian J. Mar. Sci.* **2011**, *40*, 267–274.
28. Chiadikobi, K.C.; Chiaghanam, O.I.; Omoboriowo, A.O.; Etukudoh, M.V.; Okafor, N.A. Detection of hydrocarbon reservoirs using the controlled-source electromagnetic (CSEM) method in the 'Beta' field deep water offshore Niger Delta, Nigeria. *Int. J. Sci. Emerg. Technol.* **2012**, *1*, 7–18.
29. Weiss, C.J. The fallacy of the 'shallow-water problem' in marine CSEM exploration. *Geophysics* **2007**, *72*, A93–A97. [CrossRef]
30. Booton, R.C. *Computational Methods for Electromagnetic and Microwaves*; John Wiley & Sons: New York, NY, USA, 1992.
31. Puzyrev, V.; Koldan, J.; de la Puente, J.; Houzeaux, G.; Vazquez, M.; Cela, J.M. A parallel finite-element method for three-dimensional controlled-source electromagnetic forward modelling. *Geophys. J. Int.* **2013**, *193*, 678–693. [CrossRef]
32. Key, K.; Weiss, C. Adaptive finite element modeling using unstructured grids: The 2D magnetotelluric example. *Geophysics* **2006**, *71*, G291–G299. [CrossRef]
33. Franke, A.; Börner, R.U.; Spitzer, K. Adaptive unstructured grid finite element simulation of two-dimensional magnetotelluric fields for arbitrary surface and seafloor topography. *Geophys. J. Int.* **2007**, *171*, 71–86. [CrossRef]
34. Li, Y.; Pek, J. Adaptive finite element modelling of two-dimensional magnetotelluric fields in general anisotropic media, *Geophys. J. Int.* **2008**, *175*, 942–954.
35. Bakr, SA.; Pardo, D.; Mannseth, T. Domain decomposition Fourier finite element method for the simulation of 3D marine CSEM measurements. *J. Comput. Phys.* **2013**, *255*, 456–470. [CrossRef]
36. Cox, C.S.; Constable, S.C.; Chave, A.D.; Webb, S.C. Controlled-source electromagnetic sounding of the oceanic lithosphere. *Nature* **1986**, *320*, 52–54. [CrossRef]
37. CST STUDIO SUITE Electromagnetic Field Simulation Software. Available online: https://www.3ds.com/products-services/simulia/products/cst-studio-suite/ (accessed on 15 July 2019).
38. Rasmussen, C.E.; Planck, M. *Gaussian Process in Machine Learning*; Institute of Biological Cybermatics: Tubingen, Germany, 2012.
39. Pal, M.; Dewal, S. Modelling pile capacity using Gaussian Process regression. *Comput. Geotech. J.* **2010**, *37*, 942–947. [CrossRef]
40. Petelin, D.; Grancharova, A.; Kochigan, J. Evolving Gaussian Process models for prediction of Ozone concentration in the air. *Simul. Model. Pract. Theory* **2013**, *33*, 68–80. [CrossRef]

41. Sun, A.Y.; Wang, D.; Xu, X. Monthly Stream flow forecasting using Gaussian Process Regression. *J. Hydrol.* **2013**, *511*, 72–81. [CrossRef]
42. Grbic, R.; Kurtagic, D.; Sliskovic, D. Stream water temperature prediction based on Gaussian Process Regression. *J. Expert Syst. Appl.* **2013**, *40*, 7407–7414. [CrossRef]
43. Liu, D.; Pang, J.; Zhou, J.; Peng, Y.; Pecht, M. Prognostics for state of health estimation of lithium-ion batteries based on combination Gaussian process functional regression. *Microelectron. Reliab.* **2013**, *53*, 832–839. [CrossRef]
44. Yin, F.; Zhao, Y.; Gunnarsson, F.; Gustafsson, F. Received-Signal-Strength Threshold Optimization Using Gaussian Processes. *IEEE Trans. Signal. Process.* **2017**, *65*, 2164–2177. [CrossRef]
45. Wang, H.; Gao, X.; Zhang, K.; Li, J. Fast single image super-resolution using sparse Gaussian process regression. *Signal. Process.* **2017**, *134*, 52–62. [CrossRef]
46. Sacks, J.; Welch, W.J.; Mitchell, T.J.; Wynn, H.P. Design and Analysis of Computer Experiments. *Stat. Sci.* **2002**, *4*, 409–428. [CrossRef]
47. Harari, O.; Steinberg, D.M. Optimal designs for Gaussian process models |via spectral decomposition. *J. Stat. Plan. Inference* **2014**, *154*, 87–101. [CrossRef]
48. Mukhtar, S.M.; Daud, H.; Dass, S.C. Prediction of hydrocarbon using Gaussian process for seabed logging application. *Procedia Comput. Sci.* **2015**, *72*, 225–232. [CrossRef]
49. Mohd Aris, M.N.; Daud, H.; Dass, S.C. Processing synthetic seabed logging (SBL) data using Gaussian Process regression. *J. Phys. Conf. Ser.* **2018**, *1123*, 012025. [CrossRef]
50. Aris, M.N.M.; Daud, H.; Dass, S.C. Prediction of hydrocarbon depth for seabed logging (SBL) application suing Gaussian process. *J. Phys. Conf. Ser.* **2018**, *1132*, 012075. [CrossRef]
51. Rasmussen, C.E.; Nickisch, H. Gaussian Processes for Machine Learning (GPML) Toolbox. *J. Mach. Learn. Res.* **2010**, *11*, 3011–3015.

Article

The Performance and Exhaust Emissions of a Diesel Engine Fuelled with *Calophyllum inophyllum*—Palm Biodiesel

Natalina Damanik [1], Hwai Chyuan Ong [2,*], M. Mofijur [3,*], Chong Wen Tong [2], Arridina Susan Silitonga [4,*], Abd Halim Shamsuddin [5], Abdi Hanra Sebayang [4], Teuku Meurah Indra Mahlia [3], Chin-Tsan Wang [6] and Jer-Huan Jang [7]

1 Perusahaan Listrik Negara, The Indonesia State Electricity Company, Jakarta 12160, Indonesia; natalinadamanik@gmail.com

2 Department of Mechanical Engineering, Faculty of Engineering, University of Malaya, Kuala Lumpur 50603, Malaysia; chong_wentong@um.edu.my

3 School of Information, Systems and Modelling, Faculty of Engineering and Information Technology, University of Technology Sydney, Sydney, NSW 2007, Australia; TMIndra.Mahlia@uts.edu.au

4 Department of Mechanical Engineering, Politeknik Negeri Medan, Medan 20155, Indonesia; abdisebayang@yahoo.co.id

5 Institute of Sustainable Energy, Universiti Tenaga Nasional, Kajang 43000, Selangor, Malaysia; abdhalim@uniten.edu.my

6 Department of Mechanical and Electro-Mechanical Engineering, National Ilan University, Yilan 260, Taiwan; ctwang@niu.edu.tw

7 Department of Mechanical Engineering, Ming Chi University of Technology, New Taipei City 243, Taiwan; jhjang@mail.mcut.edu.tw

* Correspondence: onghc@um.edu.my (H.C.O.); mdmofijur.rahman@uts.edu.au (M.M.); arridina@polmed.ac.id (A.S.S.); Tel.:+61-46985-1901 (M.M.)

Received: 31 July 2019; Accepted: 30 August 2019; Published: 5 September 2019

Abstract: Nowadays, increased interest among the scientific community to explore the *Calophyllum inophyllum* as alternative fuels for diesel engines is observed. This research is about using mixed *Calophyllum inophyllum*-palm oil biodiesel production and evaluation that biodiesel in a diesel engine. The *Calophyllum inophyllum*–palm oil methyl ester (CPME) is processed using the following procedure: (1) the crude *Calophyllum inophyllum* and palm oils are mixed at the same ratio of 50:50 volume %, (2) degumming, (3) acid-catalysed esterification, (4) purification, and (5) alkaline-catalysed transesterification. The results are indeed encouraging which satisfy the international standards, CPME shows the high heating value (37.9 MJ/kg) but lower kinematic viscosity (4.50 mm^2/s) due to change the fatty acid methyl ester (FAME) composition compared to *Calophyllum inophyllum* methyl ester (CIME). The average results show that the blended fuels have higher Brake Specific Fuel Consumption (BSFC) and NO_x emissions, lower Brake Thermal Efficiency (BTE), along with CO and HC emissions than diesel fuel over the entire range of speeds. Among the blends, CPME5 offered better performance compared to other fuels. It can be recommended that the CPME blend has great potential as an alternative fuel because of its excellent characteristics, better performance, and less harmful emission than CIME blends.

Keywords: *Calophyllum inophyllum* biodiesel; palm biodiesel; engine performance; exhaust emissions; alternative fuel; transesterification

1. Introduction

Petroleum derived fuels are the main source of primary energy consumption worldwide. Because of the negative impact and limited reserve of fossil fuels, scientists have focused on the new sources of energy to replace the fossil fuel [1,2]. Renewable energy sources have been proven to create less or zero-emission energy generation and can play an important role to lower fossil fuel consumption [3]. In many countries, different types of renewable energy sources including solar, wind, hydro, geothermal, bioenergy and biofuel has been introduced [4–9]. However, some renewable energy, including wind and solar, are only available for a certain time and period and therefore energy storage is required for these kinds of sources [10]. Due to this problem, researchers attempt to find other types of energy storage material that can be commercialized [11–14]. Therefore, some scientists, especially in developing countries are more interested in the energy sources that can be kept for a long period, such as bioenergy, bioethanol, and biodiesel [15–17]. Biodiesel is one renewable energy source, which can significantly lower emissions due to fossil fuel combustion that create air pollution, global warming, and acid rain [18]. Biodiesel sources include soybean oil, sunflower oil, palm oil and cottonseed oil, *Jatropha curcas oil*, mahua (*Madhuca indica*) oil, jojoba (*Simmondsia chinensis*) oil, tobacco seeds, salmon oil, tamanu (*Calophyllum inophyllum*) oil, sea mango oil (*Cerbera odollam*), and microalgae [19–22].

Palm oil has been commonly used in Malaysia and Indonesia as a biodiesel source due to its availability and favorable characteristics [23]. The productive lifetime of palm oil is around 25 years and it has to be replanted after that period [20]. Palm oil can yield methyl ester over 80%. Since 2006, the Indonesia government has paid attention to biodiesel as part of the National Security Act of Indonesia because of world crude oil price fluctuation. It is also supported because Indonesia is the largest crude palm oil (CPO) producer. However, until 2010, the Indonesia government failed to achieve biodiesel blending targets due to the increase in the world crude palm oil price and decrease in the crude oil price. As an impact, the biodiesel price has been not competitive compared to the diesel fuel price [24]. As Ong at al. [19] reported on sensitivity analysis that differences in the price of sources will have considerable impact on the life cycle cost of biodiesel by at least 79%. However, many new policies were introduced in 2014 by the Indonesian government to promote the use of biodiesel. Ong at al. [25] suggested that a financial incentive and subsidy policy should be enforced to make the price of biodiesel competitive to diesel fuel. However, based on a cost-benefit analysis (CBA), this will enhance the net benefit of palm oil plantation and biodiesel producers but will lessen the net welfare for society and the government of Indonesia. Therefore, the policy in the future will focus on reducing costs that improve the net social benefit [24].

Calophyllum inophyllum seed is an inedible oil source, which has a high oil content. Therefore, *Calophyllum inophyllum* seed is also a potential feedstock for biodiesel fuel [19] in Indonesia and Malaysia due to its abundant availability. This feedstock is a biodiverse plant that was previously known as a medicinal source due to its high antioxidant content [26]. However, *Calophyllum inophyllum* is grouped into high-acid-number feedstocks that allow biodiesel production to be equipped with special treatments, such as triple-stage transesterification, degumming, and neutralization [6]. In fact, *Calophyllum inophyllum* biodiesel has a poor oxidation stability because it has about 72.65% of unsaturated fatty acids that make this fuel unfavourable for long-term storage [27]. Excessive chemical treatment for minimizing total acid number (TAN) in oil refining may lead to a reduction of antioxidant content and oxidation stability [28]. Recently, some experiments reported that the antioxidant addition into biodiesel has improved its oxidation stability.

However, recently many studies have been reported on the fractional replacement of conventional fuel by palm and CIME. There are not many studies that have been reported on the prospect of palm and *Calophyllum inophyllum* biodiesel mixture. In this work, palm and *Calophyllum inophyllum* oil were mixed prior to the biodiesel production process and compared their performance with conventional fuel in a diesel engine. This method is believed to be able today reduce the chemical process during the acid value reduction of *Calophyllum inophyllum*–palm oil compound. Moreover, the objective of this study is also to investigate the engine performance (specifically, the Brake Specific Fuel Consumption (BSFC) and Brake Thermal Efficiency (BTE) and exhaust emission characteristics NO_x, HC, and CO emissions) of *Calophyllum inophyllum*–palm biodiesel mixture. It is expected that there is a potential for these blends to be commercialized in Indonesia and Malaysia due to the abundant supply of *Calophyllum inophyllum* seed oil and palm oil in these countries.

2. Materials and Methods

2.1. Crude Oils

Crude *Calophyllum inophyllum* oil and palm oils were purchased from a local store in Kuala Lumpur, Malaysia. The crude *Calophyllum inophyllum* and palm oils were mixed at 50:50 equal volume % in order to produce the CPME.

2.2. Production of CPME

Firstly, the blend was prepared by mixing 1 L of the crude oil from each source with 1 % of phosphoric acid (H_3PO_4, Merck Sdn. Bhd., Kuala Lumput, Malaysia) and 10 % of purified water (*v*/*v*) for 30 min. The crude oil mixture was degummed at 60 °C with an agitation speed of 800 rpm. The degumming process is essential to remove impurities and compounds (i.e., resins, proteins, phosphates, carbohydrates, and water residue). Next, acid-catalysed esterification was conducted. The details of the esterification process can be found in Silitonga et al. [29]. Molar ratio and catalyst percentage influence the esterification process of the oils [30]. In this study, it displayed the optimum molar ratio and H_2SO_4 catalyst (Merck Sdn. Bhd., Kuala Lumpur, Malaysia) concentration are 1:16.6 and 2.0 vol.%, respectively, since these parameters result in the highest esterified oil yield and fastest reaction time. According to [31], the presence of excess water can increase the formation of peroxides and increase the free fatty acid content of esterified oils. Thus, purification is crucial to remove excess water, which can be done by evaporation using a rotary evaporator, followed by the separation process with a separating funnel [32,33].

For this experiment, the esterified *Calophyllum inophyllum*–palm oil was purified by stirring the oil in a rotary evaporator (RV10 DIGITAL V IKA, Germany) at 60 °C with a stirring speed of 100 rpm for 30 min. The maximum pressure of the rotary evaporator was 7.2 MPa (72 bars). Following this, the esterified *Calophyllum inophyllum*–palm oil was poured into a separating funnel for the settling and left for 18 h. Karmakar et al. [21] also found that the high temperature of the purification process results in hydrolysis of the triglycerides, which in turn, removes water from the esterified oil.

Next, transesterification was done by mixing the esterified oils with 50% of methanol and 0.5 volume % of sodium hydroxide (KOH, Merck Sdn. Bhd., Kuala Lumpur, Malaysia) catalyst. The reaction mixture was stirred continuously in a jacketed reactor for 90 min maintaining the temperature at 60 °C. On the completion of the transesterification, the mixture was left for 4–6 h in a funnel. There are two distinct layers of liquid formed in the funnel where biodiesel was in the top and glycerol at the bottom. The glycerol was drained out from the funnel and biodiesel was washed by using sanitized water for a number of times in order to further remove impurities. The similar purification process was maintained both for the esterification and transesterification process.

2.3. Production of Methyl Ester

The CIME and palm oil methyl ester (POME) were prepared in the same manner. The crude *Calophyllum inophyllum* and palm oils were first degummed to remove impurities. The degummed oils were then esterified under the following process conditions: (1) reaction temperature; 60 °C, (2) stirring speed; 800 rpm, (3) reaction time; 60 min, (4) oil-to-methanol molar ratio; 1:16.6, and (5) H_2SO_4 catalyst concentration; 1.0 vol.%. The esterified oils were then purified to remove extraneous water present in the oils. Next, the purified *Calophyllum inophyllum* and palm oils were transesterified under the following process conditions: (1) reaction temperature; 60 °C, (2) stirring speed; 800 rpm, (3) reaction time; 90 min, (4) oil-to-methanol ratio; 1:8, and (5) catalyst- KOH with concentration; 0.5 vol.%. Likewise, the reaction mixtures were left to settle in separating funnels for 4–6 h after the transesterification process. In the final step, the CIME and POME were cleaned using sanitized water several times.

2.4. Characteristics of the CPME

The characteristics (i.e., density, kinematic viscosity (KV), flash point (FP), acid value(AV), high heating value (HHV), FAME content, and oxidation stability of the CPME and its blends were examined and compared to diesel, POME, CIME, as well as their blends. The FAME content was determined by employing a gas chromatograph–mass spectrometer (Model: GCMS-QP2010 Ultra, Shimadzu, Japan) fitted with a low-bleed GC-MS column (Model: RTX-5MS, RESTEK, Tokyo, Japan) details operating condition can be found elsewhere [34]. The temperature of the flame ionization detector and split injector was 300 °C. The biodiesels chemical and physical properties are collected from literature as a comparison.

The FAME content in per cent (%) determined by the following Equation:

$$\text{FAME} = \frac{(\sum A) - A_{EI}}{A_{EI}} \times \frac{C_{EI} \times V_{EI}}{m} \times 100 \quad (1)$$

Here, $\sum A$ is the summation of the peak areas of FAME, A_{EI} is the methyl heptadecanoate peak area, which is the internal standard, C_{EI} is the methyl heptadecanoate solution concentration in heptane (mg/mL), V_{EI} is the methyl heptadecanoate solution volume (mL) and m is the methyl ester mass (mg).

The percentage (%) of the methyl ester yield can be calculated by the following Equation:

$$Methyl\ ester\ yield = \frac{FAME \times B_{cpme}}{O_{cipo}} \times 100 \quad (2)$$

The FAME is the fatty acid methyl ester content (%), B_{cp} is the *Calophyllum inophyllum*-palm oil methyl ester weight (g) and O_{so} is the weight of the *Calophyllum inophyllum*-palm mixed oil (g).

2.5. Experimental Set-Up

Engine tests were done to study the engine performance and the characteristics of exhaust emission for CPME blends and CIME blends and the data collected compared to diesel fuel. These fuel blends were prepared in this study: (1) CPME5, (2) CPME10, (3) CIME5, and (4) CIME10. In this study, the performance parameters BSFC and BTE whereas the exhaust gases parameter NO_x, HC, and CO were measured. A single-cylinder diesel engine (Yanmar YX2500CX-A 170F, Osaka, Japan) was used to investigate the performance that set in full throttle. The engine speed varied from 1400 to 2800 rpm. A BOSCH BEA 350 gas analyser was used in order to measure the emissions. The detail of the engine test-bed and emission analyser is given in Table 1.

Table 1. Diesel engine technical specifications.

Brand	Yanmar
Model	2500CX-A 170 F
Type	1-cylinder, DI
Displacement (cc)	211
Speed (rpm)	3000
Maximum output(HP)	4.2
Cont. output (HP)	3.8
Governor System	Centrifugal weight system
Starting system	Recoil or electric
Lube oil capacity(L)	0.75
Fuel tank capacity(L)	12.5
Operational capacity (hrs.)	14

2.6. Uncertainties of the Experimental

Generally, the uncertainties of the experiment happened due to several reasons, namely: (1) instruments type and condition, (2) instruments calibration, (3) environmental conditions, and (4) procedure of experimental. To make sure the accuracy of the data between the limit, therefore the accuracy of the experimental data should be verified. Consequently, the uncertainties percentage of selected variables, namely BSFC, BTE, CO, NO_X, and HC were investigated according to the instrument's percentage uncertainties employed in the experiments. The speed accuracy, fuel consumption flowrate and time, which were ±10 rpm, ±1%, and ±0.1 s, respectively. The BSFC uncertainty was investigated by the uncertainty linearized approximation method. The details of % of uncertainties are given in Table 2.

Table 2. The percentage of uncertainties.

Measured Quantity	Measurement Range	Accuracy	Type of Instrument	Percentage Uncertainty (%)
Load	±8 Nm	±0.1 Nm	Strain gauge type load cell	±1.27
Speed	1400–2800 rpm	±1 rpm	Magnetic pickup type speed sensor	±0.1
Time	-	±0.1 s	-	±0.2
Fuel flow measurement	1–25 L/h	±0.1 L/h	Positive displacement gear wheel flow meter	±1.53
CO	0%–10% by vol.	±0.001%	Non-dispersive infrared gas sensor	±1.13
HC	0–9,999 ppm	±1 ppm	Heated flame ionization detector	±1.4
NO_X	0–5,000 ppm vol	±1 ppm vol	Electrochemical gas sensor	±1.1
BSFC	-	±0.1 L/kWh	-	±1.5
BTE	-	±0.2%	-	±1.5

3. Results and Discussion

3.1. Physicochemical Properties

The properties for POME, CIME, CPME, and their blends are given in Table 3. It is seen that the density of the CPME (880 kg/m^3) is lower than that for CIME (884 kg/m^3). The KV of the CPME was found lower than that for CIME and similar to that for POME (4.4 mm^2/s). In general, the KVs for CPME, CIME, and POME are inline with ASTM D6751 limit. The FP of CPME is 160 °C, which is above the limit of ASTM D6751 standard. The higher FP is important as it reduces the fire hazard risk, which is the main concern on fuels to handle, transport, and store [35]. However, the HHV of the CPME (37.9 MJ/kg) is found to be greater than CIME and POME (37.3 and 36.4 MJ/kg, respectively).

Table 3. Comparative physicochemical properties of the fuel sample used.

Property	Limit			Biodiesel						Biodiesel Blends					
	ASTM D6751	**EN 14214**	**Diesel**	**POME**	**CIME**	**CPME**	**CSO + WSO (Fadhil, 2017)**	**JCME (Dharma, 2016)**	**NSME + CPME (Yunus khan, 2014)**	**CIME5**	**CIME10**	**CPME5**	**CPME10**	**JCB10 (Dharma, 2016)**	**NSCPB (Yunus khan, 2014)**
Density at 15 °C (kg/m^3)	880.0	860.0–900.0	846.3	874.0	884.0	880.0	898.9	831.2	884.8	852.0	854.0	853.0	854.0	854	854.0
KV at 40 °C (mm^2/s)	1.90–6.00	3.50–5.00	2.98	4.40	4.80	4.50	3.61	3.95	4.44	3.76	4.00	3.82	4.00	3.55	3.70
FP (°C)	>130.0	Min. 101.0	80.0	246.5	179.0	160.0	246.5	84	186.5	86.0	88.0	79.9	82.0	76.5	87.5
HHV (MJ/kg)	–	35.0	45.3	36.4	37.3	37.9	36.4	40.88	39.94	43.1	42.9	44.1	43.9	42.76	44.2
AV (mg KOH/g)	<0.50	<0.50	–	0.1	0.5	0.4	0.1	0.06	0.14	0.1	0.5	0.4	0.1	0.36	0.1
Water content (%v)	Max. 0.05	-	-	0.025	0.015	0.018	-	-	-	0.015	0.0015	0.002	0.0018	-	-

3.2. Fatty Acid Methyl Ester (FAME) Composition

The FAME compositions of the CIME, POME, and CPME are summarized in Table 4. In general, all of these biodiesels have high palmitic acid content. However, the POME has a higher percentage of oleic acid, whereas the CPME has a higher percentage of antioxidants, such as methyl palmitic acid (C16:0), stearic acid (C18:0), linoleic acid ($C_{18}H_{36}O_2$), and 9-Octadecene,1-methoxy-, (E) ($C_{19}H_{38}O$) [34]. Moreover, the CPME has a high oleic acid percentage (C18:1), with a value of 52.94 wt.%, which also serves as a lubricant.

Table 4. Fatty Acid Methyl Ester (FAME) composition of *Calophyllym inophyllum* Methyl ester, CIME, Palm Oil Methyl Ester (POME), and *Ceiba Pentandra* Methyl ester (CPME).

Fatty Acid	CIME (wt.%)	POME (wt.%)	CPME (wt.%)
Lauric acid	0.10	0.10	0.10
Myristic acid	0.75	1.52	0.93
Palmitic acid	16.85	25.10	28.22
Palmitoleic acid	0.70	0.67	0.75
Stearic acid	15.57	22.46	31.99
Oleic acid	41.5	56.29	52.94
Linoleic acid	15.10	6.85	16.35
Linolenic acid	0.13	7.61	5.32
Arachidic acid	0.10	0.10	0.10

3.3. Brake Specific Fuel Consumption (BSFC)

Figure 1 shows the BSFC for diesel, CPME and CIME biodiesel blends at various engine speeds. It can be observed that all the blended fuel have higher BSFC compared to the diesel fuel except CPME5 blend. On average, biodiesel blended fuels have 16%–21% higher BSFC than diesel fuel. This finding is consistent with the literature [36–38]. Öztürk et al. [38] investigated the mixture of canola oil–hazelnut soap stock biodiesel-diesel and they found that the BSFC of blend fuel is more than the diesel fuel. The combined effects of the density, KV and HHV of the fuel caused that result [39]. During the suction stroke, biodiesel is injected on a volume basis; thus more fuels are fed inside the cylinder [40]. Consequently, more fuel is needed in order to achieve the same power because the HHV of biodiesel is lower than diesel. Among the blends, the average BSFC was highest for CIME10 blend (2.58 Ltr/kWhr) and lowest for CIME5 (2.21 Ltr/kWhr), which can be attributed by the HHV of the CIME10 blends. According to the data presented in Table 3, fuel sample CIME10 have a slightly higher heating value (43.9 MJ/kg) compared with CPME5 (43.1 MJ/kg).

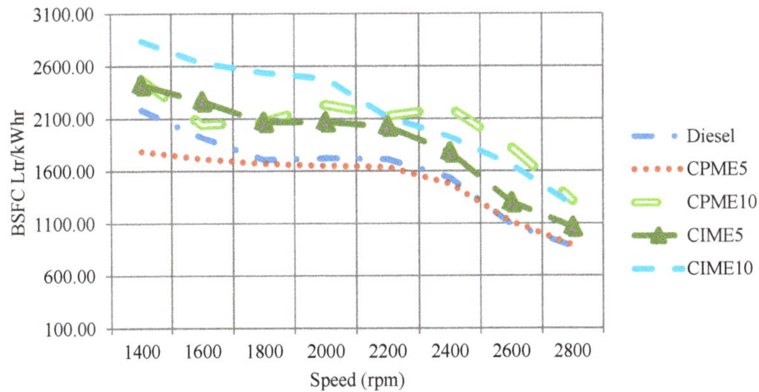

Figure 1. Changes in Brake Specific Fuel Consumption (BSFC) of diesel, CPME, and CIME blends with speeds.

3.4. Brake Thermal Efficiency (BTE)

Figure 2 shows the BTE for all fuel samples at different speeds of the engine. It is seen that the BTEs of all fuel samples used in this study increases with the speed and maximum BTE was found for diesel fuel compared to blended fuels. This can be explained by the higher heating value and lower BSFC of diesel fuel [41]. Diesel fuel showed maximum BTE followed by the CPME5, CIME5, CPME10, and CIME10 fuels. On average blended fuel lowers 1.25%–22% BTE compared to diesel fuel. The lower viscosity and higher heating value of diesel fuel, which improves the fuel atomization; thus increased the BTEs. The data obtained from the experiment are similar to the results presented by Sharma et al. [42]. They reported that the mixed *Jatropha* and Cottonseed blend produce lower BTE than diesel fuel. The reason was explained by the poor spray formation, higher viscosity, and poor ignition quality.

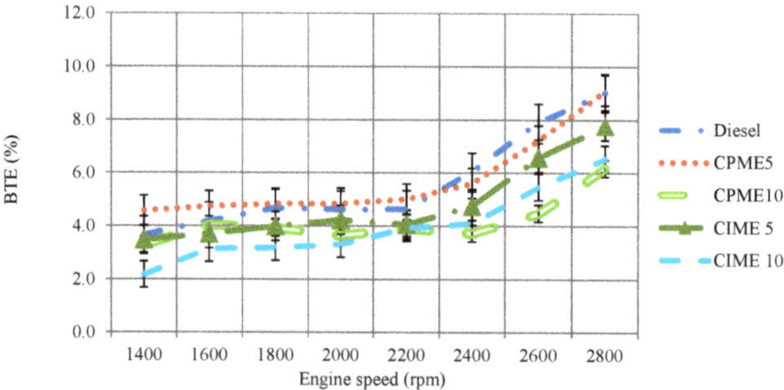

Figure 2. Changes in Brake Thermal Efficiency (BTE) of diesel, CPME, and CIME blends with speeds.

3.5. Nitrogen Oxide Emissions (NO_x) Emission

The nitrogen oxides emissions in exhaust consist of nitric oxide (NO) and nitrogen dioxide (NO_2). Figure 3 shows the NO_x emissions for diesel, and the CPME and CIME biodiesel blends at various engine speeds. It is evident that the NO_x emissions increase with an increase in engine speed. It is clear that biodiesel blended fuels give more NO_x emissions compared to diesel fuel. A similar report was found in the literature [43] for B7 and B100. The average NO_x for diesel fuel was found to be 112 ppm, which is 1.5%–29% higher than the blended fuels. This can be explained by the lean air/fuel ratio because biodiesel fuel has more inherent oxygen than diesel fuel. It has been reported that oxygenated fuel blends cause higher NO_x emissions [36]. Also, the higher KV of the biodiesel fuel leads to bigger droplets and shorter ignition delays, which affects the NO_x emission [44]. In addition, the unsaturated fatty acid content of biodiesels leads to fuels higher adiabatic flame temperature than diesel fuel, which causes higher NO_x emission [43].

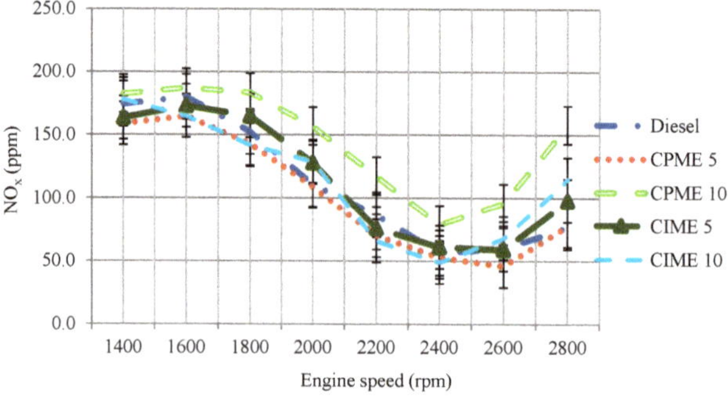

Figure 3. Changes in Nitrogen Oxide (NO_x) emissions of diesel, CPME, and CIME blends with speeds.

3.6. Carbon Monoxide (CO) Emissions

Figure 4 shows the CO emissions of all fuel samples at various engine speeds. The results indicate that the CO emissions are generally fewer for the biodiesel blends than the diesel fuel. Among the fuel samples, biodiesel fuel lowers 5% to 15% CO emission on average compared to the diesel fuel. The reason is described by the higher oxygen content of the biodiesels, which results in cleaner, better combustion [45,46]. CO is formed due to the incomplete combustion of the fuel due to insufficient oxygen or low gas temperature. As mentioned earlier, biodiesel fuel has a 12% higher oxygen content than diesel fuel, which accepts more carbon molecules to be burnt completely [36].

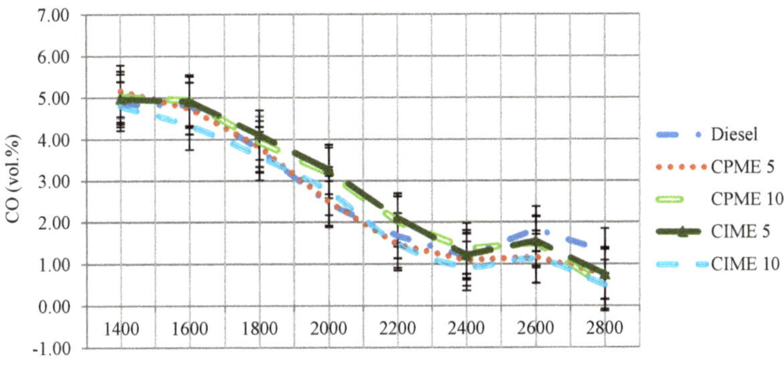

Figure 4. Changes in Carbon Monoxide (CO) emissions of diesel, CPME, and CIME blends with speeds.

3.7. Hydrocarbon (HC) Emissions

The comparison of emission among the fuel samples related to HC is presented in Figure 5. It was found that average HC emissions of blends were less than diesel. It is obvious that biodiesel blended fuel lowers HC emissions by 13%–22% than diesel fuel. The HC emissions can be reduced by the combustion quality improvement in biodiesel diesel blends due to the existence of excess oxygen atoms in biodiesel [47]. Similar results were reported by Mofijur et al. [37]. They explained that lower hydrocarbon emissions of moringa biodiesel-diesel occur because of higher oxygen contents of biodiesel fuel than diesel fuel. Also from the graph, it is seen that with increasing engine speeds, the HC emission decreases. Kegl et al. [48] presented similar results that both biodiesel and diesel fuels emit higher HC emissions when engines run at lower speeds.

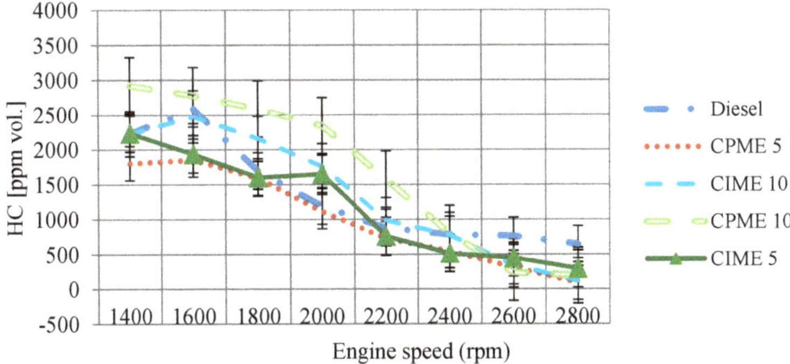

Figure 5. Changes in Hydrocarbon (HC) emissions of diesel, CPME, and CIME blends with speeds.

4. Conclusions

In this study, CPME is produced by a systematic procedure that started from crude oil mixing and ended by the transesterification process. Based on the findings, the following conclusions can be made:

1. The physicochemical properties of CPME meet ASTM D6751 and EN 14214 standards
2. The blended fuel results in lower BTE and higher BSFC compared the diesel fuel because of its higher KV, density, and lower HHV.
3. The use of blended fuel as a partial replacement of diesel significantly decreased the CO and HC emission, which is likely due to the fact that this blend promotes complete combustion whereas there is a slight increase in NO_x emissions due to higher oxygen contents.
4. Among the blends, CPME5 showed a better performance compared to the other blends.

Finally, it can be concluded the CPME blend has potential as a diesel engine alternative fuel to lower the harmful emission.

Author Contributions: Conceptualization, Methodology, results and formal analysis were initiated and wrote by N.D., A.S.S. and H.C.O.; T.M.I.M., A.H.S. (Abd Halim Shamsuddin) and C.W.T. contributed to supervision. A.H.S. (Abdi Hanra Sebayang), M.M. contributed to the Mathematical derivation and results analysis; C.-T.W. and J.-H.J. checked and improved the manuscript. All authors read and approved the final manuscript.

Funding: This research is funded by the Centre for Advanced Modeling and Geospatial Information Systems (CAMGIS), UTS under Grants 321740.2232397 and AAIBE Chair of Renewable grant no: 201801 KETTHA. The authors would like to acknowledge the University of Malaya, Kuala Lumpur for the financial support under SATU joint research scheme (ST010-2018), the Direktorat Jenderal Penguatan Riset dan Pengembangan Kementerian Riset, Teknologi dan Pendidikan Tinggi Republik Indonesia, (Grant no. 147/SP2H/LT/ DRPM/2019) and Politeknik Negeri Medan, Medan, Indonesia.

Acknowledgments: The authors would like to acknowledge the University of Technology Sydney, Australia for supporting this research.

Conflicts of Interest: The authors declare no conflict of interest.

Abbreviations

CIME	*Calophyllum inophyllum* methyl ester
CIME5	5% *Calophyllum inophyllum* methyl ester + 95% of diesel
CIME10	10% *Calophyllum inophyllum* methyl ester + 90% of diesel
CPME	*Calophyllum inophyllum*–palm oil methyl ester
CPME5	5% *Calophyllum inophyllum*–palm oil methyl ester + 95% of diesel
CPME10	10% *Calophyllum inophyllum*–palm oil methyl ester + 90% of diesel
BSFC	Brake Specific Fuel Consumption
CO	Carbon monoxides
HC	Hydrocarbon
NO_x	Nitrogen oxides

References

1. Norhasyima, R.S.; Mahlia, T.M.I. Advances in CO_2 utilization technology: A patent landscape review. *J. CO2 Util.* **2018**, *26*, 323–335. [CrossRef]
2. Mofijur, M.; Masjuki, H.H.; Kalam, M.A.; Hazrat, M.A.; Liaquat, A.M.; Shahabuddin, M.; Varman, M. Prospects of biodiesel from Jatropha in Malaysia. *Renew. Sustain. Energy Rev.* **2012**, *16*, 5007–5020. [CrossRef]
3. Anwar, M.; Rasul, M.G.; Ashwath, N.; Rahman, M.M. Optimisation of second-generation biodiesel production from Australian native stone fruit oil using response surface method. *Energies* **2018**, *11*, 2566. [CrossRef]
4. Ismail, M.S.; Moghavvemi, M.; Mahlia, T.M.I. Techno-economic analysis of an optimized photovoltaic and diesel generator hybrid power system for remote houses in a tropical climate. *Energy Convers. Manag.* **2013**, *69*, 163–173. [CrossRef]

5. Milano, J.; Ong, H.C.; Masjuki, H.H.; Silitonga, A.S.; Chen, W.-H.; Kusumo, F.; Dharma, S.; Sebayang, A.H. Optimization of biodiesel production by microwave irradiation-assisted transesterification for waste cooking oil-Calophyllum inophyllum oil via response surface methodology. *Energy Convers. Manag.* **2018**, *158*, 400–415. [CrossRef]
6. Ong, H.C.; Masjuki, H.H.; Mahlia, T.M.I.; Silitonga, A.S.; Chong, W.T.; Leong, K.Y. Optimization of biodiesel production and engine performance from high free fatty acid Calophyllum inophyllum oil in CI diesel engine. *Energy Convers. Manag.* **2014**, *81*, 30–40. [CrossRef]
7. Silitonga, A.S.; Masjuki, H.H.; Mahlia, T.M.I.; Ong, H.C.; Chong, W.T. Experimental study on performance and exhaust emissions of a diesel engine fuelled with Ceiba pentandra biodiesel blends. *Energy Convers. Manag.* **2013**, *76*, 828–836.
8. Silitonga, A.S.; Masjuki, H.H.; Ong, H.C.; Sebayang, A.H.; Dharma, S.; Kusumo, F.; Siswantoro, J.; Milano, J.; Daud, K.; Mahlia, T.M.I.; et al. Evaluation of the engine performance and exhaust emissions of biodiesel-bioethanol-diesel blends using kernel-based extreme learning machine. *Energy* **2018**, *159*, 1075–1087. [CrossRef]
9. Uddin, M.; Techato, K.; Taweekun, J.; Rahman, M.; Rasul, M.; Mahilia, T.; Ashrafur, S. An overview of recent developments in biomass pyrolysis technologies. *Energies* **2018**, *11*, 3115.
10. Mofijur, M.; Mahlia, T.M.I.; Silitonga, A.S.; Ong, H.C.; Silakhori, M.; Hasan, M.H.; Putra, N.; Rahman, S.M.A. Phase change materials (PCM) for solar energy usages and storage: An overview. *Energies* **2019**, *12*, 3167. [CrossRef]
11. Aricò, A.S.; Bruce, P.; Scrosati, B.; Tarascon, J.M.; van Schalkwijk, W. Nanostructured materials for advanced energy conversion and storage devices. *Nat. Mater.* **2005**, *4*, 366–377. [CrossRef] [PubMed]
12. Liu, C.; Li, F.; Ma, L.P.; Cheng, H.M. Advanced materials for energy storage. *Adv. Mater.* **2010**, *22*, E28–E62. [CrossRef] [PubMed]
13. Amin, M.; Putra, N.; Kosasih, E.A.; Prawiro, E.; Luanto, R.A.; Mahlia, T.M.I. Thermal properties of beeswax/graphene phase change material as energy storage for building applications. *Appl. Therm. Eng.* **2017**, *112*, 273–280. [CrossRef]
14. Mehrali, M.; Latibari, S.T.; Mehrali, M.; Mahlia, T.M.I.; Metselaar, H.S.C.; Naghavi, M.S.; Sadeghinezhad, E.; Akhiani, A.R. Preparation and characterization of palmitic acid/graphene nanoplatelets composite with remarkable thermal conductivity as a novel shape-stabilized phase change material. *Appl. Therm. Eng.* **2013**, *61*, 633–640. [CrossRef]
15. Colombo, G.; Ocampo-Duque, W.; Rinaldi, F. Challenges in bioenergy production from sugarcane mills in developing countries: A case study. *Energies* **2014**, *7*, 5874–5898. [CrossRef]
16. Chia, S.R.; Ong, H.C.; Chew, K.W.; Show, P.L.; Phang, S.-M.; Ling, T.C.; Nagarajan, D.; Lee, D.-J.; Chang, J.-S. Sustainable approaches for algae utilisation in bioenergy production. *Renew. Energy* **2018**, *129*, 838–852. [CrossRef]
17. Dharma, S.; Masjuki, H.H.; Ong, H.C.; Sebayang, A.H.; Silitonga, A.S.; Kusumo, F.; Mahlia, T.M.I. Optimization of biodiesel production process for mixed Jatropha curcas-Ceiba pentandra biodiesel using response surface methodology. *Energy Convers. Manag.* **2016**, *115*, 178–190. [CrossRef]t
18. Silitonga, A.S.; Mahlia, T.M.I.; Kusumo, F.; Dharma, S.; Sebayang, A.H.; Sembiring, R.W.; Shamsuddin, A.H. Intensification of Reutealis trisperma biodiesel production using infrared radiation: Simulation, optimisation and validation. *Renew. Energy* **2019**, *133*, 520–527. [CrossRef]
19. Ayodele, O.O.; Dawodu, F.A. Production of biodiesel from *Calophyllum inophyllum* oil using a cellulose-derived catalyst. *Biomass Bioenergy* **2014**, *70*, 239–248. [CrossRef]
20. Kusumo, F.; Silitonga, A.S.; Masjuki, H.H.; Ong, H.C.; Siswantoro, J.; Mahlia, T.M.I. Optimization of transesterification process for *Ceiba pentandra* oil: A comparative study between kernel-based extreme learning machine and artificial neural networks. *Energy* **2017**, *134*, 24–34. [CrossRef]
21. Kusumo, F.; Silitonga, A.S.; Ong, H.C.; Masjuki, H.H.; Mahlia, T.M.I. A comparative study of ultrasound and infrared transesterification of Sterculia foetida oil for biodiesel production. *Energy Sour. Part A Recovery Util. Environ. Eff.* **2017**, *39*, 1339–1346. [CrossRef]
22. Goh, B.H.H.; Ong, H.C.; Cheah, M.Y.; Chen, W.-H.; Yu, K.L.; Mahlia, T.M.I. Sustainability of direct biodiesel synthesis from microalgae biomass: A critical review. *Renew. Sustain. Energy Rev.* **2019**, *107*, 59–74. [CrossRef]

23. Hayyan, A.; Mjalli, F.S.; Hashim, M.A.; Hayyan, M.; AlNashef, I.M.; Al-Wahaibi, T.; Al-Wahaib, Y.M. A Solid organic acid catalyst for the pretreatment of low-grade crude palm oil and biodiesel production. *Int. J. Green Energy* **2014**, *11*, 129–140. [CrossRef]
24. Halim, I.F. Cost-Benefit Analysis of Biodiesel Related Policies: The Assessment of Applicability. Master's Thesis, Delft University of Technology, Delft, The Netherland, 2015.
25. Ong, H.C.; Masjuki, H.H.; Mahlia, T.M.I.; Silitonga, A.S.; Chong, W.T.; Yusaf, T. Engine performance and emissions using Jatropha curcas, Ceiba pentandra and Calophyllum inophyllum biodiesel in a CI diesel engine. *Energy* **2014**, *69*, 427–445.
26. Prasad, J.; Shrivastava, A.; Khanna, A.K.; Bhatia, G.; Awasthi, S.K.; Narender, T. Antidyslipidemic and antioxidant activity of the constituents isolated from the leaves of *Calophyllum inophyllum*. *Phytomedicine* **2012**, *19*, 1245–1249. [CrossRef] [PubMed]
27. Sahoo, P.K.; Das, L.M.; Babu, M.K.G.; Naik, S.N. Biodiesel development from high acid value polanga seed oil and performance evaluation in a CI engine. *Fuel* **2007**, *86*, 448–454. [CrossRef]
28. Pullen, J.; Saeed, K. Experimental study of the factors affecting the oxidation stability of biodiesel FAME fuels. *Fuel Process. Technol.* **2014**, *125*, 223–235. [CrossRef]
29. Silitonga, A.S.; Masjuki, H.H.; Ong, H.C.; Yusaf, T.; Kusumo, F.; Mahlia, T.M.I. Synthesis and optimization of *Hevea brasiliensis* and *Ricinus communis* as feedstock for biodiesel production: A comparative study. *Ind. Crops Prod.* **2016**, *85*, 274–286. [CrossRef]
30. Kombe, G.G. Re-esterification of high free fatty acid oils for biodiesel production. *Biofuels* **2015**, *6*, 31–36. [CrossRef]
31. Yaakob, Z.; Narayanan, B.N.; Padikkaparambil, S.; Unni, K.S.; Akbar, P.M. A review on the oxidation stability of biodiesel. *Renew. Sustain. Energy Rev.* **2014**, *35*, 136–153. [CrossRef]
32. Dinkov, R.; Hristov, G.; Stratiev, D.; Boynova Aldayri, V. Effect of commercially available antioxidants over biodiesel/diesel blends stability. *Fuel* **2009**, *88*, 732–737. [CrossRef]
33. Kirk-Othmer. *Esterification. Kirk-Othmer Encyclopedia of Chemical Technology*, 4th ed.; John Wiley & Sons, Inc.: Hoboken, NJ, USA, 2004.
34. Kumar, N.R.; Reddy, J.S.; Gopikrishna, G.; Solomon, K.A. GC-MS determination of bioactive constituent of cycas beddomei cones. *Int. J. Pharm. Bio Sci.* **2012**, *3*, 344–350.
35. Nantha Gopal, K.; Thundil Karupparaj, R. Effect of *Pongamia* biodiesel on emission and combustion characteristics of DI compression ignition engine. *Ain Shams Eng. J.* **2015**, *6*, 297–305. [CrossRef]
36. Mofijur, M.; Masjuki, H.H.; Kalam, M.A.; Atabani, A.E. Evaluation of biodiesel blending, engine performance and emissions characteristics of *Jatropha curcas* methyl ester: Malaysian perspective. *Energy* **2013**, *55*, 879–887. [CrossRef]
37. Mofijur, M.; Masjuki, H.H.; Kalam, M.A.; Atabani, A.E.; Fattah, I.M.R.; Mobarak, H.M. Comparative evaluation of performance and emission characteristics of *Moringa oleifera* and Palm oil based biodiesel in a diesel engine. *Ind. Crops Prod.* **2014**, *53*, 78–84. [CrossRef]
38. Öztürk, E. Performance, emissions, combustion and injection characteristics of a diesel engine fuelled with canola oil–hazelnut soapstock biodiesel mixture. *Fuel Process. Technol.* **2015**, *129*, 183–191. [CrossRef]
39. Mofijur, M.; Masjuki, H.H.; Kalam, M.A.; Atabani, A.E.; Arbab, M.I.; Cheng, S.F.; Gouk, S.W. Properties and use of *Moringa oleifera* biodiesel and diesel fuel blends in a multi-cylinder diesel engine. *Energy Convers. Manag.* **2014**, *82*, 169–176. [CrossRef]
40. Syed, A.; Quadri, S.A.P.; Rao, G.A.P.; Mohd, W. Experimental investigations on DI (direct injection) diesel engine operated on dual fuel mode with hydrogen and mahua oil methyl ester (MOME) as injected fuels and effects of injection opening pressure. *Appl. Therm. Eng.* **2017**, *114*, 118–129. [CrossRef]
41. Rahman, M.M.; Rasul, M.G.; Hassan, N.M.S.; Azad, A.K.; Uddin, N. Effect of small proportion of butanol additive on the performance, emission, and combustion of Australian native first- and second-generation biodiesel in a diesel engine. *Environ. Sci. Pollut. Res.* **2017**, *24*, 22402–22413. [CrossRef]
42. Sharma, L.; Grover, N.K.; Bhardwaj, M.; Kaushal, I. Comparison of engine performance of mixed Jatropha and cottonseed derived biodiesel blends with conventional diesel. *Int. J. Emerg. Technol.* **2012**, *3*, 29–32.
43. Özçelik, A.E.; Aydoğan, H.; Acaroğlu, M. Determining the performance, emission and combustion properties of camelina biodiesel blends. *Energy Convers. Manag.* **2015**, *96*, 47–57. [CrossRef]
44. Kalam, M.A.; Masjuki, H.H.; Jayed, M.H.; Liaquat, A.M. Emission and performance characteristics of an indirect ignition diesel engine fuelled with waste cooking oil. *Energy* **2011**, *36*, 397–402. [CrossRef]

45. Pinto, A.C.; Guarieiro, L.L.N.; Rezende, M.J.C.; Ribeiro, N.M.; Torres, E.A.; Lopes, W.A.; Pereira, P.A.d.; de Andrade, J.B. Biodiesel: An overview. *J. Braz. Chem. Soc.* **2005**, *16*, 1313–1330. [CrossRef]
46. Dinesha, P.; Mohanan, P. Combined effect of oxygen enrichment and exhaust gas recirculation on the performance and emissions of a diesel engine fueled with biofuel blends. *Biofuels* **2018**, *9*, 45–51. [CrossRef]
47. Man, X.J.; Cheung, C.S.; Ning, Z.; Wei, L.; Huang, Z.H. Influence of engine load and speed on regulated and unregulated emissions of a diesel engine fueled with diesel fuel blended with waste cooking oil biodiesel. *Fuel* **2016**, *180*, 41–49. [CrossRef]
48. Kegl, B. Influence of biodiesel on engine combustion and emission characteristics. *Appl. Energy* **2011**, *88*, 1803–1812. [CrossRef]

Article

A Sugarcane-Bagasse-Based Adsorbent Employed for Mitigating Eutrophication Threats and Producing Biodiesel Simultaneously

Wan Nurain Farahah Wan Basri [1,2], Hanita Daud [1], Man Kee Lam [2,3], Chin Kui Cheng [4], Wen Da Oh [5], Wen Nee Tan [6], Maizatul Shima Shaharun [1], Yin Fong Yeong [3], Ujang Paman [7], Katsuki Kusakabe [8], Evizal Abdul Kadir [9], Pau Loke Show [10] and Jun Wei Lim [1,2,*]

1 Department of Fundamental and Applied Sciences, Universiti Teknologi PETRONAS, 32610 Seri Iskandar, Perak Darul Ridzuan, Malaysia
2 Centre for Biofuel and Biochemical Research, Institute of Self-Sustainable Building, Universiti Teknologi PETRONAS, 32610 Seri Iskandar, Perak Darul Ridzuan, Malaysia
3 Department of Chemical Engineering, Universiti Teknologi PETRONAS, 32610 Seri Iskandar, Perak Darul Ridzuan, Malaysia
4 Faculty of Chemical and Natural Resources Engineering, Universiti Malaysia Pahang, Lebuhraya Tun Razak, 26300 Gambang Pahang, Malaysia
5 School of Chemical Sciences, Universiti Sains Malaysia, 11800 Penang, Malaysia
6 Chemistry Section, School of Distance Education, Universiti Sains Malaysia, 11800 Penang, Malaysia
7 Department of Agribusiness, Faculty of Agriculture, Islamic University of Riau. Jl. Kaharuddin Nasution No. 113, Pekanbaru 28284, Riau Indonesia
8 Department of Nanoscience, Sojo University, 4-22-1 Ikeda, Nishi-ku, Kumamoto 860-0082, Japan
9 Department of Informatics Engineering, Universitas Islam Riau, Jl. Kaharuddin Nasution No. 113, Marpoyan, Pekanbaru Riau 28284, Indonesia
10 Department of Chemical and Environmental Engineering, Faculty of Science and Engineering, University of Nottingham Malaysia, Jalan Broga 43500 Semenyih, Selangor, Malaysia
* Correspondence: junwei.lim@utp.edu.my; Tel.: +60-5368-7664; Fax: +60-5365-5905

Received: 25 June 2019; Accepted: 5 August 2019; Published: 28 August 2019

Abstract: Eutrophication is an inevitable phenomenon, and it has recently become an unabated threat. As a positive, the thriving microalgal biomass in eutrophic water is conventionally perceived to be loaded with myriad valuable biochemical compounds. Therefore, a sugarcane-bagasse-based adsorbent was proposed in this study to harvest the microalgal biomass for producing biodiesel. By activating the sugarcane-bagasse-based adsorbent with 1.5 M of H_2SO_4, a highest adsorption capacity of 108.9 ± 0.3 mg/g was attained. This was fundamentally due to the surface potential of the 1.5 M H_2SO_4 acid-modified sugarcane-bagasse-based adsorbent possessing the lowest surface positivity value as calculated from its point of zero charge. The adsorption capacity was then improved to 192.9 ± 0.1 mg/g by stepwise optimizing the adsorbent size to 6.7–8.0 mm, adsorption medium pH to 2–4, and adsorbent dosage to 0.4 g per 100 mL of adsorption medium. This resulted in 91.5% microalgae removal efficiency. Excellent-quality biodiesel was also obtained as reflected by the fatty acid methyl ester (FAME) profile, showing the dominant species of C16–C18 encompassing 71% of the overall FAMEs. The sustainability of harvesting microalgal biomass via an adsorption-enhanced flocculation processes was also evidenced by the potentiality to reuse the spent acid-modified adsorbent.

Keywords: eutrophication; sugarcane bagasse; adsorption; harvest; biodiesel; reusability

---

1. Introduction

"Eutrophication is the enrichment of water with nutrient salts that causes structural changes to the ecosystem, namely, the increase in production of microalgae and aquatic plants, depletion of fish species, general deterioration of water quality and other effects that reduce and preclude use". This is one of the definitions given to the eutrophic process by the Organization for Economic Cooperation and Development [1,2]. Eutrophication is a serious environmental threat since it gives rise to deterioration of water quality, and it is also one of the major impediments in achieving the quality objectives established by the Water Framework Directive in the EU as well as those in other countries [3]. Intrinsically, all water resources are subjected to a natural and slow eutrophication process. However, in recent years, the eutrophication threat has undergone very rapid progression due to the presence of various human activities, particularly the farming of cash crops in agriculture. When the eutrophication threat becomes intense, undesirable effects and environmental imbalances arise. The two most acute phenomena to stem from eutrophication are hypoxia in the deep parts of lakes (or lack of oxygen) and microalgal blooms that may produce harmful toxins. Both occurrences can plausibly bring severe devastation to the aquatic life living in the afflicted water bodies [4–8].

Malaysia and Indonesia are growing countries with steadily improving economies due to huge contributions from many agricultural sectors such as palm oil, rice and paddies, sugarcane, and other planted cash crops. Although oil palm is a major cash crop that can promote the gross domestic product (GDP), another rising cash crop in Malaysia and Indonesia is sugarcane, with total productions of 28.1 and 28.0 million metric tons, respectively, in years 2016/17 [9–12]. This gives rise to gigantic levels of agricultural waste from the juice crushing process, i.e., sugarcane bagasse which is usually left to decay on the fields. The accumulated sugarcane bagasse waste without proper management will potentially lead to the spreading of diseases that adversely affect humans, animals, and the environment. Of late, a fraction of this bagasse has been used as fuels in sugar factories, and some raw materials have been exploited in pulp and paper making. Sugarcane bagasse is a type of lignocellulosic biomass that generally consists of cellulose (50%), hemicellulose (25%), and lignin (25%) [13]. The hemicelluloses are made up of C5 and C6 sugar, while lignin comprises about one-fourth of the lignocellulose biomass. These biological polymers have hydroxyl and/or phenolic functional groups that can be chemically activated to produce materials with new properties. Accordingly, the carboxylic and hydroxyl groups can improve the capacity of adsorption via ion exchange and complexation processes. Owing to its low ash content (approximately 2.4%), sugarcane bagasse can offer many advantages when compared with other crop residues such as rice straw and wheat straw, which have about 17.5% and 11.0% ash contents, respectively [14]. In addition, sugarcane bagasse is a porous material with a relatively high fixed carbon content. Considering these advantages, sugarcane bagasse has been among the preferred waste materials chosen to be used as an adsorbent by many researchers previously to resolve issues associated with water pollution via adsorption processes [5,8,9,15,16].

Nevertheless, the application of adsorption processes to concentrate planktonic microalgal cells and subsequently flocculate and separate them from a liquid medium has been documented little of late. Accordingly, the potential use of a fabricated sugarcane-bagasse-based adsorbent to mitigate the eutrophication phenomenon served as the prime objective of this comprehensive study. The adsorptive operating conditions were initially optimized to spur the removal of microalgal biomass via adsorption-enhanced flocculation processes from the eutrophic water. In tandem with this study, the fundamental rationale describing the attachment of microalgal cells onto the surface of the fabricated sugarcane-bagasse-based adsorbent was also unveiled. Since microalgal biomass is conventionally lauded for containing myriad valuable biochemical compounds, the harvested sugarcane-bagasse-based adsorbent loaded with microalgal biomass was subsequently exploited for biodiesel production. The plausible reusability of the spent sugarcane-bagasse-based adsorbent was lastly assessed to confirm the sustainability of the fabricated sugarcane-bagasse-based adsorbent in mitigating the eutrophication threat.

2. Materials and Methods

2.1. Fabrication of Acid-Modified Sugarcane-Bagasse-Based Adsorbents

The residual sugarcane bagasse generated after juice crushing was used as a precursor in fabricating various acid-modified sugarcane-bagasse-based adsorbents. Residual sugarcane bagasse was initially amassed from the local market and cut into pieces of about 3–4 cm each. The pieces of bagasse were then boiled and washed thoroughly to remove the entrapped sugar. The wet sugar-free bagasse was later air-dried to partially remove the water content before drying to constant weight at 102–105 °C in an oven. The dried sugarcane bagasse was pulverized using a 0.5 mm blade grinder and subsequently activated using various H_2SO_4 acid concentrations: 0.1, 0.5, 1.0, 1.5, 2.0, and 2.5 M. This was achieved by individually introducing 15 g of pulverized sugarcane bagasse into 1 L of each of the H_2SO_4 acid concentrations. Each mixture was finally stirred for 24 h and washed with distilled water until a neutral pH was attained in discarded washing water prior to once again drying to constant weight at 102–105 °C in an oven. All the fabricated acid-modified sugarcane-bagasse-based adsorbents were separately kept in a desiccator ready for use.

2.2. Determination of Adsorbent Point of Zero Charge

Points of zero charge (pH_{PZC}) were initially measured from each of the fabricated acid-modified sugarcane-bagasse-based adsorbents. The pH_{PZC} is determined as the adsorption medium pH that causes the surface charge density of the adsorbent to become zero. The pH_{PZC} values of all the acid-modified sugarcane-bagasse-based adsorbents were measured following the solid addition method as described by Ofomaja and Naidoo [17] with modifications. Briefly, to a series of 100 mL Erlenmeyer flasks, 45 mL of KNO_3 solution with a concentration of 0.01 M was added to every flask. The pH in each flask was then individually adjusted from pH 2 to 9 and filled up to 50 mL using a similar KNO_3 solution. The initial pH values of all flasks were accurately recorded, and 0.1 g of specific acid-modified sugarcane-bagasse-based adsorbent was administered into each flask. All flasks were securely capped and agitated manually and sporadically for the next 2 days, allowing the charges to equilibrate before recording the final pH values from each flask. Subsequently, the differences between the initial and final pH values (ΔpH) of each flask were plotted against the respective initial pH values. The pH_{PZC} of the adsorbent was finally determined from the point of intersection of the resulting curve at $\Delta pH = 0$. Later, these procedures were repeated using other acid-modified sugarcane-bagasse-based adsorbents to determine their respective pH_{PZC} values.

2.3. Characteristics of Eutrophic Water

A water sample was collected from a eutrophic body at a site which receives discharged streams from the aquacultural activities in the vicinity (Figure 1). The in situ measurements indicated a pH of 3.8 ± 0.3, temperature of 31.6 ± 1.7 °C, and dissolved oxygen concentration of 0.1 ± 0.1 mg/L. The sample was instantly transported to the laboratory, and ex situ determination of the microalgal biomass concentration was immediately executed via gravimetric analysis. The eutrophic water was found to be loaded with 0.86 ± 0.2 g/L of microalgal biomass. The remaining sample was stored in the cold room upon reaching the laboratory at 2.0 ± 1.5 °C to minimize biological and chemical changes after isolating the sample from the eutrophic body. A predetermined volume of eutrophic water was later siphoned from the homogenized stock sample in the cold room and allowed to reach ambient temperature prior to use in the experiments outlined hereafter.

Figure 1. Eutrophic body which currently receives discharged streams from the aquacultural activities in the vicinity.

2.4. Setups for Eutrophic Water Treatment

Six 150 mL Erlenmeyer flasks were each initially filled with 100 mL of the eutrophic water sample. The sample was homogenized via vigorous shaking before the predetermined volume was transferred into each Erlenmeyer flask. A quantity of 0.5 g of acid-modified sugarcane-bagasse-based adsorbent with respective H_2SO_4 concentrations of 0.1, 0.5, 1.0, 1.5, 2.0, and 2.5 M was individually administered into each of the Erlenmeyer flasks containing a eutrophic water sample. All the mixtures were immediately adjusted to pH 3, and sufficient agitation was provided for each flask to prevent the settlement of adsorbents to the bases of all flasks. Sampling of microalgal biomasses present in the eutrophic water over time from each flask was executed to determine the residual microalgal biomass concentrations as well as for the time course studies. Samplings were also performed to determine the adsorption efficiencies (Equation (1)) and capacities (Equation (2)) of each acid-modified sugarcane-bagasse-based adsorbent loaded into each flask at equilibria as confirmed from the time course studies.

$$\text{Adsorption efficiency} = \frac{[Initial\ microalgal\ biomass] - [Microalgal\ biomass\ at\ any\ time]}{[Initial\ microalgal\ biomass]} \times 100\% \quad (1)$$

$$\text{Adsorption capacity} = \frac{Weight\ of\ adsorbed\ microalgal\ biomass}{Weight\ of\ adsorbent} \quad (2)$$

The best acid-modified sugarcane-bagasse-based adsorbent was then selected and sieved into four different sizes, namely, <4.0 mm, 5.6–6.7 mm, 6.7–8.0 mm, and >8.0 mm. A quantity of 0.5 g of each size of adsorbent was individually administered into the Erlenmeyer flasks containing the eutrophic water sample. Similar procedures to those described above were later executed in selecting the best size of acid-modified sugarcane-bagasse-based adsorbent. Afterward, stepwise optimization of the pH of the adsorption media and the adsorbent dosage were also performed to enhance the adsorptive removal of microalgal biomass from the eutrophic water. The studied ranges of the pH of adsorption media and adsorbent dosage were from pH 2 to 10 and 0.1 to 0.7 g in 100 mL of adsorption medium, respectively.

2.5. Microalgal Lipid Extraction and Transesterification into Fatty Acid Methyl Esters (FAMEs) of Biodiesel

Lipid extraction from the adsorbed microalgal biomass was accomplished following the procedures described by Bligh and Dyer [18] with modifications. Initially, the microalgal biomass adsorbed onto acid-modified sugarcane-bagasse-based adsorbent was harvested from the adsorption medium using a sieving net and dried to constant weight at 102–105 °C in an oven. The dried microalgal biomass, together with the adsorbent, was doused with water mixed with chloroform: methanol (1:2 *v*/*v*) at a ratio of 1.6:6.0 *v*/*v*. The mixture was immediately sonicated for 30 min at 40 kHz and 40 °C. For further separating the free lipids into the chloroform layer after sonication, an approximately 20% chloroform: water (1:1 *v*/*v*) solution by volume with respect to the total volume of the sonicated sample was added and mixed well manually. Centrifugation for 20 min at 5600 rpm was then employed to separate the sample solution into three layers, namely, a lipid and chloroform mixture (bottom layer), residual microalgal biomass adsorbed onto the spent sugarcane-bagasse-based adsorbent (middle layer), and a methanol and water mixture (top layer). The bottom layer was then collected by suction, and 8% chloroform by volume with respect to the total volume of the sample was added to the remaining mixture, mixed well manually, and centrifuged again to recover the residual lipids. The amassed volume of the lipid and chloroform mixture from both consecutive separation processes was lastly dried under compressed air blow to constant weight prior to the lipid yield determination (Equation (3)).

$$\text{Lipid yield} = \frac{Weight\ of\ extracted\ lipid\ from\ adsorbed\ microalgal\ biomass}{Weight\ of\ adsorbed\ microalgal\ biomass\ onto\ adsorbent - Weight\ of\ initial\ adsorbent} \times 100\% \quad (3)$$

The extracted lipids from the adsorbed microalgal biomass were subsequently transesterified into FAMEs of biodiesel following the procedures as described by Mohd-Sahib and Lim [19]. Briefly, a vortex mixer at 2000 rpm was initially employed to homogenize about 10–15 mg of extracted lipid, 1 mL of chloroform, and 2 mL of KOH in methanol (1.5 mg/mL). The sample was then capped well and transesterified for 3 h at a temperature of 60 °C. The temperature was maintained by placing the sample in a water bath until the transesterification process was completed. The FAMEs mixture of the biodiesel produced was later purified by repeatedly washing with 5 mL of distilled water followed by separation using a separating funnel until a neutral pH was measured from the aqueous phase. Next, the chloroform layer loaded with FAMEs was dried under compressed air blow to constant weight. The FAMEs mixture was finally injected into a gas chromatograph to analyze the individual FAME composition forming the biodiesel. The FAME profile obtained determines the quality of the biodiesel derived from the harvested microalgal biomass from eutrophic water. The details of the operating conditions of the employed Shimadzu brand gas chromatograph (Model GC-2010 plus) equipped with a flame ionization detector (GC-FID) can be acquired from Mohd-Sahib and Lim [19]. Ultimately, the individual FAME composition was calculated as follows:

$$\text{Composition of specific FAME species} = \frac{A_S}{A_{ISTD}} \times \frac{C_{ISTD}\ X\ V_{ISTD}}{m} \times 100\% \quad (4)$$

where A_S = Peak area of specific FAME species;
A_{ISTD} = Peak area of internal standard;
C_{ISTD} = Concentration of internal standard;
V_{ISTD} = Volume of internal standard;
m = Mass of sample.

2.6. Assessment of the Reusability of Spent Sugarcane-Bagasse-Based Adsorbent

Upon lipid extraction, the spent sugarcane-bagasse-based adsorbent or lipid-exhausted sugarcane-bagasse-based adsorbent was dried to constant weight at 102–105 °C in an oven. The dry adsorbent was then administered into 100 mL of fresh eutrophic water sample, and the optimum operational conditions to remove microalgal biomass from the adsorption medium were employed according to the first cycle, termed Cycle-1. Cycle-2 was allowed to continue until the equilibrium was attained as confirmed from the time course measurement of the residual microalgal biomass concentration. To that end, the adsorption efficiency and capacity were recorded following Equations (1) and (2), respectively. The harvested microalgal biomass adsorbed onto spent sugarcane-bagasse-based adsorbent was subsequently subjected to lipid extraction again, followed by drying of the lipid-exhausted sugarcane-bagasse-based adsorbent to constant weight at 102–105 °C. The dry adsorbent was later reused to remove microalgal biomass from fresh eutrophic water samples in Cycle-3 to Cycle-5.

3. Results and Discussion

3.1. Enhancement of Microalgal Biomass Adsorption from Eutrophic Water

The potential of various acid-modified sugarcane-bagasse-based adsorbents for removing microalgal biomass via adsorption-enhanced flocculation is presented in Table 1. In essence, the fresh negatively charged microalgal cells are hypothesized to be electrostatically attracted to the positively charged supporting materials. The acid-modified treatment mode was employed in this study in order to economically intensify the positive surface potential of the sugarcane-bagasse-based adsorbents. Increasing the H_2SO_4 acid concentration during the treatment of sugarcane-bagasse-based adsorbents progressively increased the separation of the microalgal biomass from the liquid medium, reaching the highest adsorption efficiency of 89.6% while using 1.5 M of H_2SO_4 to acid-modify the sugarcane-bagasse-based adsorbent. The adsorption efficiency and capacity were observed to gradually decrease thereafter when concentrations of H_2SO_4 higher than 1.5 M were employed to treat the sugarcane-bagasse-based adsorbents. Many studies have associated the reduction of adsorption with increasing acid concentration employed for treating adsorbents to the destruction of adsorbent structures [20,21]. Instead of agreeing with this typical rationale, the surface potentials of all fabricated acid-modified sugarcane-bagasse-based adsorbents were measured and exploited to elucidate the adsorption phenomena observed in this study. The positivity value of each adsorbent calculated by subtracting the pH of the adsorption medium from their pH_{PZC} is demonstrated in Figure 2. A mirror image trend against either the adsorption efficiency or capacity (Table 1) was obtained for the positivity values of increasingly H_2SO_4 acid-modified sugarcane-bagasse-based adsorbents. In this regard, the 1.5 M H_2SO_4 acid-modified sugarcane-bagasse-based adsorbent presented the lowest positivity value, attracting more negatively charged microalgal cells to adsorb onto the adsorbent surfaces. On the flipside, the higher positivity values of the surface potentials of other acid-modified sugarcane-bagasse-based adsorbents would attract more counter ions from the adsorption medium, forming a layer of negatively charged counter ions. This layer would shield the microalgal cells from interacting with the acid-modified sugarcane-bagasse-based adsorbents, indirectly preventing the adsorption-enhanced flocculation processes. In the case of putting the 1.5 M H_2SO_4 acid-modified sugarcane-bagasse-based adsorbent to use, the formation of this counter ions layer was inconspicuous, thereby permitting the adsorption of greater microalgal biomass than the other fabricated adsorbents.

Table 1. The performance of various acid-modified sugarcane-bagasse-based adsorbents in removing microalgal biomass from eutrophic water.

Concentration of H_2SO_4 Used for Adsorbent Modification (*M*)	0.1	0.5	1.0	1.5	2.0	2.5
Adsorption efficiency (%)	88.0 ± 0.5	88.6 ± 0.3	89.4 ± 0.8	89.6 ± 0.9	86.6 ± 0.1	85.9 ± 0.1
Adsorption capacity (mg/g)	106.9 ± 0.4	107.6 ± 0.3	108.5 ± 0.3	108.9 ± 0.3	105.0 ± 0.3	104.3 ± 0.4

To examine adsorbent sizes, the 1.5 M H_2SO_4 acid-modified sugarcane-bagasse-based adsorbent was sieved into four different sizes, namely, <4.0 mm, 5.6–6.7 mm, 6.7–8.0 mm, and >8.0 mm. The residual biomass concentrations of microalgae along with the adsorption time and the consequential adsorption capacities from the use of each adsorbent size are shown in Figure 3. Although the adsorption capacities attained by the adsorbent sizes of 6.7–8.0 and >8.0 mm were comparable, the use of 6.7–8.0 mm sized adsorbent could achieve a faster adsorption-enhanced flocculation equilibrium than the >8.0 mm sized adsorbent—specifically, 0.0576 and 0.0522 g/L h, respectively. As the targeted adsorbate in this study was a suspended microalgal biomass having a size range of 20–50 μm rather than the dissolved adsorbate, a large adsorbent size offered more macropores for accommodating microalgal cells. For comparison with dissolved adsorbate, the powdered form of adsorbents is generally preferred due to the presence of more micropores and mesopores capable of capturing the targeted dissolved adsorbate [22,23]. However, continuous increase of adsorbent size, especially beyond 8.0 mm, would culminate in the reduction of external surface area, impoverishing the frequency of contact between the adsorbent and microalgal cells. Therefore, acid-modified sugarcane-bagasse-based adsorbent with a size range of 6.7–8.0 mm was regarded as the best size for adsorption-enhanced flocculation of microalgal biomass; its adsorption capacity increased to 143.6 ± 1.7 mg/g, as compared with only 108.9 ± 0.3 mg/g using the unsieved adsorbent (Table 1). Besides this, the flocculated microalgal biomass adsorbed onto this size range of adsorbent could also be easily harvested from the liquid medium via a simple sieving net.

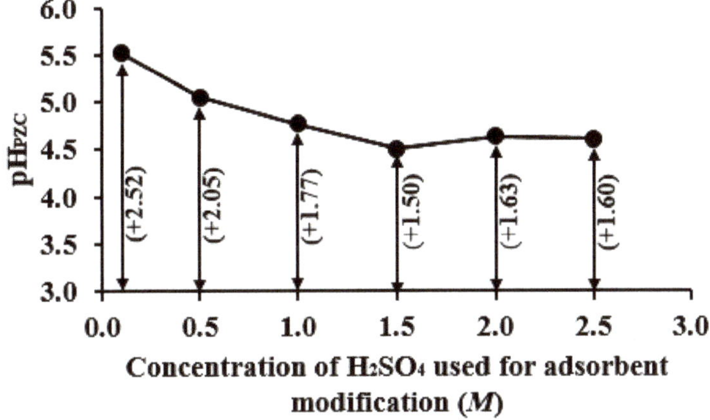

Figure 2. The point of zero charge (pH_{PZC}) values of various acid-modified sugarcane-bagasse-based adsorbents. The values in parentheses indicate the positivity of pH_{PZC} subtracted by pH 3, i.e., the pH of the adsorption medium.

The adsorption capacities were maintained in a range of 143.0–148.0 mg/g when the adsorption-enhanced flocculation processes were performed in adsorption media with pH values between 2 and 4. Increasing the pH of the adsorption medium to 6 caused a decrease in the adsorption capacity to 127 ± 4.3 mg/g. The adsorption capacities were noticed to plummet to merely 119.9 ± 0.5 and 104.3 ± 3.3 mg/g when the pH was shifted to basic in the adsorption media, namely, at pH values of 8 and 10, respectively. As the 1.5 M H_2SO_4 acid-modified sugarcane-bagasse-based adsorbent possessed a pH_{PZC} value of 4.50 (Figure 2), increasing pH value of the absorption medium above 4.50 would gradually increase the adsorbent surface potential negativity. Since microalgal cells are negatively charged, repulsion between the adsorbent and adsorbate arising from charge similarity would debilitate the adsorption-enhanced flocculation processes of the microalgal biomass. In acidic adsorption mediums, adsorption efficiencies of more than 80% could be easily attained at pH values between 2 and 4. Indeed, the pH value of the eutrophic water of 3.8 ± 0.3 also fell within this pH range, therefore safely circumventing the necessity to pretreat the eutrophic water through pH adjustment prior to adsorption-enhanced flocculation processes.

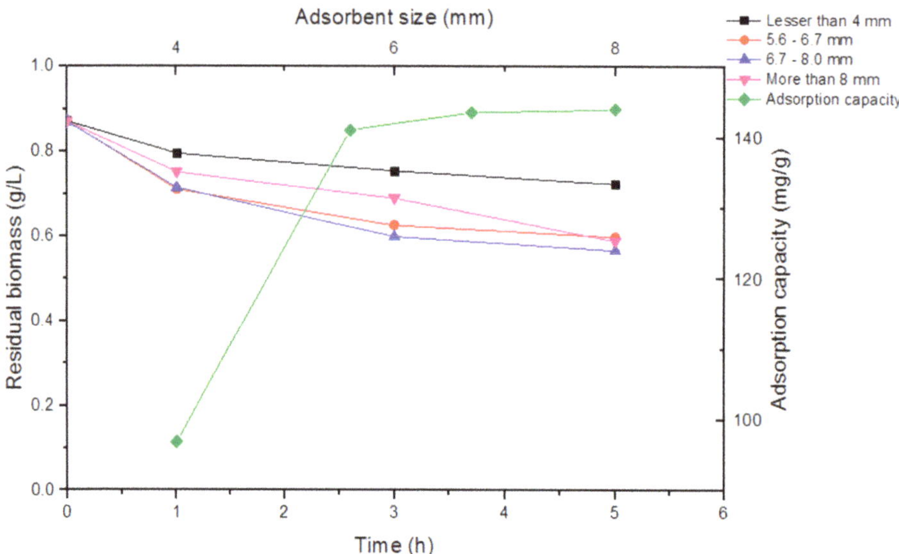

Figure 3. The time courses of residual biomass concentrations of microalgae adsorbed onto various sizes of acid-modified sugarcane-bagasse-based adsorbent and the consequent adsorption capacities from the use of each adsorbent size.

Increased amounts of adsorbent in terms of grams per 100 mL of adsorption medium increased the adsorption capacity, as revealed in Figure 4. At an adsorbent dosage of 0.1 g/100 mL, all the macropore active sites were swiftly occupied by microalgal cells, as demonstrated by the rapid attainment of adsorption-enhanced flocculation equilibrium during the early time course study. The insufficient active sites were later offset by the increasing adsorbent dosage, reaching a maximum adsorption capacity of 192.9 ± 0.1 mg/g when 0.4 g of adsorbent was introduced into 100 mL of adsorption medium, equivalent to an adsorption efficiency of 91.5%. Further increment of adsorbent dosages beyond 0.4 g/100 mL gave rise to the presence of increasing numbers of free active sites, thereby reducing the adsorption capacities. Increasing the adsorbent dosage from 0.4 to 0.7 g/100 mL slightly prompted more microalgal biomass to adsorb onto the adsorbent, steadily increasing the adsorption efficiency from 91.5% to 94.9%, respectively. This negligible rise in adsorption efficiency was due to the presence of excessive free active sites available to capture more microalgal cells from the diluted adsorption medium. As the percentage point increase was only about 3% (from 91.5% to 94.9%) but required a 75% increase in adsorbent dosage (from 0.4 to 0.7 g/100 mL), the 0.4 g/100 mL adsorbent dosage was considered ideal for executing adsorption-enhanced flocculation processes for the remediation of eutrophic water.

In summary, 1.5 M H_2SO_4 acid-modified sugarcane-bagasse-based adsorbent with a size range of 6.7–8.0 mm was employed to spur the separation of microalgal biomass from eutrophic water via adsorption-enhanced flocculation processes. The adsorption capacities achieved from sequential studies were 108.9 ± 0.3 and 143.6 ± 1.7 mg/g, respectively. Subsequently, the adsorption capacity was maintained in the range of 143.0–148.0 mg/g when adsorption-enhanced flocculation processes were carried out in adsorption media with pH values between 2 and 4. Finally, the adsorption capacity was further improved to 192.9 ± 0.1 mg/g when an adsorbent dosage of 0.4 g was introduced into 100 mL of adsorption medium. A high adsorption efficiency of microalgae of 91.5% and low residual biomass concentration of microalgae of 0.064 g/L in the adsorption medium were attained.

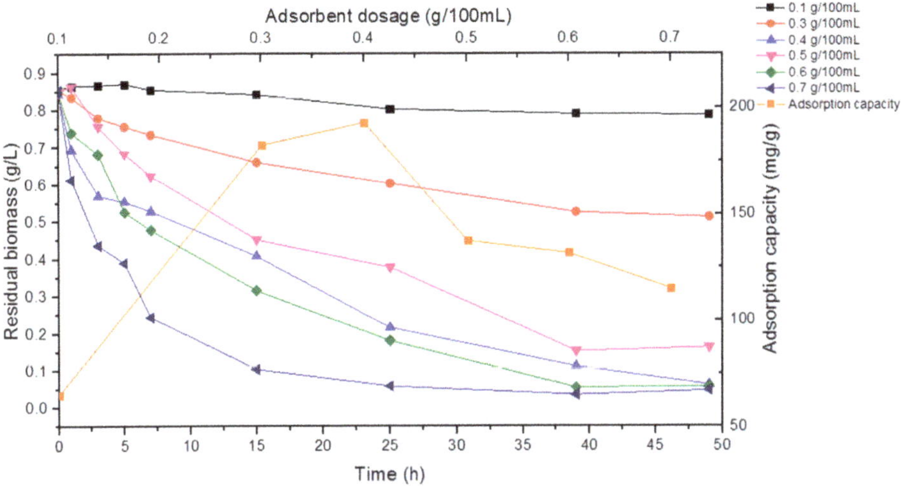

Figure 4. The time courses of residual biomass concentrations of microalgae adsorbed onto various dosages of acid-modified sugarcane-bagasse-based adsorbent and the consequent adsorption capacities from the use of each adsorbent dosage.

3.2. Biodiesel Derived from Harvested Microalgal Biomass

The microalgal biomass harvested from the eutrophic water was then sequentially subjected to extraction to obtain the lipid biocompounds and perform transesterification into a FAMEs mixture of biodiesel. The lipid yield acquired from the extraction was 30.3 ± 0.0 wt %, which was very close to the lipid yield of the control microalgal biomass of 31.0 ± 0.2 wt %. The control microalgal biomass was harvested via the centrifugation of fresh eutrophic water. With a standard deviation value of merely 0.5%, the insignificant difference between the two lipid yields showed that the adsorption-enhanced flocculation processes employed to remove the microalgal biomass from the liquid medium did not have an obvious deleterious impact on the microalgal lipid content. Finally, the FAME profile obtained from the transesterification of extracted lipids is presented in Table 2. The FAMEs are important components of biodiesel; thus, the quality of biodiesel can be concluded from the FAME profile study. There were 21 species of FAMEs identified in the biodiesel derived from microalgal biomass adsorbed onto the sugarcane-bagasse-based adsorbent. Among the FAMEs, C16 to C18 were the dominant species, making up approximately 71% of the overall FAMEs mixture. These species are also naturally found in many oil-bearing crops, e.g., soybean, sunflower seed, cottonseed, and palm oil, which are suitable for use as biodiesel [24]. According to Song and Pei [25], feedstock suited for the production of biodiesel must contain palmitic (16:0), stearic (18:0), oleic (18:1), linoleic (18:2), and linolenic (18:3) acids, which were all found in the FAME profile of the microalgal biomass adsorbed onto the sugarcane-bagasse-based adsorbent. The saturation degrees of the FAMEs mixture were calculated afterward and showed the mixture to contain 44.87% of saturated fatty acid (SFA), 32.55% of monounsaturated fatty acid (MUFA), and 22.58% of polyunsaturated fatty acid (PUFA). Biodiesel with higher levels of SFA will generally have a high cetane number and oxidative stability but poor low-temperature properties. The presence of MUFA species of palmitoleic (C16:1), oleic (C18:1), eicosenoate (C20:1), and erucate (C22:1) acids also essentially give rise to biodiesel with suitable oxidative stability, besides ameliorating cold flow [26]. The presence of high levels of PUFA, on the other hand, will offer excellent cold flow. However, this biodiesel type is easily oxidized [27]. The low degree of PUFA (<30%) and high degree of MUFA and SFA (>65%) as reported by Mohd-Sahib and Lim [19] were also attained in biodiesel derived from microalgal biomass adsorbed onto the sugarcane-bagasse-based adsorbent. According to Mohd-Sahib and Lim [19], this type of FAMEs mixture has great potential for the production of high-quality biodiesel with acceptable oxidative stability and cold flow properties.

Table 2. The fatty acid methyl ester (FAME) profile of biodiesel from microalgal biomass harvested via adsorption-enhanced flocculation processes.

Carbon Number — FAME Species	Composition in Biodiesel (%)
C14:1 — M. myristoleate	0.78
C15:0 — M. pentadecenoate	0.97
C15:1 — M. cis—10—pentadecenoate	1.25
C16:0 — M. palmitate	13.76
C16:1 — M. palmitoleate	1.49
C17:1 — M. cis—10—heptadecanoate	2.88
C18:0 — M. stearate	22.08
C18:1 — cis M. oleate	7.14
C 18:1 — (E) — M. 9—octadecanoate	7.74
C18:2 — M. linoleate	8.39
C18:2 — M. linolelaidate	6.60
C18:3 — M. y—linolenate	0.74
C20:1 — M. eicosenoate	7.04
C20:2 — M. cis—11,14—eicosadienoate	3.95
C20:3 — M. cis—11,14,17—eicosatrienoate	2.10
C21:0 — M. heneicosanoate	1.44
C22:0 — M. arachidate	0.35
C22:1 — M. erucate	4.23
C22:6 — M. cis—4,7,10,13,16,19—docosahexaer	0.80
C23:0 — M. tricosanoate	5.30
C24:0 — M. lignocerate	0.97

3.3. *Potential Reusability of Spent Sugarcane-Bagasse-Based Adsorbent*

From the point of view of sustainability, the potential to reuse spent sugarcane-bagasse-based adsorbent after the first cycle of removing microalgal biomass via adsorption-enhanced flocculation processes was evaluated. The performance results of spent sugarcane-bagasse-based adsorbent for five consecutive cycles of reuse are presented in Table 3. Cycle-1 represents the first use of virgin sugarcane-bagasse-based adsorbent upon fabrication to carry out adsorption-enhanced flocculation processes. After extracting the lipids from the microalgal biomass adsorbed onto sugarcane-bagasse-based adsorbent, the lipid-exhausted sugarcane-bagasse-based adsorbent was then used to carry out adsorption-enhanced flocculation processes again in Cycle-2, and these procedures were reiteratively repeated until Cycle-5. The adsorption efficiency diminished about 10% in Cycle-2 when compared to Cycle-1, though the adsorption capacity was still maintained above 100 mg/g. This could be undoubtedly rationalized by the presence of vacant and unexploited active sites left unoccupied after Cycle-1. When the same adsorbent was employed for the third time, parts of the loose cellulosic materials were noticed to inevitably detach from the sugarcane-bagasse-based adsorbent. Accordingly, the total weight of the lipid-exhausted sugarcane-bagasse-based adsorbent plunged about 20% to merely 0.33 g in Cycle-3. This could be due to continuous mechanical abrasion among the adsorbent solids, stemming from the agitation provided during the adsorption-enhanced flocculation and lipid extraction processes. As a result, the adsorption efficiency and capacity also dropped to lowest values of 38.2% and 59.8 mg/g, respectively, in this cycle. As the material structure of the remaining lipid-exhausted sugarcane-bagasse-based adsorbent was more firm and stable after Cycle-3, the adsorption efficiency was observed to improve to about 48% in Cycle-4, later plateauing in Cycle-5. As the total weight of the lipid-exhausted sugarcane-bagasse-based adsorbent did not differ much in Cycle-4 and Cycle-5, the adsorption capacities were measured in the range of approximately 70–80 mg/g in these two cycles.

Table 3. Performance of spent sugarcane-bagasse-based adsorbent for reiterative removal of microalgal biomass via adsorption-enhanced flocculation processes.

Reusability (Cycle)	Cycle-1	Cycle-2	Cycle-3	Cycle-4	Cycle-5
Adsorption efficiency (%)	91.5 ± 1.1	77.5 ± 1.3	38.2 ± 1.5	48.8 ± 8.3	44.7 ± 3.0
Adsorption capacity (mg/g)	192.9 ± 0.1	129.0 ± 4.0	59.8 ± 2.2	82.5 ± 0.1	70.3 ± 1.0
Lipid-exhausted adsorbent (g)	0.42 ± 0.00	0.4 ± 0.03	0.33 ± 0.03	0.32 ± 0.01	0.29 ± 0.01

4. Conclusions

Acid-modified sugarcane-bagasse-based adsorbent was successfully employed to remove microalgal biomass from eutrophic water via adsorption-enhanced flocculation processes. By activating the adsorbent with only 1.5 M of H_2SO_4, a microalgal biomass adsorption capacity of 108.9 ± 0.3 mg/g was achieved at equilibrium. This is due to 1.5 M H_2SO_4 acid-modified sugarcane-bagasse-based adsorbent having the lowest surface positivity value among the adsorbents tested, minimizing negative counter ion formation whilst maximizing the negatively charged microalgal cell interaction. In enhancing microalgal biomass separation from eutrophic water, the employment of a 6.7–8.0 mm adsorbent size resulted in an increase of the adsorption capacity to 143.6 ± 1.7 mg/g. Further optimizing the adsorbent dosage permitted the adsorption capacity to reach 192.9 ± 0.1 mg/g with a dosage of 0.4 g of acid-modified adsorbent in 100 mL of adsorption medium. This was equivalent to a 91.5% microalgae removal efficiency from eutrophic water. The harvested microalgal biomass also produced excellent-quality biodiesel, as manifested by the high levels of C16–C18 components (71%) in the FAME profile. The biodiesel quality was also proven by the low degree of PUFA (22.58%) and high degree of MUFA (32.55%) and SFA (44.87%). From the sustainability viewpoint, the spent acid-modified adsorbent also could be reused immediately after lipid extraction from the adsorbed microalgal biomass without the necessity to regenerate.

Author Contributions: Conceptualization, M.K.L. and J.W.L.; methodology, W.N.F.W.B. and H.D.; validation, C.K.C., W.D.O. and W.N.T.; formal analysis, M.S.S. and W.N.F.W.B.; resources, K.K. and E.A.K.; data curation, U.P.; writing—original draft preparation, W.N.F.W.B.; writing—review and editing, P.L.S. and J.W.L.; visualization, Y.F.Y.; supervision, J.W.L.; project administration, J.W.L. and M.K.L.; funding acquisition, J.W.L.

Funding: Funding from Ministry of Education Malaysia through HICoE awarded to the Centre for Biofuel and Biochemical Research, Universiti Teknologi PETRONAS.

Acknowledgments: The financial supports from International Grant—Universitas Islam Riau (UIR), Pekanbaru, Indonesia with the cost center 015ME0-039 and Universiti Teknologi PETRONAS via YUTP-FRG with the cost center 0153AA-E48 are gratefully acknowledged. Funding from Ministry of Education Malaysia through HICoE awarded to the Centre for Biofuel and Biochemical Research, Universiti Teknologi PETRONAS is as well duly acknowledged.

Conflicts of Interest: All authors declare that they have no conflict of interest.

References

1. Altmann, J.; Rehfeld, D.; Träder, K.; Sperlich, A.; Jekel, M. Combination of granular activated carbon adsorption and deep-bed filtration as a single advanced wastewater treatment step for organic micropollutant and phosphorus removal. *Water Res.* **2016**, *92*, 131–139. [CrossRef] [PubMed]
2. Azmi, N.B.; Bashir, M.J.; Sethupathi, S.; Wei, L.J.; Aun, N.C. Stabilized landfill leachate treatment by sugarcane bagasse derived activated carbon for removal of color, COD and NH3-N–optimization of preparation conditions by RSM. *J. Environ. Chem. Eng.* **2015**, *3*, 1287–1294. [CrossRef]
3. Barros, A.I.; Gonçalves, A.L.; Simões, M.; Pires, J.C. Harvesting techniques applied to microalgae: A review. *Renew. Sustain. Energy Rev.* **2015**, *41*, 1489–1500. [CrossRef]
4. Daliry, S.; Hallajsani, A.; Mohammadi Roshandeh, J.; Nouri, H.; Golzary, A. Investigation of optimal condition for Chlorella vulgaris microalgae growth. *Glob. J. Environ. Sci. Manag.* **2017**, *3*, 217–230.

5. Meinel, F.; Zietzschmann, F.; Ruhl, A.S.; Sperlich, A.; Jekel, M. The benefits of powdered activated carbon recirculation for micropollutant removal in advanced wastewater treatment. *Water Res.* **2016**, *91*, 97–103. [CrossRef]
6. Gerardo, M.L.; Van Den Hende, S.; Vervaeren, H.; Coward, T.; Skill, S.C. Harvesting of microalgae within a biorefinery approach: A review of the developments and case studies from pilot-plants. *Algal Res.* **2015**, *11*, 248–262. [CrossRef]
7. International Energy Agency. *Key Worlds Energy Statistics 2014*; OECD Publishing: Paris, France, 2014.
8. Liu, J.; Zhu, Y.; Tao, Y.; Zhang, Y.; Li, A.; Li, T.; Zhang, C. Freshwater microalgae harvested via flocculation induced by pH decrease. *Biotechnol. Biofuels* **2013**, *6*, 98. [CrossRef]
9. Mohanta, D.; Ahmaruzzaman, M. Bio-inspired adsorption of arsenite and fluoride from aqueous solutions using activated carbon@ SnO 2 nanocomposites: Isotherms, kinetics, thermodynamics, cost estimation and regeneration studies. *J. Environ. Chem. Eng.* **2018**, *6*, 356–366. [CrossRef]
10. Rashid, N.; Rehman MS, U.; Sadiq, M.; Mahmood, T.; Han, J.I. Current status, issues and developments in microalgae derived biodiesel production. *Renew. Sustain. Energy Rev.* **2014**, *40*, 760–778. [CrossRef]
11. Milledge, J.J.; Heaven, S. A review of the harvesting of micro-algae for biofuel production. *Rev. Env. Sci. Bio Technol.* **2013**, *12*, 165–178. [CrossRef]
12. Olkiewicz, M.; Fortuny, A.; Stüber, F.; Fabregat, A.; Font, J.; Bengoa, C. Evaluation of different sludges from WWTP as a potential source for biodiesel production. *Procedia Eng.* **2012**, *42*, 634–643. [CrossRef]
13. Su, S.; Liu, Q.; Liu, J.; Zhang, H.; Li, R.; Jing, X.; Wang, J. Functionalized Sugarcane Bagasse for U (VI) Adsorption from Acid and Alkaline Conditions. *Sci. Rep.* **2018**, *8*, 793. [CrossRef] [PubMed]
14. Zhao, Y.; Jiang, C.; Yang, L.; Liu, N. Adsorption of Lactobacillus acidophilus on attapulgite: Kinetics and thermodynamics and survival in simulated gastrointestinal conditions. *LWT Food Sci. Technol.* **2017**, *78*, 189–197. [CrossRef]
15. Kharat, D.S. Adsorption of Reactive Blue 19 Dye by Sugarcane Bagasse and the Proposed Modelling. *Curr. Env. Eng.* **2018**, *5*, 155–165. [CrossRef]
16. Shehzad, A.; Bashir, M.J.; Sethupathi, S.; Lim, J.W. An overview of heavily polluted landfill leachate treatment using food waste as an alternative and renewable source of activated carbon. *Process Saf. Environ. Prot.* **2015**, *98*, 309–318. [CrossRef]
17. Ofomaja, A.; Naidoo, E.; Modise, S. Removal of copper (II) from aqueous solution by pine and base modified pine cone powder as biosorbent. *J. Hazard. Mater.* **2009**, *168*, 909–917. [CrossRef] [PubMed]
18. Bligh, E.G.; Dyer, W.J. A rapid method of total lipid extraction and purification. *Can. J. Biochem. Physiol.* **1959**, *37*, 911–917. [CrossRef] [PubMed]
19. Mohd-Sahib, A.A.; Lim, J.W.; Lam, M.K.; Uemura, Y.; Isa, M.H.; Ho, C.D.; Rosli, S.S. Lipid for biodiesel production from attached growth Chlorella vulgaris biomass cultivating in fluidized bed bioreactor packed with polyurethane foam material. *Bioresour. Technol.* **2017**, *239*, 127–136. [CrossRef]
20. Argun, M.E.; Dursun, S. Removal of heavy metal ions using chemically modified adsorbents. *J. Int. Environ. Appl. Sci.* **2006**, *1*, 27–40.
21. Duan, W.; Xu, X.; Ling, C.; Xu, G.; Su, S. Preparation of acid-modified-attapulgite/$Al_2(SO_4)_3$ adsorbent for enhanced removal of dom in WWTP secondary effluent. *Fresenius Environ. Bull.* **2016**, 4637–4644.
22. Shehzad, A.; Bashir, M.J.; Sethupathi, S.; Lim, J.W. An insight into the remediation of highly contaminated landfill leachate using sea mango based activated bio-char: Optimization, isothermal and kinetic studies. *Desalin. Water Treat.* **2016**, *57*, 22244–22257. [CrossRef]
23. Leong, K.Y.; See, S.; Lim, J.W.; Bashir, M.J.; Ng, C.A.; Tham, L. Effect of process variables interaction on simultaneous adsorption of phenol and 4-chlorophenol: Statistical modeling and optimization using RSM. *Appl. Water Sci.* **2017**, *7*, 2009–2020. [CrossRef]
24. Lam, M.K.; Lee, K.T. Catalytic transesterification of high viscosity crude microalgae lipid to biodiesel: Effect of co-solvent. *Fuel Process. Technol.* **2013**, *110*, 242–248. [CrossRef]
25. Song, M.; Pei, H.; Hu, W.; Ma, G. Evaluation of the potential of 10 microalgal strains for biodiesel production. *Biores. Technol.* **2013**, *141*, 245–251. [CrossRef] [PubMed]

26. Hoekman, S.K.; Broch, A.; Robbins, C.; Ceniceros, E.; Natarajan, M. Review of biodiesel composition, properties, and specifications. *Renew. Sustain. Energy Rev.* **2012**, *16*, 143–169. [CrossRef]
27. Hu, Q.; Sommerfeld, M.; Jarvis, E.; Ghirardi, M.; Posewitz, M.; Seibert, M.; Darzins, A. Microalgal triacylglycerols as feedstocks for biofuel production: Perspectives and advances. *Plant J.* **2008**, *54*, 621–639. [CrossRef]

Article

Thermal Analysis of Nigerian Oil Palm Biomass with Sachet-Water Plastic Wastes for Sustainable Production of Biofuel

Bello Salman [1], Mei Yin Ong [1], Saifuddin Nomanbhay [1,*], Arshad Adam Salema [2], Revathy Sankaran [3] and Pau Loke Show [4,*]

1 Institute of Sustainable Energy (ISE), Universiti Tenaga Nasional (The Energy University), Jalan IKRAM-UNITEN, Kajang 43000, Malaysia

2 School of Engineering, Monash University Malaysia, Jalan Lagoon Selatan, Bandar Sunway 46150, Malaysia

3 Institute of Biological Sciences, Faculty of Science, University of Malaya, Kuala Lumpur 50603, Malaysia

4 Department of Chemical and Environment Engineering, Faculty of Science and Engineering, Jalan Broga Semenyih 43500, Malaysia

* Correspondence: saifuddin@uniten.edu.my (S.N.); pauloke.show@nottingham.edu.my (P.L.S.); Tel.: +603-8921-7285 (S.N.); +603-8924-8605 (P.L.S.)

Received: 17 June 2019; Accepted: 8 July 2019; Published: 23 July 2019

Abstract: Nigeria, being the world's largest importer of diesel-powered gen-sets, is expected to invest in bio-fuels in the future. Hence, it is important to examine the thermal properties and synergy of wastes for potential downstream resource utilization. In this study, thermal conversion as a route to reduce the exploding volume of wastes from sachet-water plastic (SWP) and oil palm empty fruit bunch (OPEFB) biomass was studied. Thermogravimetric (TGA) and subsequent differential scanning calorimeter (DSC) was used for the analysis. The effect of heating rate at 20 °C min^{-1} causes the increase of activation energy of the decomposition in the first-stage across all the blends (0.96 and 16.29 kJ mol^{-1}). A similar phenomenon was seen when the heating rate was increased from 10 to 20 °C min^{-1} in the second-stage of decomposition. Overall, based on this study on the synergistic effects during the process, it can be deduced that co-pyrolysis can be an effective waste for energy platform.

Keywords: sachet-water plastic waste; oil palm empty fruit bunch; TGA-DSC analysis; activation energy; physio-thermal analysis; co-pyrolysis

1. Introduction

Concern about the growing demand for energy, with emphasis on the developing economies, has prompted the urgent calls to implement renewable energy planning and advancement in reducing solid waste by utilizing it for energy production. The use of conventional energy sources is largely responsible for increasing CO_2 emissions in the atmosphere [1]. Thus, it is regarded as the underlying cause of greenhouse gas emissions and global warming. It was reported that the African energy-related CO_2 emissions are projected to increase by about 40% by the year 2030, with Nigeria contributing significantly to this growth [2].

According to the Power Africa Fact Sheet in Nigeria, only 45% of the nation has access to the national power grid [3]. This electricity mix percentage is mostly generated using 80% natural gas and 20% hydropower [4]. Another electrification method is the use of diesel-powered generator sets (gen-sets). Self-generation of electricity became necessary to avoid blackouts, especially in the densely-populated northern region of Nigeria. Furthermore, it is estimated that demand for energy in Nigeria will grow by at least 500% by 2035. However, the current trajectory of the electricity supply will increase by around just 1% in the same period [5]. The current situation of the power generation requires a new approach to impede the growing threats of CO_2 emissions from gen-sets. In this context, alternative energy generation from local biomass could be one of the most viable solutions to instantly minimize the intake of fossil fuels to reduce environmental complications [6–8].

Energy from waste, mainly lignocellulosic, is advantageous due to its widely recognized social, economic, environmental and renewable properties [9]. At the same time, plastic waste poses an ecological dilemma due to its long life-time, and symbolizes an essential element of waste management [10]. According to Ben-Iwo et al. and Stoler, biomass and sachet water plastics produced from agricultural and industrial activities in Nigeria are estimated to be nearly 144 million ton y^{-1} and 70–100 million ton y^{-1}, respectively [11,12]. Sachet water, primarily known as 'pure water', is symbolic to the sub-Saharan region of Africa, sold in mechanically-sealed 500 mL plastic sleeves at a unit price of not more than 0.10 US $. This water is regarded as a multibillion-dollar industry. The primary concern related to this water industry is the generation of plastic waste that continues to be one of the most significant threats to the region because of clogging gutters, causing routine flooding and exposing residents to a variety of health risks. In Nigeria, waste from plastics comprises about 65% of national solid waste streams. Therefore, in this study, the thermal and kinetic behavior of plastic waste will be examined, and the possible application of this waste for energy generation and as a platform for waste management will be explored.

Several methods, including gasification, hydrothermal liquefaction, combustion and pyrolysis, have been explored to treat the waste biomass [13–15]. Among these, pyrolysis involves the thermochemical decomposition of hydrocarbon or organic materials (usually biomass) at elevated temperatures in the absence of oxygen. Through the pyrolysis process, biomass will be converted into energetic products, such as bio-oil, syngas and bio-char. Researchers have explored a possible solution by co-pyrolysis of plastic waste with biomass [16]. The co-pyrolysis with plastics is preferred due to its ability to balance the ratios of carbon, oxygen, and hydrogen in the bio-oil derived from biomass pyrolysis [17]. In general, plastics are known to have high hydrogen compounds which make them suitable, and the potential substrate to improve the quality of the bio-oil product. Previous investigations claim apparent interactions and synergistic effects between biomass and plastics in the co-pyrolysis [18]. The inherent individual characteristics observed were due to the lower thermal stability of biomass being caused by the presence of plastics. These contradictions may have been caused by significant variations in composition, the origins of biomass, and complexity of chemical reactions during pyrolysis [19]. Thus, the need for increased research on the thermal and physio-chemical anatomization of biomass and secondary waste resources to predict the downstream product, as well as the associated environmental impacts, is necessary.

To the author's knowledge, no data is available on the pyrolysis of Nigerian oil palm empty fruit bunch (OPEFB), sachet-water plastic wastes (SWP) and their blends. In view of the large amount of OPEFB and SWP produced in Nigeria, it is necessary that a study on waste-to-value added product should be undertaken. Further, the results are expected to provide useful information for individuals and institutions who are interested in using Nigerian biomass and plastics for thermochemical conversion. Therefore, the objective of the study is to determine the thermal behavioral properties and synergy (if any) between OPEFB and SWP wastes as a potential application in renewable biofuel and biomaterials.

2. Materials and Methods

2.1. Biomass Samples

The OPEFB samples were provided by the Nigeria Institute for Oil Palm Research (NIFOR), Nigeria and SWP (Low Density Polyethylene) were acquired from Deezor Pharmaceuticals Limited, Nigeria. The samples were dried in an oven at 105 °C for 24 h and then ground in Fristch Pulverisette (model 19) to the size of 250 μm and further to 200 μm size in a Fristch Pulverisette (model 16).

The biomass-plastic (OPEFB and SWP) blends were prepared with different weight percentages (wt.%) of 20%, 40%, 50%, 60%, 80%, and 100%. However, for DSC analysis, the blends of 10, 20 and 30 wt.% of biomasses to plastic blends were used. The ratio of SWP to OPEFB was capped at maximum of 30% (w/w) to retain OPEFB as the major component of the sample based on results from TGA analysis of this work. To make certain of the blending homogeneity, samples were put through vortex shaker (IKA Vortex 3) for two minutes at 2500 rpm.

2.2. Elemental and Proximate Analysis

The ultimate/elemental analysis of the sample was performed using a CHNS/O analyzer (2400 model by PerkinElmer, Waltham, MA, USA), followed ASTM D5373-93 method. The volatile matter, fixed carbon and ash content of the sample (proximate analysis), however, were determined according to ASTM E 897-82 and ASTM D 1102-84 method. All experiments were conducted in triplicated and averaged values are reported in Table 1.

Table 1. Ultimate and proximate composition of samples.

Component	Method	Composition (wt.%) OPEFB	SWP
Ultimate analysis			
Carbon (C)	Elemental analyzer	54.40	86.93
Hydrogen (H)	Elemental analyzer	7.64	16.54
Oxygen (O)	By difference	36.44	1.39
Nitrogen (N)	Elemental analyzer	1.04	0.09
Sulphur (S)	Elemental analyzer	0.48	0.12
Proximate analysis			
Volatile matter	ASTM E 897-82	81.4	99.6
Fixed carbon	By difference	18.6	0.0
Ash	ASTM D 1102-84	4.6	0.4

2.3. Thermal Analysis Using Thermogravimetric Analysis (TGA) and Differential Scanning Calorimetry (DSC)

Pyrolysis of original and blend samples (~4 mg) was carried out in a programmable TG analyzer (DSC-TGA Q Series instrument and SDT Q600 thermal analyser, manufactured by TA Instrument, New Castle, DE, USA) from room temperature to 800 °C and at two heating rates (10 and 20 °C min^{-1}). Isothermal DSC measurements can be successfully applied for information about the heat capacity as a function of temperature during phase transitions, autoxidation, thermal decomposition and adsorption of different kinds of fuels [20]. For DSC (DSC823 manufactured by Mettler Toledo) experiments, the samples were heated from 25 °C to 600 °C at a heating rate of 15 °C min^{-1} to measure the heat flow during pyrolysis. For all experiments (TGA/DSC), nitrogen was used as an inert carrier gas with a flow rate of 50 mL min^{-1}. The instrument continuously recorded TG and DTA data which were used to analyze its thermal characteristics and to calculate the kinetic parameters. All tests were performed in triplicates to ensure reproducibility.

2.4. Kinetic Reaction

Several models have been used in kinetic analysis [21–23]. The Coats-Redfern is one of the most widely-used approach to obtain the kinetic parameters of carbonaceous materials [24]. It is a model-fitting method to determine the activation energy, pre-exponential factor and reaction order from a single measurement of thermogravimetric [25]. In general, biomass pyrolysis can be characterized using an infinite number of reaction mechanisms, described by the n-order law:

$$\frac{d\alpha}{dt} = k(T)f(\alpha) \tag{1}$$

$$f(\alpha) = (1-\alpha)^n \tag{2}$$

where $k(T)$ is the reaction rate constant, α indicates the amount of conversion or the fractional weight loss (Equation (3)), and n is the reaction order.

$$\alpha = \frac{m_i - m_o}{m_i - m_f} \tag{3}$$

where m_i, m_o and m_f are the initial mass, the current mass at time 't' and the final mass of the sample respectively. Note that α value is always between 0 and 1.

The reaction rate constant, $k(T)$ is a function of temperature, T with a unit of K and can be expressed as (Equation (4)) based on Arrhenius relationship.

$$k(T) = Ae^{-\frac{E_a}{RT}} \tag{4}$$

where A symbolizes the pre-exponential factor (min^{-1}). On the other hand, E_a indicates the activation energy of the decomposition reaction (kJ mol^{-1}) and R is the universal gas constant (8.314 J mol^{-1} K^{-1}). By substituting (Equation (1)) and (Equation (3)) into (Equation (4)), the kinetic equation for the sample decomposition is expressed as follow:

$$\frac{\mathrm{d}\alpha}{\mathrm{dt}} = Ae^{-\frac{E_a}{RT}}(1-\alpha)^n \tag{5}$$

For the non-isothermal case (at constant heating rate, β), however, the above equation can be further modified to:

$$\frac{\mathrm{d}\alpha}{\mathrm{dT}} \cdot \frac{\mathrm{dT}}{\mathrm{dt}} = Ae^{-\frac{E_a}{RT}}(1-\alpha)^n \tag{6}$$

As

$$\beta = \frac{\mathrm{dT}}{\mathrm{dt}} \tag{7}$$

Hence, the final kinetic equation in non-isothermal TG experiments is:

$$\frac{\mathrm{d}\alpha}{\mathrm{dT}} = \frac{A}{\beta}e^{-\frac{E_a}{RT}}(1-\alpha)^n \tag{8}$$

According to Coats and Redfern method, Equation (8) was then rearranged, integrated and finally expressed as:

$$ln\left\{\frac{-ln(1-\alpha)}{T^2}\right\} = ln\left\{\psi\left(1 - \frac{2RT}{E_a}\right)\right\} - \frac{E_a}{RT} \tag{9}$$

where $\psi = \frac{AR}{\beta E_a}$ and the reaction is assumed to be first-order.

By assuming $\frac{2RT}{E_a} \ll 1$, $\ln\left\{\psi\left(1 - \frac{2RT}{E_a}\right)\right\} \approx \ln(\psi)$ [26]. Thus, Equation (9) can be further modified to:

$$ln\left\{\frac{-ln(1-\alpha)}{T^2}\right\} = ln(\psi) - \frac{E_a}{RT} \tag{10}$$

By plotting this equation, the activation energy and the pre-exponential factor can then be determined.

3. Results and Discussion

3.1. Characteristic properties of OPEFB and SWP

In Table 1, the proximate analysis shows that the volatile matter (OPEFB: 81.4%; SWP: 99.6%) and fixed carbon (OPEFB: 18.6%; SWP: 0%) yielded high-level reactivity and volatility benefits which were appropriate for production of liquid fuels [27]. Nonetheless, the values in this study differ slightly, i.e., by around 2%, from the experimental analysis of OPEFB reported by Oyedun et al. but are comparable to those in other agricultural residues [27]. This could be attributed to the erratic nature of biomass due to species of plant being used, the age of those plants, or the climate of the plantation. Interestingly, crop origin did slightly affect the (O) element property in biomass material that is in the scope of 40 wt.% to 44 wt.%. Additionally, the ash content was at 4.6%, which was more consistent with the results from [28,29]. Therefore, this low value of ash is expected to have a positive effect on heating value since high-level ash quotas in biomass are known to lead to unfavorable yield conditions in terms of liquid by-product in fast pyrolysis [30].

The pure samples (OPEFB: SWP) contained traces of Carbon (C), 54.4: 86.93%, Oxygen (O) 36.44: 1.39% and Hydrogen (H) 7.64: 16.54% respectively. The plastic material had higher C and lower O contents, and therefore lower O/C ratio. On the other hand, the OPEFB biomass had higher sulphur contents compared to the SWP plastics. The content of sulphur (0.48%) in OPEFB is slightly higher than in the SWP (0.12%) of plastic wastes. It has been reported that the typical amount of sulphur in oil palm empty fruit bunch biomass materials is in the range of <0.1 to 0.68% [27,29]. Thus, the 0.48 wt.% value determined in this study is consistent, irrespective of the geographical location. The trace of sulphur in plastic could be because waste plastic contains some contamination, e.g., in the pigment used to impart color in the plastic material [31].

3.2. Thermal Characteristics

Figures 1 and 2 show the TG and DTG curves of OPEFB, SWP, and their blends at a heating rate of 10 and 20 °C min^{-1} in a nitrogen environment. The pyrolysis process in the present study was differentiated into three main stages; (a) drying stage (30–120 °C), (b) pyrolysis stage (120–500 °C), and (c) char stage (500–800 °C).

The initial weight loss that occurred during the drying stage from room temperature to about 120 °C is due to the removal of free and bound moisture content. SWP showed almost no mass loss in the drying stage as compared to OPEFB (which suffered about 7.1 wt.% mass loss in this region) [32]. The drying stage can also be detected from the presence of the first small peak in the DTG curves (Figures 1B and 2B). A similar phenomenon was also reported by Xu et al. in their work on pine sawdust with and without polyvinylidene [19].

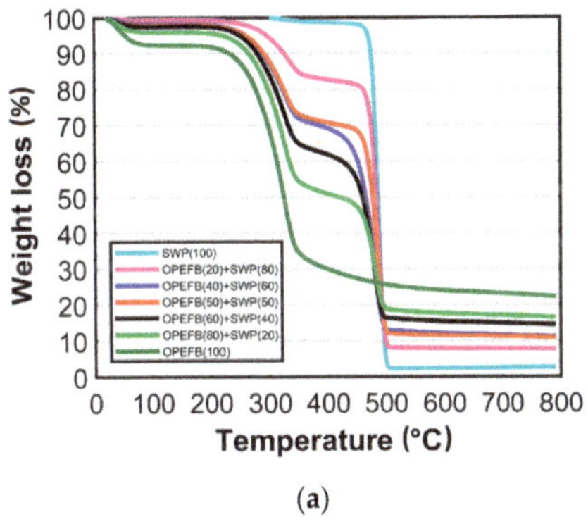

(a)

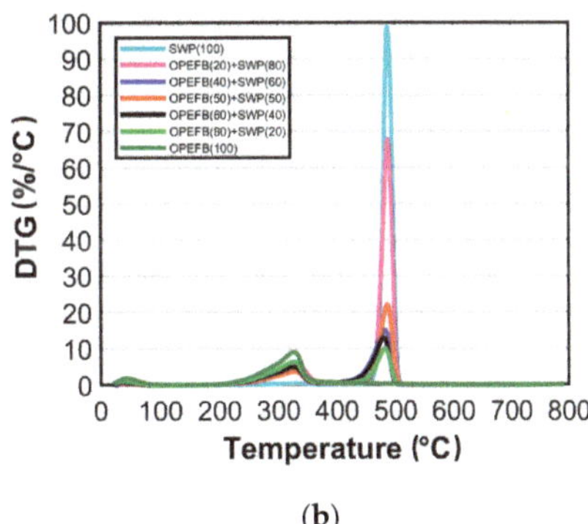

(b)

Figure 1. (**a**) TG, and (**b**) DTG curves for OPEFB and SWP samples and their blends at heating rate of 10 °C min^{-1}.

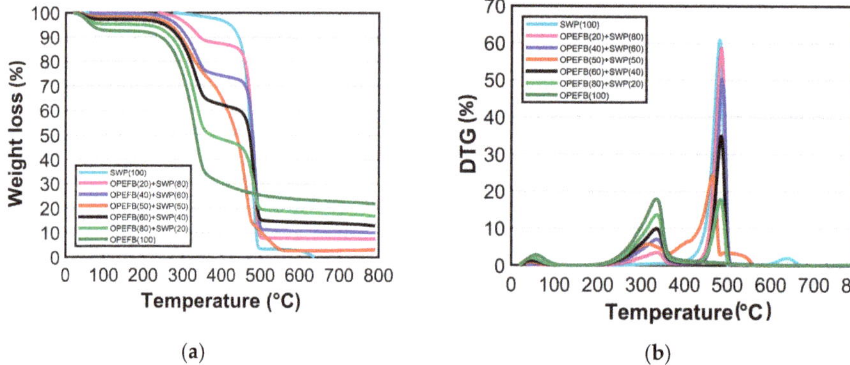

Figure 2. (**a**) TG, and (**b**) DTG curves for OPEFB and SWP samples and their blends at heating rates of 20 °C min^{-1}.

The pyrolysis stage follows immediately after the moisture was released in two stages; (a) pyrolysis of light volatization between 120 and 220 °C. At this stage, the chemical structure of the polymer and biomass starts to depolymerize and soften though without any loss in mass [16]. The next significant weight loss occurred during (b), the main pyrolysis in the temperature range between 220 °C and 500 °C due to the removal of heavy organic compounds. The pyrolysis of original OPEFB and SWP showed only one major peak in the central pyrolysis region. The total weight loss in this region for OPEFB and SWP was about 62 wt.% and 94 wt.%, respectively. Basically, during this stage, the biomass gets converted into volatiles which contain condensable and non-condensable gases. Original SWP plastic (100%) depicted a single peak at both heating rates (10 and 20 °C min^{-1}) in this region, as evident in Figures 1A and 2A. The major weight loss (~94 wt.%) for SWP started at 456 °C (at 10 °C min^{-1}) and 398 °C (at 20 °C min^{-1}) and completed at 505 °C (at 10 °C min^{-1}) and 497 °C (at 20 °C min^{-1}). This thermal behavior of plastic was similar to that in previous studies [27,33]. The thermal decomposition of plastic is vastly different among biomass materials due to different chemical bonds. The degradation mechanisms of polymers are due to the rapid primary hemolytic scission and intra-molecular hydrogen transfer in macro-radical forming oligomers at low temperatures, i.e., around 300 °C [34]. This random scission and de-polymerization occur at intermediate temperatures of 500 °C, further b-scission, and intra-molecular hydrogen transfer to form ethylbenzene, toluene, and a-methylstyrene at ~600 °C [34]. In theory, lignocellulosics such as oil palm empty fruit bunch is acknowledged to contain significant amounts of cellulose, hemicellulose and lignin [35], and the devolatilization has been shown to primarily correspond to the degradation of the components occurring in this region of significant weight loss [36]. Within this temperature range, the devolatilization stages of OPEFB can be divided into the lower temperature stage between, 195 °C and 300 °C, representing hemicellulose decomposition, and the intermediate temperature stage, 270–370 °C representing cellulose decomposition. It was reported that the significant devolatilization of biomass is because of the thermal formation of more stable and stronger bonds which replace the weaker bonds in the biomass structure [37]. The findings of the present study are in agreement with those obtained by previous researchers [38,39].

Some interesting facts were revealed when the OPEFB biomass was blended with SWP at different weight percentage ratios. Obvious differences in the pyrolysis of the original OPEFB/SWP from blend were noticed in the central pyrolysis region. The blends showed two peaks, with the DTG peak height and positions depicting the reactivity of the materials [40]. For instance, with the increase in biomass weight percentage in the blend, the reactivity and the mass loss rate of blends decreased significantly, depicting some degree of interaction between the two samples. Thus, it seems that the decomposition of biomass is, to some extent, affected by the presence of plastic and vice versa. According to Fang et al., the pattern of the fractured surface characteristic of OPEFB may provide substantial information about the adhesion and interfacial compatibility between the OPEFB fiber and SWP during co-pyrolysis [41]. The fractured surfaces in OPEFB also implies the potential of the higher hygroscopic behavior of the material. Hence, simultaneous degradation may be observed during the co-pyrolysis process, and thus, might re-orientate the chemical and thermal features of the co-pyrolysis.

Furthermore, it is reported that plastic starts softening at around 365 °C, but does not decompose completely [32]. In this study, based on Figures 1A and 2A, it can be observed that there was a slight drop in the weight loss of the SWP, which indicated that within the heating range there was an effect on the heat and mass transfer process. The peaks in this stage might overlap due to the simultaneous devolatilization of both OPEFB and SWP materials, resulting in an amplified synergetic effect. Also, a gap exists between the 2nd and 3rd peaks of the DTG curves (Figures 1B and 2B), which is similar to the results obtained by Oyedun et al. [27]. Worthy of note in Figure 2B is the fact that the OPEFB (50) +SWP (50) blend and SWP (100) depicted tails in the second peak at temperatures beyond 500 °C. This feature is different to the pyrolysis under 10 °C min^{-1} shown in Figure 1B. This might be due to the uneven pyrolysis of the constituents. In many cases, this is difficult to avoid due to the preparation procedures of the samples.

Table 2 displays the initial and final temperatures, peak temperatures and total mass loss in the main pyrolysis stage for biomasses and their blends. A shift in the pyrolysis range and peak temperature was observed due to biomass with SWP blending. A similar observation was reported in a previous work [42], and can be attributed to the thermal resistance between the reacting and evolved species which occur at higher heating rates. The initial degradation temperature in Table 2 shows the minimum temperature where the feedstock starts to decompose, which is also important because it gives information about the minimum ignition temperature required to decompose the material. For instance, the initial degradation temperature of OPEFB was 225 °C, as reported by Abdullah and Gerhauser [28]. However, in the present study, it was about 216 °C, a slightly lower than the values reported by previous authors. This could be due to the difference in the analysis method such as particle size at 250–355 μm; furthermore, a 100 mL min^{-1} nitrogen flowrate was applied in the work of Abdullah and Gerhauser [28]. On the other hand, the initial degradation temperature for SWP was in close agreement with the study of Banat and Fares [43].

Table 2. Properties of active pyrolysis zone at different heating rates.

Component	Heating Rate (°C min^{-1})	Pyrolysis Range (°C)	Peak Temperature (°C)	Total Mass Loss (%)
OPEFB: SWP				
100:0	10	219–380	328	59.98
	20	216–364	335	58.4
80:20	10	220–502	337 & 484	72.29
	20	220–497	332 & 483	74.5
60:40	10	232–498	339 & 480	77.33
	20	220–502	337 & 485	81.5
50:50	10	243–503	331 & 485	83.40
	20	237–552	304 & 460	93.16
40:60	10	235–503	329 & 481	83.96
	20	237–502	338 & 483	85.7
20:80	10	262–501	329 & 486	89.14
	20	273–501	329 & 484	90.75
0:100	10	464–507	486	94.45
	20	411–497	480	92.55

The last stage is called char or the carbonization stage, in which the carbon-rich residual is formed due to the relatively slow degradation of lignin. In this study, slow and steady weight loss could be observed at temperatures of around 500 °C. Typically, the degradation of lignin starts from the very early stage, i.e., from 300 °C and continues until 800 °C [44]. In this study, the carbonization stage of the lignin component of biomass OPEFB takes place within a temperature range of approximately 375–800 °C. The mass loss in this stage was about 22 wt.% for OPEFB. Meanwhile, the final residue from the carbonization stage of SWP was about 5 wt.% at 500 °C and completely melted at 650 °C. The leftover char residues at the end of pyrolysis (~700 °C) were about OPEFB20+SWP80 (17.2 &17.7 wt.%), OPEFB40 + SWP60 (14.9 & 13.8 wt.%), OPEFB50+SWP50 (11.3 & 2.8 wt.%), OPEFB60+SWP40 (11.3 & 10.5 wt.%), OPEFB80+SWP20 (7.8 & 7.6 wt.%) at 10 and 20 °C min^{-1} respectively. Among the blends, the char residues decreased in mass loss with increasing portions of OPEFB with OPEFB80+SWP20 showing the highest degradation rate. These variations further suggest a synergetic effect among the biomass and plastic mixtures in that the chemical components of the plastic are acting as catalyst [38,45]. This gives new insights into the behavior of co-adding plastics in the co-pyrolysis of biomass which have not been widely discussed in previous studies. However, the effect of the heating rate on the char residues is less pronounced, except in the case of OPEFB50+SWP50 at 20 °C min^{-1}, which could be attributed to mixing errors.

3.3. TGA Kinetic Analysis

The kinetics parameters of OPEFB, SWP and blends co-pyrolysis were determined using Coat Redfern's (CR) Method. The activation energy, E_a was estimated from the slope of -(E_a/RT) by a linear fit of the experimental points. By substituting this value back into the CR Equation (Equation (10)) gives the pre-exponential factor. The peak temperature (T_p) of each reaction, as revealed in Figures 1 and 2, is to be expected within a small range around the local maximum of mass loss rate. Listed in Table 3 are the calculated values for the kinetic parameters, including activation energy, E_a, pre-exponential factor, A, and reaction rate constant at peak temperature, of all the samples at the applied heating rates of 10 and 20 °C. In all cases, the value of R^2, (correlation coefficient) of the fitting straight line was above 0.90; this indicates that the corresponding non-isothermal model-fitting is in good agreement with the pyrolysis analysis and kinetics [46]. Furthermore, a comparison of the selected heating rates on pyrolysis of blends on solid-state kinetics data is presented. There is limited evidence in the literature of the effects of different heating rates on the kinetics of thermal decomposition to describe the devolatilization process in co-pyrolysis of biomass and plastics.

From Table 3, the thermal decomposition of OPEFB and SWP can be described by a single step reaction, while their blends showed two consecutives first and second step reactions. In view of the results of the kinetic parameters of the isolated samples tabulated in Table 3, it may be seen that a pure sample of SWP at varying heating rates (10 and 20 °C min^{-1}) has the least conversion and slower reactive characteristics due to its higher (E_a) value of 346.93 and 234.36 kJ mol^{-1}. Nonetheless the pre-exponential factor which expresses the probability of colliding molecules resulting in a reaction revealed 2.95×10^{23} min^{-1} and 9.13×10^{15} min^{-1} respectively. It was also expected that higher heating rates reduce the complex energy required to decompose the polymer atoms. This confirms that plastics decompose at higher temperatures (480–486 °C). As to the isolated OPEFB sample the (E_a) values of 46.83 and 44.21 kJ mol^{-1} was revealed at 10 and 20 °C min^{-1} and decomposes at a much lower temperature range (227–334 °C), compared to isolated SWP. Thus, a similar trend could be deduced regarding the effect of heating rate on their activation energies. Nonetheless, the effect of the heating rate is more apparent in the degradation mechanism of solid-state reactions of the SWP. The kinetics parameters in Table 3 also show that the temperature and energy needed to decompose the blends is higher with a higher weight percentage of SWP or plastic. Meanwhile, Nyakuma examined decomposition of pelletized oil palm empty fruit bunch through thermogravimetric analyzer and calculated kinetic parameters in the range from 36.60 kJ mol^{-1} to 233.92 kJ mol^{-1} through *Popescu* method [47]. Also, the apparent (E_a) and (A) obtained with the CR method employed in this study were in accordance with different biomass as demonstrated by the E_a values for oil palm empty fruit bunch, 50.37 kJ mol^{-1} [27] and those of other biomass species including almond and hazelnut shells (11.2–254.4 kJ mol^{-1}) and (40.3–144.9 kJ mol^{-1}) [48]. Similarly, in comparison with the polystyrene parameters reported in the literature, the activation energy obtained in this study at 10 and 20 °C min^{-1} falls within the reported ~213.78 kJ mol^{-1} [49] to 253.69 kJ mol^{-1} [50].

Table 3. Kinetics parameters of samples at different heating rates.

Reaction	Heating Rate (°C min^{-1})	T_p (°C)	R^2	E_a (kJ mol^{-1})	A (min^{-1})	Reaction Rate Constant at T_p (min^{-1})	
						Average	Standard Deviation
OPEFB 100	10	327	0.9821	46.83	1.29×10^3	0.15	0.0386
	20	334	0.9655	44.21	1.18×10^3		
OPEFB 80, SWP 20-First stage	10	331	0.9934	48.42	1.10×10^3	0.10	0.0332
	20	332	0.9894	49.97	2.84×10^3		
OPEFB 80, SWP 20-Second stage	10	486	0.9467	75.14	3.05×10^4	4.22	4.01
	20	484	0.9659	72.33	8.05×10^5		
OPEFB 60, SWP 40-First stage	10	320	0.9948	49.89	1.10×10^3	0.07	0.027
	20	334	0.9926	50.85	2.34×10^3		
OPEFB 60, SWP 40-Second stage	10	480	0.9523	93.25	7.03×10^5	0.42	0.1763
	20	484	0.9281	123.65	2.01×10^8		
OPEFB 50, SWP 50-First stage	10	327	0.9934	45.81	3.28×10^2	0.04	0.0107
	20	316	0.9910	49.28	1.29×10^3		
OPEFB 50, SWP 50-Second stage	10	485	0.9627	183.00	1.72×10^{12}	0.54	0.1202
	20	464	0.9862	112.16	5.89×10^7		
OPEFB 40, SWP 60-First stage	10	326	0.9748	45.46	3.11×10^2	0.06	0.0223
	20	334	0.9992	61.75	1.61×10^4		
OPEFB 40, SWP 60-Second stage	10	479	0.9587	217.67	4.30×10^7	0.40	0.4002
	20	484	0.9463	195.99	2.66×10^{13}		
OPEFB 20, SWP 80-First stage	10	327	0.9066	40.08	3.91×10^1	0.02	0.0092
	20	335	0.9683	55.97	1.99×10^3		
OPEFB 20, SWP 80–Second stage	10	481	0.9395	240.26	1.55×10^{16}	0.62	0.2648
	20	482	0.9393	229.86	7.02×10^{15}		
SWP 100	10	486	0.9092	346.93	2.95×10^{23}	0.71	0.3111
	20	480	0.9540	234.36	1.83×10^{16}		

In the case of kinetic results of the blended mixtures generally as expected, the (E_a) manifested a significant increase alongside the corresponding (A) with an increasing percentage of SWP in blends. The experimental data shown in Table 3 revealed that the values obtained for (E_a) and (A) of the SWP/OPEFB blends are relatively different from those of the individual materials. This means that a synergetic effect is observed in the SWP/OPEFB blends that might have an overlapping degradation temperature, which creates the opportunity for free radicals from biomass pyrolysis to participate in reactions of plastic decomposition. For example, the activation energies of 80 and 60% composition of SWP were much higher than the value obtained from the pyrolysis of pure samples, and presented obvious changes with the increase of plastics weight percentage. However, the value of the activation energy and pre-exponential factor decreases when the weight percentage of SWP in the blend decreases. Similar results have been reported in previous studies [17,33]; however, interestingly, the addition of SWP in the OPEFB biomass revealed a negligible effect on the activation energy at the first reaction order of decomposition.

However, at the second stage, an appreciable increase in action energy was revealed. At higher compositions of SWP 80% E_a was 40.08 kJ mol^{-1} and 240.26 kJ mol^{-1} at heating rate of 10 °C min^{-1} in the first and second stages, respectively. A similar trend was observed for the heating rate of 20 °C min^{-1}, where the activation energy of 60% SWP composition was 61.75 kJ mol^{-1} and 195.99 kJ mol^{-1} in the first and second stages, correspondingly. On the other hand, the kinetics parameters in Table 3, remarkably show that less energy is required to decompose the blends (below 50%), depicting a favorable synergistic effect between lower mass percentages blends of SWP lowering the activation energy. This technically supports the discrepancy that in co-pyrolysis, the ratio of feed is the most significant variable in product yield and economics [16,51].

Overall, a higher heating rate (20 °C min^{-1}) will lead to (average 10%) an increase in the activation energy of the decomposition in the first stage across all the blends in the range of 0.96 kJ mol^{-1} and 16.29 kJ mol^{-1}. Remarkably, varying the heating rate from 10 to 20 °C min^{-1} in the second stage of decomposition reaction shows, as in the previous case, an increase in activation energy, with one exception at OPEFB50:SWP50, while with a 50% SWP in blend, a higher heating rate of 20 °C min^{-1} favors a decline in the activation energy from 183.00 to 112.16 kJ mol^{-1}. This can be explained by the aforementioned phenomenon whereby the increase in the heating rate on isolated OPEFB and SWP result in decreasing activation energy. In this case, the homogeneous blend of equal samples followed the same reaction pathway correspondingly result in much lower activation energy. It can also be concluded that a significant difference can be seen in the distribution of the thermal and elemental composition.

It can also be concluded that significant differences can be seen in the distribution of the thermal and elemental composition. Furthermore, the SWP can be identified as low density polyethylene (LDPE) from the elemental composition and the pattern of thermal degradation [52]. The SWP presented a heat release rate curve typical for intermediately thin non-charring materials of LDPE family [53]. The burning time was very short, with a steady increase in heat release rate after ignition up to the peak heat release rate at the end of combustion. Furthermore, the heat release rate pattern showed hardly any shoulder, with the pyrolysis involving nearly all the LDPE material at once due to the complete melting of the polymer [54], this interpretation was confirmed by visual observation of the thermogravimetric curves in Figures 1 and 2.

3.4. DSC Analysis

Based on the findings from the TGA analysis, lower portions of SWP in blends do not lead to a significant increase in the energy required to decompose the co-pyrolysis system. Therefore, this section examines further the enthalpy of reaction at 15 °C of lower blends (10, 20 and 30 wt.%) of SWP. Differential scanning calorimetry (DSC) is widely employed to assess the heat evolution and spontaneous combustion of carbonaceous materials, due to its sensitive analyses at relative temperatures [55]. The DSC heat evolution curve stipulates quantitative data of the heat flow, enabling the kinetics analysis during thermal heating. There are several DSC based methods for the kinetic analysis including ASTM methods (Kissinger– Akahira–Sunose and Ozawa–Flynn–Wall) [25] and the Roger & Morris method [56]. The ASTM method relies on measuring the exothermic peak temperature at various heating rate, thereby resulting in the kinetics reflecting the main region of combustion. The profiles of the heat evolution rate of samples (OPEFB-100%, OPEFB-90%: SWP-10%, OPEFB-80%: SWP-20%, OPEFB-70%: SWP-30% and SWP-100%), thermally decomposed from 25 °C to 600 °C at a fixed heating rate of 15 °C/min, are illustrated in Figure 3. The single heating rate DSC test at 15 °C min^{-1} is the best for the application of ASTM approach, which has the advantage of reducing the required time, compared other approaches [55]. DSC analysis of pure OPEFB and SWP samples registered four and two exotherms, respectively, while each of the blended samples registered three stages. The exothermic effect differs from one material to another depending on the chemical composition and decomposition behavior of the material. The maximum temperature of the exotherms in the first, second, third and fourth regions were seen at 87–122 °C, 240–251 °C, 339–340 °C, and 465–491 °C, respectively. The first region corresponds to moisture release region, and it was found between 42 °C and 131 °C for pure OPEFB sample and between 95 °C and 128 °C for pure SWP sample. The second thermal zone was determined to be between 147 °C and 292 °C for OPEFB and between 152 and 332 °C for sample comprising 70% OPEFB and 30% SWP blends. The third thermal zone was determined to be between 315 and 353 °C for pure OPEFB, 287 and 355 °C, 295 and 354 °C for OPEFB: SWP blended ratios 90:10 and 80:20, respectively. This reflects the heat evolution of the representative samples, in which similar pattern was indicated by the TG and DTG curves in Figures 1 and 2. The melting of the various crystals led to several relatively broad, endothermic and exothermic peaks. The shape of the peak mirrors the size and weight distribution of the crystals and are among the characteristics of a material. The peak region marked on the experimental thermogram corresponds to the region of some physical or physicochemical processes and is dependent on internal factors, such as the structural nature of the material, quantity and thermal conductivity of the sample and external factors such as crucible shape and material, heating rate, the position of the thermocouples [57].

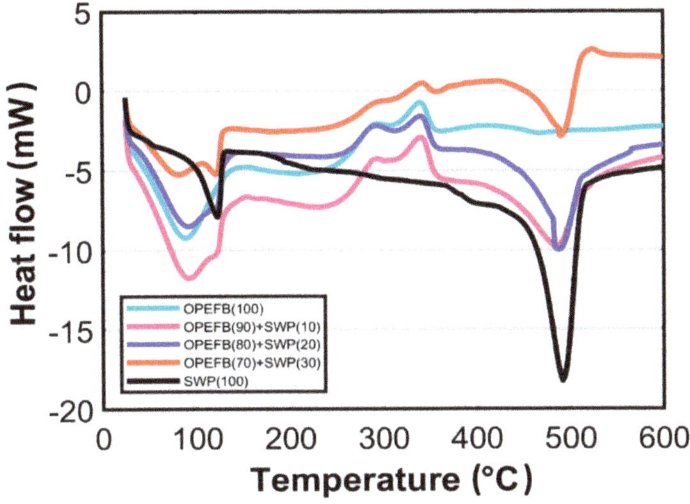

Figure 3. DSC evolutions of sample species evaluated (15 °C min^{-1}).

The enthalpy of a material has a direct correlation with the material's heating value, i.e., the ratio of the enthalpy of complete combustion to its mass [58]. It is an important thermal parameter in the reactor design and predicting efficiency of bioenergy applications. The temperatures and internal energies (enthalpy) corresponding to the beginning, maximum, and the end of the exothermic reactions are summarized in Table 4. The energy of a phase transition was calculated directly from the thermogram of the investigated material from the experimental function according to the kinetics.

Table 4 displays the heat released (J g^{-1}) through the reaction progress; it is not surprising that with increasing the SWP ratios, a shift towards lower H values was observed during the first stage decomposition. Also, isolated SWP sample exhibited lower values of heat released (117.39 J g^{-1}) as compared to the OPEFB (147.69 J g^{-1}) at the first stage of decomposition. It is no surprise that a high amount of heat was recorded during this stage. For the combustion reaction to proceed, energy is needed to overcome the tightly bonded cellulose and hemicellulose to lignin. The results were in agreement to some extent with previous studies [27,59]. Then, the (H) value for isolated OPEFB declined through the four decomposition stages to 18 J g^{-1}. Typically, it is expected for the (E_a) and (H) energies to progressively decrease due to the biomass composition mechanism as each of the components can decompose by parallel exothermic processes. Meanwhile, there was a significant further increase in the (H) of isolated SWP from 117.39 to 1030.19 J g^{-1} in its first and forth stages of decomposition, which is generally attributed to β-scissors reaction and in agreement with the TG studies. The net enthalpy of pyrolysis for the blended samples studied in this work were 398.18 J g^{-1} (OPEFB, 90: SWP 10); 365.32 J g^{-1} (OPEFB, 80: SWP, 20); 317.91 J g^{-1} (OPEFB, 70: SWP, 30). This can clearly be seen to be higher than that of pure biomass (276.82 J g^{-1}) and much lower than for pure SWP plastic (1147.58 J g^{-1}). The reason for this was that the heat flow of the blended samples at the final stage of decomposition were zero, which would indicate that no degradation reactions had taken place. However, through a comparison of the devolatization stage with the findings of TGA results, it could be observed that enthalpy (40.72) for OPEFB100% was similar to those of TGA. But a significant gap was observed for the case of SWP100%. Nonetheless, blends showed proof of a synergistic effect, and support the TGA findings in which a lower percentage weight of SWP in the co-pyrolysis mixture does not significantly boost energy required for the decomposition in both of the first stages of delovatization (84.42 J g^{-1} for OPEFB90:SWP10, 59.83 J g^{-1} for OPEFB80:SWP20, and 88.97 J g^{-1} for OPEFB70:SWP30) and second stage decomposition (158.21 J g^{-1} for OPEFB90:SWP10, 161.04 J g^{-1} for OPEFB80:SWP20, and 145.4 J g^{-1} for OPEFB70:SWP30).

Table 4. DSC results of the samples.

Sample (%)	T_{Onset} (°C)	T_{Peak} (°C)	T_{Endset} (°C)	Enthalpy, H (J g^{-1})
First stage				
OPEFB, 100	42.06	87.64	131.32	147.69
OPEFB, 90: SWP 10	45.21	91.04	130.31	155.55
OPEFB, 80: SWP, 20	42.04	90.42	146.40	144.45
OPEFB, 70: SWP, 30	89.66	120.77	127.50	83.54
SWP, 100	95.30	122.15	122.15	117.39
Second stage				
OPEFB, 100	147.23	240.96	292.32	88.41
OPEFB, 90: SWP, 10	-	-	-	84.42
OPEFB, 80: SWP, 20	-	-	-	59.83
OPEFB, 70: SWP, 30	152.08	251.87	332.97	88.97
SWP, 100	-	-	-	-
Third stage				
OPEFB, 100	315.25	340.20	353.66	22.28
OPEFB, 90: SWP, 10	287.90	339.55	355.90	158.21
OPEFB, 80: SWP, 20	295.60	339.89	354.82	161.04
OPEFB, 70: SWP, 30	-	-	-	145.40
SWP, 100	-	-	-	-
Forth stage				
OPEFB, 100	437.11	465.29	520.42	18.44
OPEFB, 90: SWP, 10	423.74	484.73	517.18	-
OPEFB, 80: SWP, 20	476.52	487.21	515.89	-
OPEFB, 70: SWP, 30	476.17	489.27	510.80	-
SWP, 100	455.68	491.56	511.48	1030.19

4. Conclusions

This study examined the thermal decomposition behavior and kinetics of Nigerian native oil palm empty fruit bunch, sachet-water plastic wastes and their blends under non-isothermal inert conditions at different heating rates. The findings show that the co-pyrolysis of OPEFB biomass and SWP plastic mixture exhibits a diverse pyrolysis reactivities at different temperatures. The co-pyrolysis of OPEFB and SWP clearly showed synergic effects due to the difference in thermal behavior and kinetics parameters. In short, the results show that co-pyrolysis can be an effective method to dispose of wastes (biomass and sachet water, SWP) and convert them into useful energy. The experimental data obtained with a model-free method would be helpful in the design and development of energy systems for sustainable waste utilization for energy.

Author Contributions: Conceptualization, S.N. and B.S.; methodology, B.S. and A.A.S.; validation and investigation, B.S. and M.Y.O.; resources, B.S.; writing—original draft preparation, B.S. and M.Y.O.; writing—review and editing, S.N., R.S., and P.L.S.; supervision, S.N. and A.A.S.; funding acquisition, S.N.

Funding: This work was funded by TNB Seed Fund (code: U-TR-RD-18-11). A note of appreciation to iRMC UNITEN for the financial support through publication fund BOLD 2025 (RJO10436494).

Acknowledgments: The authors would also like to acknowledge Monash University Malaysia and Universiti Tenaga Nasional (UNITEN) for the facilities support.

Conflicts of Interest: The authors declare no conflict of interest. Besides, the funders had no role in the design of the study; in the collection, analyses, or interpretation of data; in the writing of the manuscript, or in the decision to publish the results.

References

1. Norhasyima, R.S.; Mahlia, T.M.I. Advances in CO_2 utilization technology: A patent landscape review. *J. CO2 Util.* **2018**, *26*, 323–335. [CrossRef]
2. IEA. *Excerpt from CO_2 Emissions from Fuel Combustion*; IEA: Paris, France, 2015.
3. Power Africa. *Nigeria Power Africa Fact Sheet*; USAID: Washington, DC, USA, 2019.
4. Aliyu, A.S.; Ramli, A.T.; Saleh, M.A. Nigeria electricity crisis: Power generation capacity expansion and environmental ramifications. *Energy* **2013**, *61*, 354–367. [CrossRef]
5. Salman, B.; Neshaeimoghaddam, H. An Evaluation of The Nigeria Electricity Sector Post Privatisation. *J. Energy Environ.* **2017**, *9*, 33–37.
6. Financie, R.; Moniruzzaman, M.; Uemura, Y. Enhanced enzymatic delignification of oil palm biomass with ionic liquid pretreatment. *Biochem. Eng. J.* **2016**, *110*, 1–7. [CrossRef]
7. Bustos, G.; Arcos, U.; Vecino, X.; Cruz, J.M.; Moldes, A.B. Recycled Lactobacillus pentosus biomass can regenerate biosurfactants after various fermentative and extractive cycles. *Biochem. Eng. J.* **2018**, *132*, 191–195. [CrossRef]
8. Martínez-Patiño, J.C.; Ruiz, E.; Cara, C.; Romero, I.; Castro, E. Advanced bioethanol production from olive tree biomass using different bioconversion schemes. *Biochem. Eng. J.* **2018**, *137*, 172–181. [CrossRef]
9. Uddin, M.N.; Techato, K.; Taweekun, J.; Mofijur, M.; Rasul, M.G.; Mahlia, T.M.I.; Ashrafur, S.M. An Overview of Recent Developments in Biomass Pyrolysis Technologies. *Energies* **2018**, *11*, 3115. [CrossRef]
10. Rocha, E.P.A.; Sermyagina, E.; Vakkilainen, E.; Colodette, J.L.; de Oliveira, I.M.; Cardoso, M. Kinetics of pyrolysis of some biomasses widely available in Brazil. *J. Therm. Anal. Calorim.* **2017**, *130*, 1445–1454. [CrossRef]
11. Ben-Iwo, J.; Manovic, V.; Longhurst, P. Biomass resources and biofuels potential for the production of transportation fuels in Nigeria. *Renew. Sustain. Energy Rev.* **2016**, *63*, 172–192. [CrossRef]
12. Stoler, J. From curiosity to commodity: A review of the evolution of sachet drinking water in West Africa. *Wiley Interdiscip. Rev. Water* **2017**, *4*, e1206. [CrossRef]
13. De Caprariis, B.; Bavasso, I.; Bracciale, M.P.; Damizia, M.; De Filippis, P.; Scarsella, M. Enhanced bio-crude yield and quality by reductive hydrothermal liquefaction of oak wood biomass: Effect of iron addition. *J. Anal. Appl. Pyrolysis* **2019**, *139*, 123–130. [CrossRef]
14. Chianese, S.; Fail, S.; Binder, M.; Rauch, R.; Hofbauer, H.; Molino, A.; Blasi, A.; Musmarra, D. Experimental investigations of hydrogen production from CO catalytic conversion of tar rich syngas by biomass gasification. *Catal. Today* **2016**, *277*, 182–191. [CrossRef]
15. Nomanbhay, S.; Salman, B.; Hussain, R.; Ong, M.Y. Microwave pyrolysis of lignocellulosic biomass—A contribution to power Africa. *Energy Sustain. Soc.* **2017**, *7*, 23. [CrossRef]
16. Abnisa, F.; Wan Daud, W.M.A. A review on co-pyrolysis of biomass: An optional technique to obtain a high-grade pyrolysis oil. *Energy Convers. Manag.* **2014**, *87*, 71–85. [CrossRef]
17. Xiong, S.; Zhuo, J.; Zhou, H.; Pang, R.; Yao, Q. Study on the co-pyrolysis of high density polyethylene and potato blends using thermogravimetric analyzer and tubular furnace. *J. Anal. Appl. Pyrolysis* **2015**, *112*, 66–73. [CrossRef]
18. Chen, L.; Wang, S.; Meng, H.; Wu, Z.; Zhao, J. Synergistic effect on thermal behavior and char morphology analysis during co-pyrolysis of paulownia wood blended with different plastics waste. *Appl. Therm. Eng.* **2017**, *111*, 834–846. [CrossRef]
19. Xu, Z.-X.; Zhang, C.-X.; He, Z.-X.; Wang, Q. Pyrolysis Characteristic and kinetics of Polyvinylidene fluoride with and without Pine Sawdust. *J. Anal. Appl. Pyrolysis* **2017**, *123*, 402–408. [CrossRef]
20. Shen, J.; Igathinathane, C.; Yu, M.; Pothula, A.K. Biomass pyrolysis and combustion integral and differential reaction heats with temperatures using thermogravimetric analysis/differential scanning calorimetry. *Bioresour. Technol.* **2015**, *185*, 89–98. [CrossRef]
21. Milivojević, A.; Ćorović, M.; Carević, M.; Banjanac, K.; Vujisić, L.; Veličković, D.; Bezbradica, D. Highly efficient enzymatic acetylation of flavonoids: Development of solvent-free process and kinetic evaluation. *Biochem. Eng. J.* **2017**, *128*, 106–115. [CrossRef]
22. Costa, R.S.; Vinga, S. Control analysis of the impact of allosteric regulation mechanism in a Escherichia coli kinetic model: Application to serine production. *Biochem. Eng. J.* **2016**, *110*, 59–70. [CrossRef]

23. Martínez, I.; Santos, V.E.; Garcìa-Ochoa, F. Metabolic kinetic model for dibenzothiophene desulfurization through 4S pathway using intracellular compound concentrations. *Biochem. Eng. J.* **2017**, *117*, 89–96. [CrossRef]
24. Coats, A.W.; Redfern, J.P. Kinetic Parameters from Thermogravimetric Data. *Nature* **1964**, *201*, 68–69. [CrossRef]
25. Mishra, R.K.; Mohanty, K. Pyrolysis kinetics and thermal behavior of waste sawdust biomass using thermogravimetric analysis. *Bioresour. Technol.* **2018**, *251*, 63–74. [CrossRef] [PubMed]
26. Samaržija-Jovanović, S.; Jovanović, V.; Marković, G.; Marinović-Cincović, M.; Budinski-Simendić, J.; Janković, B. Ethylene–Propylene–Diene Rubber-Based Nanoblends: Preparation, Characterization and Applications. In *Rubber Nano Blends*; Springer: Cham, Switzerland, 2017; pp. 281–349.
27. Oyedun, A.O.; Tee, C.Z.; Hanson, S.; Hui, C.W. Thermogravimetric analysis of the pyrolysis characteristics and kinetics of plastics and biomass blends. *Fuel Process. Technol.* **2014**, *128*, 471–481. [CrossRef]
28. Abdullah, N.; Gerhauser, H. Bio-oil derived from empty fruit bunches. *Fuel* **2008**, *87*, 2606–2613. [CrossRef]
29. Sulaiman, F.; Abdullah, N. Optimum conditions for maximising pyrolysis liquids of oil palm empty fruit bunches. *Energy* **2011**, *36*, 2352–2359. [CrossRef]
30. Yang, J.; Rizkiana, J.; Widayatno, W.B.; Karnjanakom, S.; Kaewpanha, M.; Hao, X.; Abudula, A.; Guan, G. Fast co-pyrolysis of low density polyethylene and biomass residue for oil production. *Energy Convers. Manag.* **2016**, *120*, 422–429. [CrossRef]
31. Lawson, O.E.; Lawson, E.O. Physico-Chemical Parameters and Heavy Metal Contents of Water from the Mangrove Swamps of Lagos Lagoon, Lagos, Nigeria. *Adv. Biol. Res.* **2011**, *5*, 8–21.
32. Han, B.; Chen, Y.; Wu, Y.; Hua, D.; Chen, Z.; Feng, W.; Yang, M.; Xie, Q. Co-pyrolysis behaviors and kinetics of plastics–biomass blends through thermogravimetric analysis. *J. Therm. Anal. Calorim.* **2014**, *115*, 227–235. [CrossRef]
33. Uzun, B.B.; Yaman, E. Pyrolysis kinetics of walnut shell and waste polyolefins using thermogravimetric analysis. *J. Energy Inst.* **2017**, *90*, 825–837. [CrossRef]
34. Bartoli, M.; Rosi, L.; Frediani, M.; Undri, A.; Frediani, P. Depolymerization of polystyrene at reduced pressure through a microwave assisted pyrolysis. *J. Anal. Appl. Pyrolysis* **2015**, *113*, 281–287. [CrossRef]
35. Chang, S.H. An overview of empty fruit bunch from oil palm as feedstock for bio-oil production. *Biomass Bioenergy* **2014**, *62*, 174–181. [CrossRef]
36. Ro, D.; Kim, Y.-M.; Lee, I.-G.; Jae, J.; Jung, S.-C.; Kim, S.C.; Park, Y.-K. Bench scale catalytic fast pyrolysis of empty fruit bunches over low cost catalysts and HZSM-5 using a fixed bed reactor. *J. Clean. Prod.* **2018**, *176*, 298–303. [CrossRef]
37. Chandrasekaran, A.; Ramachandran, S.; Subbiah, S. Determination of kinetic parameters in the pyrolysis operation and thermal behavior of Prosopis juliflora using thermogravimetric analysis. *Bioresour. Technol.* **2017**, *233*, 413–422. [CrossRef] [PubMed]
38. Jin, Q.; Wang, X.; Li, S.; Mikulčić, H.; Bešenić, T.; Deng, S.; Vujanović, M.; Tan, H.; Kumfer, B.M. Synergistic effects during co-pyrolysis of biomass and plastic: Gas, tar, soot, char products and thermogravimetric study. *J. Energy Inst.* **2019**, *92*, 108–117. [CrossRef]
39. Kai, X.; Li, R.; Yang, T.; Shen, S.; Ji, Q.; Zhang, T. Study on the co-pyrolysis of rice straw and high density polyethylene blends using TG-FTIR-MS. *Energy Convers. Manag.* **2017**, *146*, 20–33. [CrossRef]
40. Chong, Y.Y.; Thangalazhy-Gopakumar, S.; Gan, S.; Ng, H.K.; Lee, L.Y.; Adhikari, S. Kinetics and Mechanisms for Copyrolysis of Palm Empty Fruit Bunch Fiber (EFBF) with Palm Oil Mill Effluent (POME) Sludge. *Energy Fuels* **2017**, *31*, 8217–8227. [CrossRef]
41. Fang, T.W.; Asyikin, N.; Syasya, N.S.; Shawkataly, A.K.H.; Kassim, M.H.M.; Syakir, M.I. Water Absorption and Thickness Swelling of Oil Palm Empty Fruit Bunch (OPEFB) and Seaweed Composite for Soil Erosion Mitigation. *J. Phys. Sci.* **2017**, *28*, 1–17. [CrossRef]
42. Khalid, M.; Ratnam, C.T.; Luqman, C.A.; Salmiaton, A.; Choong, T.S.Y.; Jalaludin, H. Thermal and Dynamic Mechanical Behavior of Cellulose- and Oil Palm Empty Fruit Bunch (OPEFB)-Filled Polypropylene Biocomposites. *Polym.-Plast. Technol. Eng.* **2009**, *48*, 1244–1251. [CrossRef]
43. Banat, R.; Fares, M.M. Thermo-Gravimetric Stability of High Density Polyethylene Composite Filled with Olive Shell Flour. *Am. J. Ploym. Sci.* **2015**, *5*, 65–74.
44. Yahiaoui, M.; Hadoun, H.; Toumert, I.; Hassani, A. Determination of kinetic parameters of Phlomis bovei de Noé using thermogravimetric analysis. *Bioresour. Technol.* **2015**, *196*, 441–447. [CrossRef] [PubMed]

45. Wang, X.; Ma, D.; Jin, Q.; Deng, S.; Stančin, H.; Tan, H.; Mikulčić, H. Synergistic effects of biomass and polyurethane co-pyrolysis on the yield, reactivity, and heating value of biochar at high temperatures. *Fuel Process. Technol.* **2019**, *194*, 106127. [CrossRef]
46. Romero Millán, L.M.; Sierra Vargas, F.E.; Nzihou, A. Kinetic Analysis of Tropical Lignocellulosic Agrowaste Pyrolysis. *BioEnergy Res.* **2017**, *10*, 832–845. [CrossRef]
47. Nyakuma, B.B. Kinetic analysis of oil palm empty fruit bunch (OPEFB) pellets as feedstock for pyrolysis. *PeerJ PrePrints* **2015**, *3*, e1150v1.
48. Wang, S.; Dai, G.; Yang, H.; Luo, Z. Lignocellulosic biomass pyrolysis mechanism: A state-of-the-art review. *Prog. Energy Combust. Sci.* **2017**, *62*, 33–86. [CrossRef]
49. Zhang, Y.; Chen, P.; Liu, S.; Peng, P.; Min, M.; Cheng, Y.; Anderson, E.; Zhou, N.; Fan, L.; Liu, C.; et al. Effects of feedstock characteristics on microwave-assisted pyrolysis—A review. *Bioresour. Technol.* **2017**, *230*, 143–151. [CrossRef]
50. Uzoejinwa, B.B.; He, X.; Wang, S.; El-Fatah Abomohra, A.; Hu, Y.; Wang, Q. Co-pyrolysis of biomass and waste plastics as a thermochemical conversion technology for high-grade biofuel production: Recent progress and future directions elsewhere worldwide. *Energy Convers. Manag.* **2018**, *163*, 468–492. [CrossRef]
51. Hassan, H.; Lim, J.K.; Hameed, B.H. Recent progress on biomass co-pyrolysis conversion into high-quality bio-oil. *Bioresour. Technol.* **2016**, *221*, 645–655. [CrossRef]
52. Soganciogly, M.; Yel, E.; Ahmetli, G. Pyrolysis of waste high density polyethylene (HDPE) and low density polyethylene (LDPE) plastics and production of epoxy composites with their pyrolysis chars. *J. Clean. Prod.* **2017**, *165*, 369–381. [CrossRef]
53. Scarfato, P.; Incarnato, L.; Di Maio, L.; Dittrich, B.; Schartel, B. Influence of a novel organo-silylated clay on the morphology, thermal and burning behavior of low density polyethylene composites. *Compos. Part B Eng.* **2016**, *98*, 444–452. [CrossRef]
54. Zattini, G.; Leonardi, C.; Mazzocchetti, L.; Cavazzoni, M.; Montanari, I.; Tosi, C.; Benelli, T.; Giorgini, L. Pyrolysis of Low-Density Polyethylene. In *Sustainable Design and Manufacturing 2017*; SDM 2017; Smart Innovation, Systems and Technologies; Campana, G., Howlett, R., Setchi, R.C.B., Eds.; Springer: Cham, Switzerland, 2017; pp. 480–490.
55. Li, B.; Chen, G.; Zhang, H.; Sheng, C. Development of non-isothermal TGA–DSC for kinetics analysis of low temperature coal oxidation prior to ignition. *Fuel* **2014**, *118*, 385–391. [CrossRef]
56. Ozbas, K.E.; Kök, M.V.; Hicyilmaz, C. DSC study of the combustion properties of turkish coals. *J. Therm. Anal. Calorim.* **2003**, *71*, 849–856. [CrossRef]
57. Mohan, D.J.; Kullová, L. A study on the relationship between preparation condition and properties/performance of polyamide TFC membrane by IR, DSC, TGA, and SEM techniques. *Desalin. Water Treat.* **2013**, *51*, 586–596. [CrossRef]
58. Boycheva, S.; Zgureva, D.; Vassilev, V. Kinetic and thermodynamic studies on the thermal behaviour of fly ash from lignite coals. *Fuel* **2013**, *108*, 639–646. [CrossRef]
59. Zhou, L.; Zou, H.; Wang, Y.; Le, Z.; Liu, Z.; Adesina, A.A. Effect of potassium on thermogravimetric behavior and co-pyrolytic kinetics of wood biomass and low density polyethylene. *Renew. Energy* **2017**, *102*, 134–141. [CrossRef]

Article

Fabrication of Green Superhydrophobic/Superoleophilic Wood Flour for Efficient Oil Separation from Water

Xuefei Tan [1,2], Deli Zang [3], Haiqun Qi [1], Feng Liu [3], Guoliang Cao [4] and Shih-Hsin Ho [4,*]

1 Heilongjiang Institute of Technology, College of Materials and Chemical Engineering, Harbin 150050, China
2 Dalian SEM Bio-Engineering Technology Co., Ltd., Dalian 116620, China
3 Key Laboratory of Bio-based Material Science and Technology, Ministry of Education, Northeast Forestry University, Harbin 150040, China
4 State Key Laboratory of Urban Water Resource and Environment, School of Environment, Harbin Institute of Technology, Harbin 150090, China
* Correspondence: stephen6949@hit.edu.cn or stephen6949@msn.com

Received: 15 May 2019; Accepted: 19 June 2019; Published: 2 July 2019

Abstract: The removal of oil from waste water is gaining increasing attention. In this study, a novel synthesis method of green superhydrophobic/superoleophilic wood flour is proposed using the deposition of nano–zinc oxide (nZnO) aggregated on the fiber surface and the subsequent hydrophobic modification of octadecanoic acid. The as-prepared wood flour displayed great superhydrophobicity and synchronous superoleophilicity properties with the water contact angle (WCA) of 156° and oil contact angle (OCA) of 0° for diesel oil. Furthermore, the as-prepared wood flour possessed an excellent stability, probably due to the strong adhesion of nZnO, which aggregates to the fiber surface of wood flour with the action of glutinous polystyrene. The maximum adsorption capacity of as-prepared wood flour was 20.81 g/g for engine oil, which showed that the as-prepared wood flour is a potential candidate as an efficient oil adsorbent in the field of water-oil separation. Moreover, it has good chemical steadiness and environmental durability. Taken together, all the information acquired from this research could be valuable in evaluating the potential of as-prepared wood flour as a competitive and sustainable oil-water separation material.

Keywords: wood flour; oil adsorption; superhydrophobic; superoleophilic; oil-water separation; sustainable material

---

1. Introduction

With the rapid development of modern industry, the growing crisis of global water pollution is severely affecting the environment [1–5]. A representative case is the explosion at British Petroleum's Deepwater Horizon oil rig in 2010, which resulted in the loss of life and property, as well as the spillage of huge amounts of oil into the ocean [6]. To date, various materials and approaches have been adopted to remove spilled oil from water bodies, including physical diffusion [7], activated carbon [8], oil containment booms [9], exfoliated graphite [10], waste barley straw [11,12] and membranes [13,14]. Nevertheless, these traditional techniques have certain deficiencies, such as being time-consuming, economically infeasible, environmentally damaging and non-renewable. Therefore, the exploration and use of novel green materials, which can effectively separate oil contaminants from water, is highly desirable. This is not only important for environmental protection, but also for sustainable urban development.

In recent years, a number of self-cleaning plant surfaces have drawn significant research attention. The superhydrophobic nature of lotus leaves is a good example of this [15–18]. In general, the water contact angle is a critical index used to characterize the wettability of a surface. Due to their remarkable waterproofing performance, superhydrophobic materials have been prepared through various approaches, involving a sol-gel process [19], vapor phase deposition [20], chemical etching [21,22], surface fluorination [23] and electrospinning [24]. To date, a number of advanced materials with superhydrophobic and superoleophilic properties have been synthesized and applied to the disposal of spilled oil [25–29]. For example, Wang et al. [30] developed a facile electrochemical deposition method to prepare a novel functional micro-nano hierarchical structured copper mesh film with special superhydrophobic and superoleophilic characteristics for the effective removal of oil from water. Zhang and Seeger [31] successfully synthesized superhydrophobic and superoleophilic polyester textiles using silicone nanofilaments using one-step growth, which could be used for oil/water separation. Cortese et al. [32] described a convenient approach to fabricate cotton textiles with superhydrophobic and superoleophilic properties using plasma-enhanced chemical vapor deposition. Yue et al. [33] developed a kind of superhydrophobic cellulose/LDH (layered double hydroxide) membrane and applied it in an open oil/water two-phase system. Li et al. [34] successfully synthesized superhydrophobic/superoleophilic cotton fabrics combined with polyvinylsilsesquioxanes polymer and nano–Al_2O_3 particles.

Wood flour is one of the most common forms of agricultural waste, and when burnt, can cause severe environmental problems, such as dust or air pollution. There is an urgent need to reuse waste wood flour in a sustainable way, which could alleviate these problems. Wood flour has the advantages of having a low density, being bio-degradable, eco-friendly and low-cost [35]. However, the synthesis of superhydrophobic and superoleophilic materials using waste wood flour has received little research attention. In this work, wood flour was evenly coated with a layer of octadecanoic acid-modified nZnO aggregates with the assistance of glutinous polystyrene, which enabled it to acquire an excellent performance with regards to both superhydrophobicity and superoleophilicity. The superhydrophobic/superoleophilic wood flour with a WCA of 156° and OCA of 0° was prepared for an efficient oil removal from contaminated wastewater. Regardless of the oil or organic solvent, the maximum oil adsorption capacity of the as-prepared wood flour was almost three-fold higher than the pristine wood flour. The adsorption efficiency ranged from 98% to 100%. After the adsorbed oils were removed using acetone, the superhydrophobic/superoleophilic wood flour could be reused multiple times with a good chemical stability and environmental durability.

2. Materials and Methods

2.1. Materials and Chemicals

The wood flour used in this study was obtained from the Larch species (*Larix gmelini*), provided by the Carpentry Laboratory of Northeast Forestry University, China. Hydrogen peroxide, sodium hydroxide, hydrochloric acid, octadecanoic acid and anhydrous ethanol were provided by Tianjin Kaitong Chemical Reagent Co., Ltd., Tianjin, China. Zinc nitrate was purchased from Tianjin Hengxing Chemical Reagent Manufacturing Co., Ltd., Tianjin, China. Polystyrene (Mw = 101,900) was purchased from Shanghai xibao Biological Technology Co., Ltd., Shanghai, China. Methylbenzene was obtained from Tianli Chemical Co., Ltd., Guangdong, China. The gasoline, crude oil, diesel, engine oil, chloroform, n–hexane and toluene were purchased from Tianjin Hengxing Chemical Reagent Manufacturing Co., Ltd., Tianjin, China. All chemical reagents were of analytical grade, and used without further purification.

2.2. Pretreatment of Wood Flour

Raw wood flour was first sifted through a filter screen with a uniform grading smaller than 74 μm. Then, 1.5 g wood flour was dipped into 7 mL hydrogen peroxide (30%) and 200 mL sodium hydroxide aqueous solution (0.5 wt.%) at room temperature under constant stirring for 14 h. After that, hydrochloric acid (6 mol/L) was added to adjust the pH within the range of 6.5–7.0. Finally, the wood flour was washed several times using ultrapure water and dried at 50 °C for 3 h until the weight became constant.

2.3. Preparation of nZnO Particles

2.4 g sodium hydroxide and 200 mL ultrapure water were added to a 250 mL round-bottom flask in a water bath with constant magnetic stirring at 70 °C for 10 min. Then, 2.0 g of zinc nitrate was added to the sodium hydroxide solution and continuously stirred for 8 h. The solution was placed at room temperature for some time. Next, the non-reacted reagents or by-products were removed using ultrapure water and anhydrous ethanol. The final solution was dried for 5 h in a vacuum oven at 60 °C. Finally, the particles of nZnO were obtained as white powder.

2.4. Fabrication of the Green Superhydrophobic/Superoleophilic Wood Flour

0.1 g wood flour and 0.1 g nZnO particles were mixed and immersed in a beaker containing 10 mL ethanolic octadecanoic acid solution (5%, m/v). The mixture was maintained at 70 °C for 5 h, washed with anhydrous ethanol and dried in an oven at 60 °C for 4 h to obtain octadecanoic acid modified-nZnO nanoparticles coated on the wood flour surface. Afterwards, the modified-nZnO wood flour composites were mixed with 10 mL (2%, m/v) of polystyrene methylbenzene solution. Finally, the superhydrophobic/superoleophilic wood flour was dried at 50 °C for 2 h until the weight became constant.

The mechanism of the interaction of nanoparticles, coated by octadecanoic acid with wood flour fibers, was as follows. Hydroxyl groups on the surfaces of ZnO particles interacted with the hydroxyl groups on the surfaces of wood flour fibers through the formation of hydrogen bonds, thus making the ZnO particles homogeneously cover the wood flour surface. Octadecanoic acid chemically reacted with ZnO to generate modified-ZnO particles, as shown in Figure 1b,c. Polystyrene reagent served as a binder to cover the surfaces of the modified-ZnO particles and wood flour fibers, which resulted in a firm adhesion of nanoparticles onto the fiber surface. Figure 1a shows the generation of ZnO aggregates on the sample surface. In fact, the ZnO nanoparticles on the surface of the final product existed as ZnO aggregates.

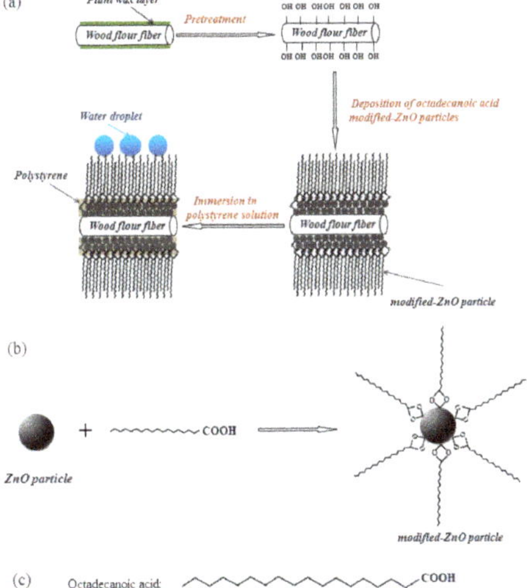

Figure 1. (**a**) Synthesis route for the preparation of superhydrophobic/superoleophilic wood flour. (**b**) Graphical representation for the modification of the ZnO particle with octadecanoic acid. (**c**) Chemical structure of octadecanoic acid.

2.5. Separation of Oil/Water Mixtures

0.5 g of as-prepared wood flour was added to a beaker containing a 150 mL mixture of water and diesel oil, which was dyed red using Sudan III in order to easily and clearly observe the phenomenon. After the adsorption, the red wood flour was recovered on the surface of the water.

The oil adsorption capability was defined using Equation (1):

$$Q = (m2 - m1)/m1 \tag{1}$$

where Q is the oil adsorption capability (g/g); $m2$ is the weight of the wood flour after adsorption (g), and $m1$ is the initial weight of the wood flour before adsorption (g).

2.6. Characterization of the Green Superhydrophobic/Superoleophilic Wood Flour

2.6.1. The Surface Morphologies of Wood Flour Sample

Scanning electron microscopy (SEM, FEI QUANTA200, Hillsborough, Oregon, USA) was used to examine the surface morphologies of the pristine and superhydrophobic/superoleophilic wood flours under the condition that all specimens were pre-coated with a layer of gold.

2.6.2. The Chemical Compositions of the Wood Flour Samples

The chemical compositions of the wood flour samples were investigated using Fourier transform infrared spectroscopy (FT-IR, Magna-IR 560, Nicolet, Madison, Wisconsin, USA), X-ray photoelectron spectrometry (XPS, PHI Thermo Fisher Scientific, Waltham, MA, USA) and energy-dispersive X-ray analysis (EDX, Quantax 70, Billerica, MA, USA). The FT-IR spectra of the wood flour samples were obtained by direct transmittance using the potassium bromide (KBr) pellet technique. For each sample, the wavenumber was measured within the range of 500–3400 cm^{-1}, and the spectrum was accumulated from a total of 32 co-added scans at a spectral resolution of 4 cm^{-1}. The preparation of the samples for the FT-IR measurement was performed by fully grinding the mixture of 2 mg wood flour and 200 mg spectroscopic grade KBr powder, which were pressed into a pellet with a diameter of 15 mm. Before analysis, the background spectrum of pure KBr was recorded. For the infrared spectroscopic analysis, the pellets were analyzed directly.

2.6.3. The Water/Oil Contact Angle

A contact angle instrument (CA-A, Hitachi, Tokyo, Japan) was used to measure the water contact angle (WCA) and oil contact angle (OCA). For these measurements, 5 μL of ultrapure water or an oil droplet was dropped at five different locations under ambient conditions. The values of WCA and OCA were obtained as averages of five repeated measurements.

3. Results and Discussion

3.1. Morphology of the Green Superhydrophobic/Superoleophilic Wood Flour

It has been widely reported that the superhydrophobic property of certain materials is primarily due to their dualistic micro/nano surface structures [36]. As a consequence, it is vital to characterize the surface morphologies of samples before and after the treatments. The morphologies of pristine and superhydrophobic/superoleophilic wood flours were characterized using SEM at different magnifications, as shown in Figure 2. According to Figure 2a–d, after the treatment with octadecanoic acid modified-nZnO particles and polystyrene, the integrity of the fiber-like structure of the wood flour survived, which means that the structures of fibers were the same for both the pristine and as-prepared wood flours. In addition, compared with the pristine wood flour, the as-prepared superhydrophobic/superoleophilic wood flour possessed a rougher surface because of the deposition of nZnO aggregates, which had an average diameter of approximately 80 nm on the surface of the fibers (Figure 2b,d). A hierarchical structure is thus presented on the surface of the superhydrophobic/superoleophilic wood flour, which included the micron fibers and abundant nanoparticles on each fiber. Due to this structure, the as-prepared wood flour achieved enough roughness to bring about the co-existing features of superhydrophobicity and superoleophilicity. The results showed that the presence of nZnO aggregates played a prominent role in the preparation of the superhydrophobic/superoleophilic wood flour. On this basis, and in combination with the modification of an octadecanoic acid (low-surface energy material), a large amount of air could be trapped in the cavities and interspaces in the as-obtained wood flour surface. Then, as soon as a water droplet was dripped onto the surface of a sample, it would come into contact with the trapped air and bounce off without any residue. The superhydrophobicity of wood flour was due to the combination of trapped air and low-surface energy material, which prevented the adsorption of water at the interface.

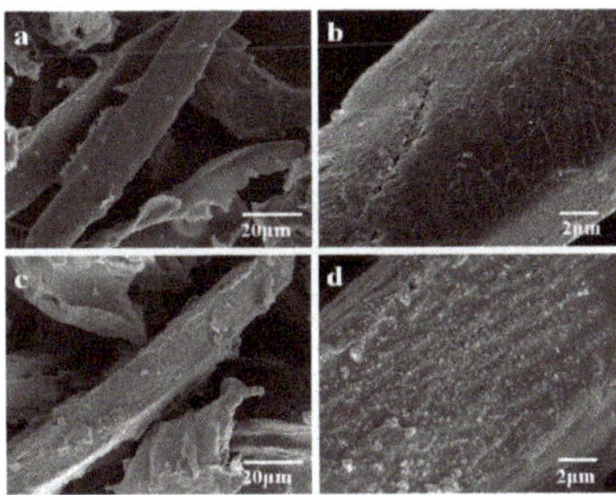

Figure 2. SEM images of pristine wood flour. (**a**,**b**) without nZnO aggregates and superhydrophobic/superoleophilic wood flour, and (**c**,**d**) with nZnO aggregates at low and high magnifications.

3.2. Surface Wettability of the Green Superhydrophobic/Superoleophilic Wood Flour

In order to verify the superhydrophobicity and superoleophilicity of the as-obtained wood flour sample, its surface wettability was investigated by measuring the values of the water/oil contact angles using a contact angle device at room temperature. As shown in Figure 3a, the water contact angle of pristine wood flour was almost 0°, indicating that abundant hydroxyl groups were present on the fiber surfaces. With regards to the sample treated with octadecanoic acid modified-ZnO, the water contact angle reached 120° (Figure 3c) and therefore had a hydrophobic surface. Furthermore, after coating with octadecanoic acid modified-nZnO and polystyrene (Figure 3b), the water droplet on the as-prepared wood flour became spherical, and the water contact angle became as high as 156°. Additionally, the scattering of the contact angle along the sample was 5°, showing that the as-prepared wood flour had excellent superhydrophobic properties. In addition, the oil wettability of the as-prepared wood flour surface was also examined using oil contact angle measurements. When the oil droplets fell onto the surface of the resulting product, all the oil droplets (diesel oil, gasoline and kerosene) were immediately adsorbed onto the treated wood flour at the interface (see Figure 3d), indicating that the oil contact angles of the as-prepared wood flour surface were 0°. A hierarchical structure is presented on the surface of the superhydrophobic/superoleophilic wood flour, including the micron fibers and abundant nanoparticles on each fiber. Due to this structure, the as-prepared wood flour achieved enough roughness to bring about the co-existing features of superoleophilicity. For the same oil or organic solvent, the maximum oil adsorption capacity of the as-prepared wood flour was almost three times that of the pristine wood flour, indicating the enhancement of superoleophilicity. In short, the dual basic properties of the as-prepared wood flour with superhydrophobicity and superoleophilicity were clearly confirmed.

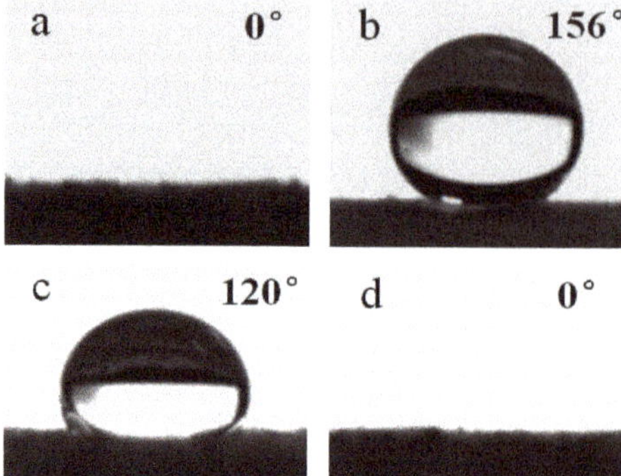

Figure 3. Images of a liquid droplet on different surfaces: typical photograph of a 5 μL water droplet on the surfaces of (**a**) pristine wood flour and (**b**) superhydrophobic/superoleophilic wood flour; (**c**) a water droplet on a wood flour surface treated with octadecanoic acid; (**d**) an oil droplet on a superhydrophobic/superoleophilic wood flour surface.

On the basis of the theoretical principle of surface wettability, the wettability of a material is a synergistic effect of the chemical composition and surface morphology [37]. The combination of the deposition of numerous nZnO aggregates and the surface decoration through low surface energy octadecanoic acid was required to produce the features of superhydrophobicity and superoleophilicity of wood flour. As a result, when a water droplet was dropped onto the as-prepared wood flour, it bounced off, leaving almost no trace of water. Overall, a novel wood flour with a very good superhydrophobic and superoleophilic performance was successfully obtained in this study.

3.3. Chemical Composition Analysis of the Green Superhydrophobic/Superoleophilic Wood Flour

In order to prove the generation of octadecanoic acid modified-nZnO and polystyrene molecules on the as-prepared wood flour's surface, FT-IR, XPS and EDX were employed to investigate the chemical components of as-prepared wood flour.

The typical FT-IR spectra of the wood flour, coated with octadecanoic acid modified-nZnO and the as-prepared wood flour, are presented in Figure 4. In both the spectra, the absorption peaks at 2954 cm^{-1} and 1398 cm^{-1} stemmed from asymmetrical stretching vibrations and symmetrical bending vibrations of $-CH_3$, respectively. In addition, the peaks at 2916 cm^{-1} and 2846 cm^{-1} were attributed to asymmetrical stretching vibrations and symmetrical stretching vibrations of $-CH_2$, respectively. All four characteristic peaks confirmed that long alkyl-chains existed on the surface of the wood flour coated with modified-nZnO. The absorption peaks at 1537 cm^{-1} were due to –COOH– stretching vibrations, whereas those at 1464 cm^{-1} were due to –COOH– bending vibrations and were induced by the $CH_3(CH_2)_{16}COO-$ groups. The presence of octadecanoic acid on both samples' surfaces was evidenced by the characteristic peaks present in both the spectra. In the high frequency region of Figure 4b, the bands at 3024 cm^{-1} and 3060 cm^{-1} were assigned to the C–H stretching vibrations of benzene ring groups, which were introduced by polystyrene. In the low frequency region of Figure 4b, two typical bands at 696 cm^{-1} and 746 cm^{-1} were due to the C–H bending vibrations of benzene ring groups of polystyrene. Moreover, compared to Figure 4a, in accordance with the characteristic absorption peaks of the benzene rings from polystyrene, the bands at 1599 cm^{-1} and 1452 cm^{-1} were visible in Figure 4b. Based on these observations, it was confirmed that nZnO was successfully functionalized by octadecanoic acid, and that polystyrene was successfully used as an adhesive agent to stick octadecanoic acid-modified nZnO to the surface of the as-prepared wood flour.

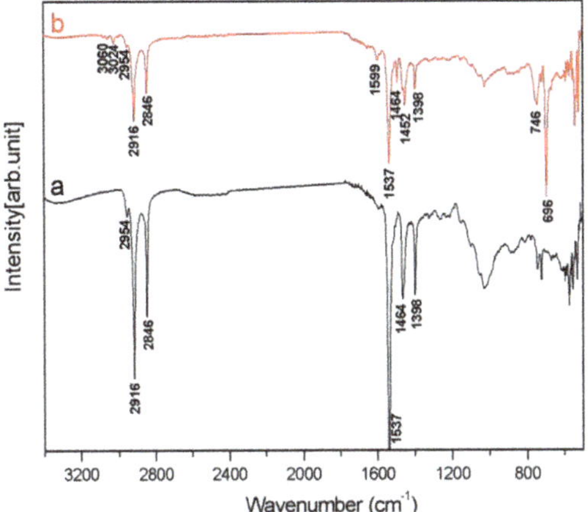

Figure 4. FT-IR spectra of (**a**) wood flour coated with octadecanoic acid modified-nZnO and (**b**) as-prepared superhydrophobic/superoleophilic wood flour.

The XPS spectra of the pristine and superhydrophobic/superoleophilic wood flours are shown in Figure 5. For the pristine wood flour (Figure 5a), peaks corresponding to C 1 s and O 1 s were observed. In comparison, the XPS spectra of the as-prepared wood flour contained a new Zn 2p peak, which accounted for the generation of ZnO. The oxygen content would change when nZnO was introduced into the surface of the wood flour. Based on the XPS spectra, the oxygen intensity in Figure 5a was higher than that in Figure 5b. Therefore, the changes in the C/O proportion were calculated based on the changes in the C and O intensities. In addition, the accurate C/O proportion changed from 65/35% to 83/13%, which resulted in an increase in the C content. In summary, it is inferred that the successful grafting of ZnO with octadecanoic acid was carried out in this study.

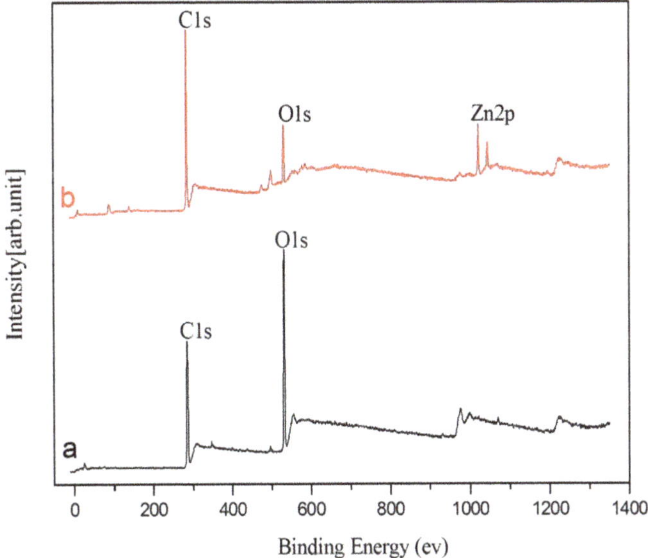

Figure 5. XPS spectra of the (**a**) pristine wood flour and (**b**) as-prepared wood flour.

In addition to the characterizations of FT-IR and XPS, the elemental composition of the as-prepared wood flour was analyzed using EDX, and the results are shown in Figure 6. The carbon (C) peak and oxygen (O) peak observed in both the spectra are attributed to the sustainable wooden material. Notably, in comparison with the pristine wood flour, there was a new peak of zinc (Zn) in the as-prepared material, which provided clear evidence of the existence of nZnO on the surface of the superhydrophobic/superoleophilic wood flour.

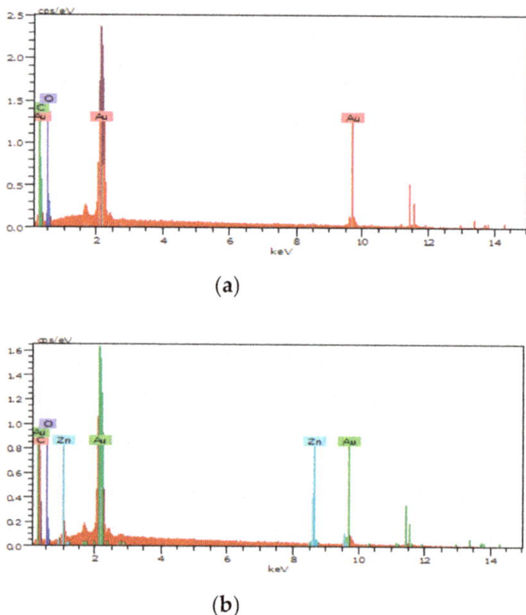

Figure 6. EDX spectra of the (**a**) pristine wood flour and (**b**) as-prepared superhydrophobic/superoleophilic wood flour.

3.4. Steadiness and Durability Analysis of the Proposed Superhydrophobic/Superoleophilic Wood Flour

In order to enhance the economic feasibility of the proposed superhydrophobic/superoleophilic wood flour, its environmental durability and chemical steadiness were analyzed. The chemical steadiness was assessed by recording the changes in the contact angles of corrosion solutions on the as-prepared wood flour (see Figure 7a). The as-prepared wood flour reacted with the aqueous solutions at various pH values ranging from 0 to 14 at room temperature for 24 h, and the changes in its contact angles were detected. From Figure 7a, the water contact angles on the sample surface changed within a very narrow range and remained higher than 150° (152–156°). Meanwhile, the oil contact angle remained at 0°. As such, the great superhydrophobicity and superoleophilicity of the as-prepared wood flour was retained in the corrosive solutions. Moreover, the environmental durability was examined under ambient conditions over 150 days. During the storage period, there were no apparent changes in the values of the contact angles of the as-prepared wood flour, showing that the superhydrophobic/superoleophilic wood flour possessed a remarkable environmental durability (Figure 7b). It is therefore concluded that the superhydrophobic/superoleophilic wood flour prepared in this study has a favorable chemical steadiness and environmental durability, which are due to the viscosity of the polystyrene present on the surface of the as-prepared wood flour fibers.

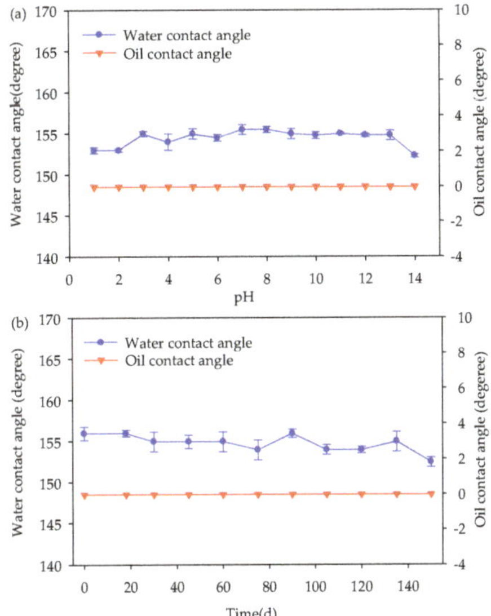

Figure 7. (**a**) The contact angles of the superhydrophobic/superoleophilic wood flour with different pHs of aqueous solution; (**b**) the contact angles of the superhydrophobic/superoleophilic wood flour for the storage evaluation in an ambient environment.

3.5. Application of the Green Superhydrophobic/Superoleophilic Wood Flour in Water-Oil Separation

In view of its dramatic chemical steadiness and environmental durability, the as-prepared wood flour has a great potential for application to oil-containing wastewaters. Figure 8 illustrates the procedure for water-oil separation using the superhydrophobic/superoleophilic wood flour as an oil sorbent. A certain amount of as-prepared wood flour was placed in the mixture of water and diesel oil, which was dyed red using Sudan III for ease of observations. As a result, the superhydrophobic/superoleophilic wood flour could effectively adsorb the diesel oil, while completely repelling water. The red diesel oil was fully adsorbed by the as-prepared wood flour within a few seconds. The transparent region on the water surface indicated that the diesel oil was effectively separated from water. After the adsorption, the red wood flour was recovered from the surface of the water.

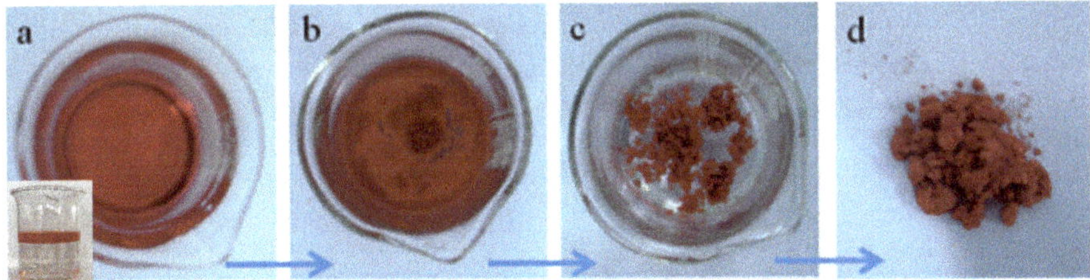

Figure 8. Photographs of the procedure for water-oil separation using the superhydrophobic/superoleophilic wood flour as an oil sorbent: (**a**) water and diesel oil mixture (diesel oil was dyed with Sudan III); (**b**) right after the addition of the as-prepared wood flour; (**c**) after a few seconds, the wood flour filled with red diesel oil floated on the surface of the water; and (**d**) the red oil-adsorbing wood flour recycled from the surface of the water.

The adsorption capacities of the pristine and as-prepared wood flours for oils and organic solvents are shown in Figure 9a.

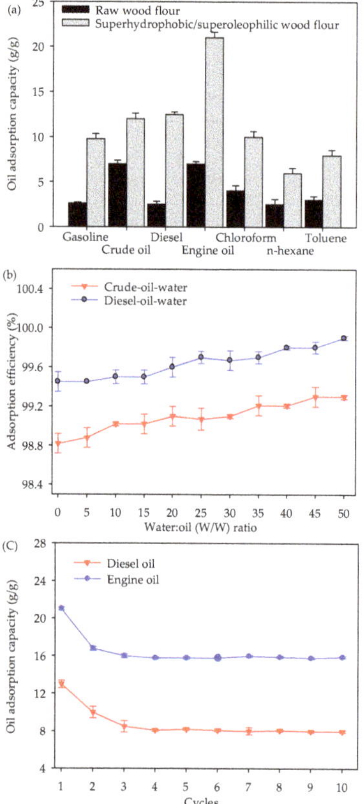

Figure 9. (**a**) The maximum adsorption capacities of the pristine and superhydrophobic/superoleophilic wood flours for various oils and organic solvents at room temperature. (**b**) The variation in adsorption efficiency (%) for diesel oil and crude oil with various mass ratios of water-to-oil. (**c**) The reusability of the superhydrophobic/superoleophilic wood flour for diesel oil and engine oil applications.

For the same oil or organic solvent, the maximum oil adsorption capacity of the as-prepared wood flour was almost three times that of the pristine wood flour, which may be due to the pretreatment of the wood flour and the composites' coating on the surface of the as-prepared wood flour. As such, the oil sorption capacity of the superhydrophobic/superoleophilic wood flour was greatly enhanced. In addition, the oil adsorption efficiency of the as-prepared superhydrophobic/superoleophilic wood flour was examined. The variation in the adsorption efficiency (%) for diesel oil and crude oil for various mass ratios of water-to-oil is shown in Figure 9b. The adsorption efficiency ranged from 98% to 100%, which was mainly potentially due to the wastage of oil with a high viscosity decreasing with the increase in the proportion of water.

The reusability of the proposed material is an important issue with regards to its practical applications. After the adsorbed oils were removed using acetone, the superhydrophobic/superoleophilic wood flour could be reused for many cycles. As seen in Figure 9c, the sorption capacity of the reused wood flour for both diesel and crude oil decreased to 83% and 77% for the second and third cycles, respectively, which was probably due to the trace residual oils left in the fibers of the as-prepared wood flour. The oil adsorption capacity of the superhydrophobic/superoleophilic wood flour remained almost unchanged after three cycles, exhibiting a good reusability. Moreover, after several cycles, the as-prepared wood flour can still be used as an environment-friendly and low-cost material with satisfactory removal efficiencies for diesel and crude oils. Finally, a comparison of the oil adsorption capacities between the proposed superhydrophobic/superoleophilic wood flour and the bio-materials reported in the literature is presented in Table 1. It can be seen that the green superhydrophobic/superoleophilic wood flour had a relatively high oil adsorption capacity, demonstrating that the as-prepared wood flour has a great potential for being used as an oil adsorbent due to its excellent superhydrophobic/superoleophilic characteristics.

Table 1. A comparison of the oil adsorption capacities of various adsorbents.

Sorbent	Adsorption Capacity (g Crude Oil/g Fiber)	Reference
As-prepared wood flour	16.89	This study
Acetylated peat moss	8.00	[2]
Sphagnum Dill	5.80	[38]
Black rice husk ash	6.22	[39]
Chitosan based polyacrylamide	2.30	[40]
Carbonized rice husks	6.00	[41]
Recycled wool-based nonwoven material (RNWM)	11.50	[42]
Superhydrophobic/superoleophilic polyurethane sponge	44.00	[43]

4. Conclusions

In this work, green superhydrophobic/superoleophilic wood flour with a WCA of 156° and OCA of 0° was prepared for an efficient oil removal from contaminated wastewater. The water-resistance and oil-adsorption properties of the as-prepared superhydrophobic/superoleophilic wood flour were comprehensively ascribed to the synergistic effect of octadecanoic acid modified-nZnO on the micron-fiber surface and glutinous polystyrene employed to attach the modified-nZnO aggregates to the surface of fibers. In addition, the as-prepared wood flour could effectively adsorb oil, while completely repelling water, demonstrating that it has a great potential for use as an effective oil adsorbent from wastewater. For the same oil or organic solvent, the maximum oil adsorption efficiency of the proposed wood flour was almost three times that of the pristine wood flour. The adsorption efficiency ranged from 98% to 100% for the proposed wood flour. After the adsorbed oils were removed using acetone, the green superhydrophobic/superoleophilic wood flour could be reused many times. Notably, the superior chemical steadiness and environmental durability of the proposed wood flour with environment-friendly characteristics add to its commercial feasibility.

Author Contributions: S.-H.H. conceived designed the experiments. D.Z. performed the experiments. X.T. performed the experiments and wrote the paper. H.Q. and G.C. analyzed the data. F.L. contributed in materials preparation and their surface analysis. All the experiments were performed under the supervision of S.-H.H. All authors read and approved the final manuscript.

Funding: This research was funded by Heilongjiang Institute of Technology Doctoral Research Fund (2017BJ31).

Conflicts of Interest: Declare conflicts of interest or state "The authors declare no conflict of interest."

References

1. Araji, M.T.; Shakour, S.A. Realizing the Environmental Impact of Soft Materials: Criteria for Utilization and Design Specification. *Mater. Des.* **2013**, *43*, 560–571. [CrossRef]
2. Cojocaru, C.; Macoveanu, M.; Cretescu, I. Peat-Based Sorbents for the Removal of Oil Spills from Water Surface: Application of Artificial Neural Network Modeling. *Colloids Surf. Phys. Eng. Asp.* **2011**, *384*, 675–684. [CrossRef]
3. Feng, B.; Xu, X.; Xu, W.; Zhou, G.; Hu, J.; Wang, Y.; Bao, Z. Self-Assembled 3D ACF–RGO–tio2 Composite as Efficient and Recyclable Spongy Adsorbent for Organic Dye Removal. *Mater. Des.* **2015**, *83*, 522–527. [CrossRef]
4. Li, H.; Liu, L.; Yang, F. Oleophilic Polyurethane Foams for Oil Spill Cleanup. In Proceedings of the 2013 International Symposium on Environmental Science and Technology (2013 ISEST), Dalian, China, 4–7 June 2013; Quan, X., Ed.; Elsevier Science Bv: Amsterdam, The Netherlands, 2013; Volume 18, pp. 528–533.

5. Xu, L.; Li, J.; Brenning, A. A Comparative Study of Different Classification Techniques for Marine Oil Spill Identification Using RADARSAT-1 Imagery. *Remote Sens. Environ.* **2014**, *141*, 14–23. [CrossRef]
6. Al-Majed, A.A.; Adebayo, A.R.; Hossain, M.E. A Sustainable Approach to Controlling Oil Spills. *J. Environ. Manag.* **2012**, *113*, 213–227. [CrossRef] [PubMed]
7. Yu, X.; Luo, Z.; Li, H.; Gan, D. The Diffusion Process of the Dense Medium and Its Effects on the Beneficiation of 0–6 Mm Oil Shale in the Continuous Vibrating Air Dense Medium Fluidized Bed. *Fuel* **2018**, *234*, 371–383. [CrossRef]
8. Chang, W.; Zhao, D.; Lian, J.; Han, S.; Xue, Y. Regeneration of Waste Rolling Oil via Inorganic Flocculation-Adsorption Process. *Pet. Sci. Technol.* **2019**, *37*, 837–844. [CrossRef]
9. Wang, E.; Wang, H.; Liu, Z.; Yuan, R.; Zhu, Y. One-Step Fabrication of a Nickel Foam-Based Superhydrophobic and Superoleophilic Box for Continuous Oil–Water Separation. *J. Mater. Sci.* **2015**, *50*, 4707–4716. [CrossRef]
10. Annunciado, T.R.; Sydenstricker, T.H.D.; Amico, S.C. Experimental Investigation of Various Vegetable Fibers as Sorbent Materials for Oil Spills. *Mar. Pollut. Bull.* **2005**, *50*, 1340–1346. [CrossRef]
11. Nam, H.; Choi, W.; Genuino, D.A.; Capareda, S.C. Development of Rice Straw Activated Carbon and Its Utilizations. *J. Environ. Chem. Eng.* **2018**, *6*, 5221–5229. [CrossRef]
12. Wang, C.; Yao, T.; Wu, J.; Ma, C.; Fan, Z.; Wang, Z.; Cheng, Y.; Lin, Q.; Yang, B. Facile Approach in Fabricating Superhydrophobic and Superoleophilic Surface for Water and Oil Mixture Separation. *ACS Appl. Mater. Interfaces* **2009**, *1*, 2613–2617. [CrossRef]
13. Obaid, M.; Barakat, N.A.M.; Fadali, O.A.; Motlak, M.; Almajid, A.A.; Khalil, K.A. Effective and Reusable Oil/Water Separation Membranes Based on Modified Polysulfone Electrospun Nanofiber Mats. *Chem. Eng. J.* **2015**, *259*, 449–456. [CrossRef]
14. Yang, Y.; Wang, H.; Li, J.; He, B.; Wang, T.; Liao, S. Novel Functionalized Nano-tio2 Loading Electrocatalytic Membrane for Oily Wastewater Treatment. *Environ. Sci. Technol.* **2012**, *46*, 6815–6821. [CrossRef] [PubMed]
15. Latthe, S.S.; Sutar, R.S.; Kodag, V.S.; Bhosale, A.K.; Kumar, A.M.; Kumar Sadasivuni, K.; Xing, R.; Liu, S. Self-Cleaning Superhydrophobic Coatings: Potential Industrial Applications. *Prog. Org. Coat.* **2019**, *128*, 52–58. [CrossRef]
16. Gou, X.; Guo, Z. Superhydrophobic Plant Leaves: The Variation in Surface Morphologies and Wettability during the Vegetation Period. *Langmuir* **2019**, *35*, 1047–1053. [CrossRef] [PubMed]
17. Bai, H.; Zhang, L.; Gu, D. Micrometer-Sized Spherulites as Building Blocks for Lotus Leaf-like Superhydrophobic Coatings. *Appl. Surf. Sci.* **2018**, *459*, 54–62. [CrossRef]
18. Zang, D.; Liu, F.; Zhang, M.; Niu, X.; Gao, Z.; Wang, C. Superhydrophobic Coating on Fiberglass Cloth for Selective Removal of Oil from Water. *Chem. Eng. J.* **2015**, *262*, 210–216. [CrossRef]
19. Li, C.; Ma, R.; Du, A.; Fan, Y.; Zhao, X.; Cao, X. One-Step Fabrication of Bionic Superhydrophobic Coating on Galvanised Steel with Excellent Corrosion Resistance. *J. Alloys Compd.* **2019**, *786*, 272–283. [CrossRef]
20. Rezaei, S.; Manoucheri, I.; Moradian, R.; Pourabbas, B. One-Step Chemical Vapor Deposition and Modification of Silica Nanoparticles at the Lowest Possible Temperature and Superhydrophobic Surface Fabrication. *Chem. Eng. J.* **2014**, *252*, 11–16. [CrossRef]
21. Liao, R.; Zuo, Z.; Guo, C.; Yuan, Y.; Zhuang, A. Fabrication of Superhydrophobic Surface on Aluminum by Continuous Chemical Etching and Its Anti-Icing Property. *Appl. Surf. Sci.* **2014**, *317*, 701–709. [CrossRef]
22. Yuan, Z.; Wang, X.; Bin, J.; Peng, C.; Xing, S.; Wang, M.; Xiao, J.; Zeng, J.; Xie, Y.; Xiao, X.; et al. A Novel Fabrication of a Superhydrophobic Surface with Highly Similar Hierarchical Structure of the Lotus Leaf on a Copper Sheet. *Appl. Surf. Sci.* **2013**, *285*, 205–210. [CrossRef]
23. Minařík, M.; Wrzecionko, E.; Minařík, A.; Grulich, O.; Smolka, P.; Musilová, L.; Junkar, I.; Primc, G.; Ptošková, B.; Mozetič, M.; et al. Preparation of Hierarchically Structured Polystyrene Surfaces with Superhydrophobic Properties by Plasma-Assisted Fluorination. *Coatings* **2019**, *9*, 201. [CrossRef]
24. Liu, Z.; Tang, Y.; Zhao, K.; Zhang, Q. Superhydrophobic sio2 Micro/Nanofibrous Membranes with Porous Surface Prepared by Freeze Electrospinning for Oil Adsorption. *Colloids Surf. Phys. Eng. Asp.* **2019**, *568*, 356–361. [CrossRef]
25. Liu, F.; Ma, M.; Zang, D.; Gao, Z.; Wang, C. Fabrication of Superhydrophobic/Superoleophilic Cotton for Application in the Field of Water/Oil Separation. *Carbohydr. Polym.* **2014**, *103*, 480–487. [CrossRef] [PubMed]
26. Li, H.; Wang, X.; He, Y.; Peng, L. Facile Preparation of Fluorine-Free Superhydrophobic/Superoleophilic Paper via Layer-by-Layer Deposition for Self-Cleaning and Oil/Water Separation. *Cellulose* **2019**, *26*, 2055–2074. [CrossRef]

27. Wang, Q.; Li, Q.; Yasir Akram, M.; Ali, S.; Nie, J.; Zhu, X. Decomposable Polyvinyl Alcohol-Based Super-Hydrophobic Three-Dimensional Porous Material for Effective Water/Oil Separation. *Langmuir* **2018**, *34*, 15700–15707. [CrossRef] [PubMed]
28. Cao, G.; Zhang, W.; Jia, Z.; Liu, F.; Yang, H.; Yu, Q.; Wang, Y.; Di, X.; Wang, C.; Ho, S.-H. Dually Prewetted Underwater Superoleophobic and under Oil Superhydrophobic Fabric for Successive Separation of Light Oil/Water/Heavy Oil Three-Phase Mixtures. *ACS Appl. Mater. Interfaces* **2017**, *9*, 36368–36376. [CrossRef] [PubMed]
29. Cao, G.; Wang, Y.; Wang, C.; Ho, S.-H. A Dually Prewetted Membrane for Continuous Filtration of Water-in-Light Oil, Oil-in-Water, and Water-in-Heavy Oil Multiphase Emulsion Mixtures. *J. Mater. Chem. A* **2019**, *7*, 11305–11313. [CrossRef]
30. Wang, S.; Song, Y.; Jiang, L. Microscale and Nanoscale Hierarchical Structured Mesh Films with Superhydrophobic and Superoleophilic Properties Induced by Long-Chain Fatty Acids. *Nanotechnology* **2006**, *18*, 1–5. [CrossRef]
31. Zhang, J.; Seeger, S. Polyester Materials with Superwetting Silicone Nanofilaments for Oil/Water Separation and Selective Oil Absorption. *Adv. Funct. Mater.* **2011**, *21*, 4699–4704. [CrossRef]
32. Cortese, B.; Caschera, D.; Federici, F.; Ingo, G.M.; Gigli, G. Superhydrophobic Fabrics for Oil-Water Separation through a Diamond like Carbon (DLC) Coating. *J. Mater. Chem. A* **2014**, *2*, 6781–6789. [CrossRef]
33. Yue, X.; Li, J.; Zhang, T.; Qiu, F.; Yang, D.; Xue, M. In Situ One-Step Fabrication of Durable Superhydrophobic-Superoleophilic Cellulose/LDH Membrane with Hierarchical Structure for Efficiency Oil/Water Separation. *Chem. Eng. J.* **2017**, *328*, 117–123. [CrossRef]
34. Li, G.; Mai, Z.; Shu, X.; Chen, D.; Liu, M.; Xu, W. Superhydrophobic/Superoleophilic Cotton Fabrics Treated with Hybrid Coatings for Oil/Water Separation. *Adv. Compos. Hybrid Mater.* **2019**, *15*, 254–265. [CrossRef]
35. Almaadeed, M.A.; Kahraman, R.; Noorunnisa Khanam, P.; Madi, N. Date Palm Wood Flour/Glass Fibre Reinforced Hybrid Composites of Recycled Polypropylene: Mechanical and Thermal Properties. *Mater. Des.* **2012**, *42*, 289–294. [CrossRef]
36. Boinovich, L.; Emelyanenko, A. Principles of Design of Superhydrophobic Coatings by Deposition from Dispersions. *Langmuir* **2009**, *25*, 2907–2912. [CrossRef] [PubMed]
37. Sommers, A.D.; Brest, T.J.; Eid, K.F. Topography-Based Surface Tension Gradients to Facilitate Water Droplet Movement on Laser-Etched Copper Substrates. *Langmuir* **2013**, *29*, 12043–12050. [CrossRef] [PubMed]
38. Olga, V.R.; Darina, V.I.; Alexandr, A.I.; Alexandra, A.O. Cleanup of Water Surface from Oil Spills Using Natural Sorbent Materials. *Procedia Chem.* **2014**, *10*, 145–150. [CrossRef]
39. Vlaev, L.; Petkov, P.; Dimitrov, A.; Genieva, S. Cleanup of Water Polluted with Crude Oil or Diesel Fuel Using Rice Husks Ash. *J. Taiwan Inst. Chem. Eng.* **2011**, *42*, 957–964. [CrossRef]
40. Sokker, H.H.; El-Sawy, N.M.; Hassan, M.A.; El-Anadouli, B.E. Adsorption of Crude Oil from Aqueous Solution by Hydrogel of Chitosan Based Polyacrylamide Prepared by Radiation Induced Graft Polymerization. *J. Hazard. Mater.* **2011**, *190*, 359–365. [CrossRef]
41. Angelova, D.; Uzunov, I.; Uzunova, S.; Gigova, A.; Minchev, L. Kinetics of Oil and Oil Products Adsorption by Carbonized Rice Husks. *Chem. Eng. J.* **2011**, *172*, 306–311. [CrossRef]
42. Radetić, M.M.; Jocić, D.M.; Jovančić, P.M.; Petrović, Z.L.; Thomas, H.F. Recycled Wool-Based Nonwoven Material as an Oil Sorbent. *Environ. Sci. Technol.* **2003**, *37*, 1008–1012. [CrossRef] [PubMed]
43. Li, B.; Liu, X.; Zhang, X.; Zou, J.; Chai, W.; Lou, Y. Rapid Adsorption for Oil Using Superhydrophobic and Superoleophilic Polyurethane Sponge. *J. Chem. Technol. Biotechnol.* **2015**, *90*, 2106–2112. [CrossRef]

Review

A Review on Insights for Green Production of Unconventional Protein and Energy Sources Derived from the Larval Biomass of Black Soldier Fly

Sabrina Hasnol [1], Kunlanan Kiatkittipong [2,*], Worapon Kiatkittipong [3], Chung Yiin Wong [1], Cheng Seong Khe [4], Man Kee Lam [5], Pau Loke Show [6], Wen Da Oh [7], Thiam Leng Chew [8] and Jun Wei Lim [1,*]

1 Department of Fundamental and Applied Sciences, HICoE-Centre for Biofuel and Biochemical Research, Institute of Self-Sustainable Building, Universiti Teknologi PETRONAS, Seri Iskandar 32610, Perak Darul Ridzuan, Malaysia; sabrinahasnol@gmail.com (S.H.); johnsonwcy@gmail.com (C.Y.W.)
2 Department of Chemical Engineering, Faculty of Engineering, King Mongkut's Institute of Technology Ladkrabang, Bangkok 10520, Thailand
3 Department of Chemical Engineering, Faculty of Engineering and Industrial Technology, Silpakorn University, Nakhon Pathom 73000, Thailand; KIATKITTIPONG_W@su.ac.th
4 Department of Fundamental and Applied Sciences, Centre of Innovative Nanostructures and Nanodevices (COINN), Institute of Autonomous Systems, Universiti Teknologi PETRONAS, Seri Iskandar 32610, Perak Darul Ridzuan, Malaysia; chengseong.khe@utp.edu.my
5 Department of Chemical Engineering, HICoE-Centre for Biofuel and Biochemical Research, Institute of Self-Sustainable Building, Universiti Teknologi PETRONAS, Seri Iskandar 32610, Perak Darul Ridzuan, Malaysia; lam.mankee@utp.edu.my
6 Department of Chemical and Environmental Engineering, University of Nottingham Malaysia, Broga Road, Semenyih 43500, Selangor Darul Ehsan, Malaysia; PauLoke.Show@nottingham.edu.my
7 School of Chemical Sciences, Universiti Sains Malaysia, Penang 11800, Malaysia; ohwenda@usm.my
8 Department of Chemical Engineering, CO_2 Research Centre (COSRES), Institute of Contaminant Management, Universiti Teknologi PETRONAS, Seri Iskandar 32610, Perak Darul Ridzuan, Malaysia; thiamleng.chew@utp.edu.my
* Correspondence: kunlanan.kia@kmitl.ac.th (K.K.); junwei.lim@utp.edu.my (J.W.L.); Tel.: +60-5368-7664 (J.W.L.)

Received: 3 January 2020; Accepted: 20 February 2020; Published: 28 April 2020

Abstract: The purpose of this review is to reveal the lipid and protein contents in black soldier fly larvae (BSFL) for the sustainable production of protein and energy sources. It has been observed from studies in the literature that the larval lipid and protein contents vary with the rearing conditions as well as the downstream processing employed. The homogenous, heterogenous and microbial-treated substrates via fermentation are used to rear BSFL and are compared in this review for the simultaneous production of larval protein and biodiesel. Moreover, the best moisture content and the aeration rate of larval feeding substrates are also reported in this review to enhance the growth of BSFL. As the downstream process after harvesting starts with larval inactivation, various related methods have also been reviewed in relation to its impact on the quality/quantity of larval protein and lipids. Subsequently, the other downstream processes, namely, extraction and transesterification to biodiesel, are finally epitomized from the literature to provide a comprehensive review for the production of unconventional protein and lipid sources from BSFL feedstock. Incontrovertibly, the review accentuates the great potential use of BSFL biomass as a green source of protein and lipids for energy production in the form of biodiesel. The traditional protein and energy sources, preponderantly fishmeal, are unsustainable naturally, pressingly calling for immediate substitutions to cater for the rising demands. Accordingly, this review stresses the benefits of using BSFL biomass in detailing its production from upstream all the way to downstream processes which are green and economical at the same time.

Keywords: black soldier fly larvae; protein; lipid; biodiesel; substrate; transesterification

1. Introduction

Fossil fuel holds the position of being the main source of energy consumed in the world. According to the World Energy Forum, the reserves of fossil-based oil, gas and coal, used mainly in the transportation, agriculture, domestic and industrial sectors, will be exhausted in less than a decade. As this main source of energy is rapidly diminishing at an alarming rate, it has accelerated the demands to find an alternative source that serves the same functions. This has lead researchers to consider renewable energy, offering not only improved energy security, but also a chance for the planet to reduce carbon emissions while providing much cleaner air. This in turn will permit the future generation to have a more sustainable green footing in regard to the environment. According to Barnwal and Sharma [1], fuels that are of biological origin, originating from vegetable oils, alcohol, biomass and biogas, are some of the alternatives presented from these past few years as sustainable fuels. Some of these fuels can be used directly, while others may need further modification before the fuels can be used. Biodiesel, one of the alternative fuels that originates from vegetable oils, animal fats and microorganisms such as microalgae, yeast, bacteria and fungi, shows promising results in becoming the main source of energy. For maximum yield, a transesterification process is carried out on the glyceride of the oily sources with alcohol in the presence of a catalyst to form fatty acid alkyl esters and glycerol [2]. However, biodiesel has challenges in implementation due to its high cost and limited availability of resources rising from the food versus fuel issue [3]. This is because the sources were limited to plant and animal feedstock, thereby competing with a food source needed for consumption. Microorganisms then became a new interest in synthesizing biofuels, making microbes such as bacteria, fungi and microalgae the next generation of biodiesel [4]. It was determined that microalgae contained the highest lipid content, over 75% measured relative to dry biomass weight [5]. However, this new source has led to the other problems such as extensive time consumption of medium preparation and intensive energy requirement for harvesting as microalgae are more buoyant and difficult to settle [6].

Thus, to generate biodiesel in a more favorable condition, researchers have suggested to derive the sustainable fuel from insects. Fuels derived from insects through insect farming allow several biochemical products and byproducts to be obtained, including proteins. Biodiesel production from insects has become more favorable since it has been found that insect breeding is economically and environmentally viable. Certain species of insects can easily degrade organic matter, converting organic waste into insect biomass. Insect breeding space is not large compared to the large land areas required for crops such as soybeans or to the large water footprint required for microalgae production. This new alternative has become more feasible, especially for countries with limited space and highly populated areas that need to devote their land for food-source production [7]. Insect larvae can accumulate lipids as their fat body and are able to stimulate the metabolic reserves needed, especially during their immature stages such as larva, pupa and nymph. Insects possesses a nutrient storage system that is used in the metamorphosis process, a structure called the "fat body". This structure is able to accumulate the lipids in the body as fat, which is used as an energy reserve and plays a role in the intermediary metabolism. From the research work conducted by Leong et al. [8], the *Hermetia illucens* larvae, or the black soldier fly larvae (BSFL), has become the ideal candidate in biodiesel production during its larval stage because the adult of the fly has been reported to be missing the mouthparts to feed and relies on food reserves, unlike common houseflies. This means that the black soldier fly is not a vector that can transmit diseases or parasites when feeding. Thus, this species of fly is not considered as a harmful pest, feeding on only kitchen waste, spoiled feed and manure. Recently, this fly, which can be commonly found in poultry- and pig-rearing units, has been found to be able to reduce unpleasant smells as it feeds on the manure or compost, efficiently reducing the polluting compounds from manures and compost. Undesirable bacteria are also reduced by the modification of the bacterial microflora by the BSFL during feeding. The BSFL is a sustainable source for biodiesel production, as the chemical composition of this species is able to accumulate fat, depending on its feeding medium during its rearing process. Upon the lipid extraction for biodiesel production, the residual is a protein-rich larval biomass and can be used as the animal feed to replace fishmeal, which is not sustainable for the long term. Various research studies have been conducted on employing the BSFL biomass as the animal feed for farming of land animals as well as for aquaculture. Figure 1 presents the flow of the present review, starting from the BSFL substrate preparations all the way until the conditions for larval biodiesel production.

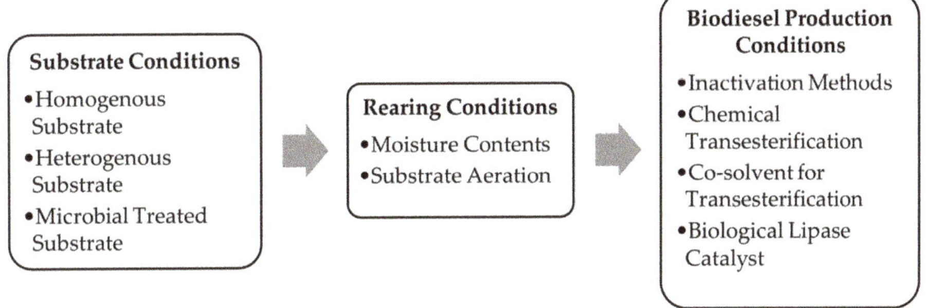

Figure 1. Flow of review, encompassing the larval substrate, rearing and biodiesel production conditions.

2. Homogenous Substrate

With a wide dietary range [9], BSFL has been evaluated for the precision and easy incorporation in formulating its diets that allow sufficient amount of lipid for biodiesel and protein production. Studies had been conducted on feeding BSFL with two different types of feeding mediums, namely, homogenous and heterogenous substrates. The homogenous substrates contain only one kind of organic matter, while the heterogenous substrates incorporate a mixture of two different types of organic matters or more before feeding the BSFL. For the homogenous substrates, there are various type of mediums that have been used to feed BSFL in order to assess its lipid content, biodiesel yield and protein content. Those single substrates are manure, animal feed/food, waste and nutritional meal.

Manure is basically an organic matter originating from the feces of animals, mainly used to fertilize crops. Different type of animals have different consumption of feeds in their diets, affecting the nutritional content of their manure. According to Li et al. [10], the use of cattle manure to feed BSFL would generate the extracted lipid content of 38.2 g, yielding 29.9% amount of fat. The biodiesel produced was 35.6 g and the BSFL that was fed with the cattle manure was able to yield 93% of biodiesel. When pig manure was used, the amount of lipid produced was 60.4 g with the yield of 29.1%, while the biodiesel produced was 57.7 g with the yield of 96%. The amount of lipid produced when chicken manure was used however gave the amount of 98.5 g with the yield of 30.1% and the biodiesel production of 91.4 g with yield of 93%. According to this study, the BSFL fat-based biodiesel fuel properties were comparable to a crop-based fuel, rapeseed. With the amount of crude fat as well as biofuel yield from the transesterification process, the results from this study show that BSFL fat has the potential as a feedstock in biodiesel production.

Other studies were also conducted by Newton et al. [11] in comparing the lipid and protein contents between poultry manure and swine manure. According to their studies, it was found that the lipid content of BSFL was slightly higher when fed with poultry manure, with the yield of 34.8%, while BSFL was able to yield 28% when fed with swine manure. The protein content of BSFL was higher when it was fed with swine manure, with yield of 43.2%, than when it was fed with poultry manure (42.1%). This study showed that BSFL contained a high concentration of oil that would yield as much energy as the methane fermentation that used the same type of manure. The difference in the

BSFL lipid and protein contents when reared by different types of manure reflects how the variation of diet affects the lipid and protein concentration, as it was tested that the other nutrients, except for phosphorus, can be found in slightly higher concentrations when fed with poultry manure. With its high level of oil in the BSFL, it would be likely best to not use BSFL as a bulk protein supplement for animal feedstock, but instead to use it as the potential energy source. According to the study conducted by Lalander et al. [9], poultry manure was fed to BSFL and the crude protein content obtained was 22.8%. It could be deduced that the development of the BSFL growth was dependent on the concentration of the protein of the BSFL. When the feed provides the BSFL with enough protein to accumulate, it will be used as part of the its development, making it consume less energy from its lipid content. However, it will result in a much smaller larva. In the same study, poultry feed gave the protein content of 17.3% and dog food gave 33.9%.

Lalander et al. [9] also investigated the effects on the concentration of crude protein of BSFL when they were fed waste materials. When food waste from local restaurants was used, 22.2% was obtained. Abattoir sheep waste gave 56.3%, human feces gave 35.5%, dewatered wastewater sludge gave 16.9%, sewage sludge gave 31.5% and the digested sludge gave the protein content of 14.7%. According to the protein conversion ratio, pure abattoir waste can have the potential to obtain a higher protein ratio if more carbon was added to allow nutrients in the substrates of the waste to be balanced. The nitrogen content of the waste can also be improved with added carbon as it allows the BSFL to utilize the protein content in a much higher usage during its development. The sludge may have low protein content as it has too few volatile solids. The feeding rate that was regulated to dry matter in this investigation was affected. Human feces has a high ratio, and this may be due to its biomass conversion ratio.

Another type of homogenous substrate was flour protein, as carried out by Arango Gutiérrez et al. [12], which contains proteinic ingredients and high digestibility that has the qualities that make it suitable for providing the right nutritional value in the animal's feed. According to the analysis, it is found that when the flour protein was fed to the BSFL, the larvae had the lipid content of 18.82% with the protein content of 36.98%. This research shows that the feed has potential ingredients to provide energy content.

3. Heterogenous Substrate

Lately, it has been found that the oxidation from the fiber of plants or crops is important factor that contributes greatly towards the metabolic activity of the BSFL. As reported by Li et al. [13], the fibers that exist provide the black soldier fly larvae the sufficient materials and energy required for life activities. Therefore, a balanced nutrition is required in the BSFL diet to ensure that the total conversion efficiency is enhanced; this will, in turn, assist the black soldier fly larvae's digestion of the materials. With a better nutrient balance, a higher yield is due to the synergy of the biological growth established being highly positive. According to Wu Li et al. [14], when corn cob residue was soaked in restaurant wastewater at the optimal soaking condition of 75 °C for 5 h, 23.34% lipid content was able to be produced from the BSFL. The restaurant wastewater was used to soak the corn because of its acidification properties, allowing cellulose hydrolysis which allows the lignocellulose of the corn to degrade easily. Different concentrations of xylose and glucose of a fibrous plant or crops in the BSFL feed greatly influences the insect's dry weight and the lipid content [13]. With xylose being the most abundant carbohydrate derived from fibers, especially corn, it became of great importance to extract the xylose to be able to produce lipid. Without any time lag, BSFL is able to consume both xylose and glucose of a plant, easily transforming it to lipid. When a standard feed with a mixture of 8% of glucose was added, 34.31% lipid was able to be yielded from the BSFL. If 6% xylose was mixed with the standard feed, 34.60% lipid was able to be yielded. This shows that both xylose and glucose are able to yield a good amount of lipid. Thus, when 0.3% glucose and 0.8% xylose were used in the mixture with standard feed, the lipid yield became 33.78%. On the other hand, 97.3% of the glucose and 93.8% of the xylose were able to be successfully converted to lipid as the dynamics changed between the three substrates within only 14 days. Another study showed that the types of substrate that are usually fed to the BSFL have a high concentration of cellulose, hemicellulose and lignin, as the animal's main diet consists of crops or plants. The BSFL do possess guts with microbiome symbioses that are able to digest the cellulose consumed. With the right enzymes available in the BSFL, the cellulose, hemicellulose and lignin can be degraded. The main challenge, however, is when a feed with a high amount of crude fiber is used as the main diet of the BSFL, such as dairy manure. More energy is required to break down the cellulose of the fiber materials, thus reducing the lipid yield for biodiesel production. Therefore, a lower fiber content, such as chicken manure, is used together with the dairy manure as co-digestion with different ratios for the BSFL. The study conducted by Rehman et al. [15] evaluated the performance of the BSFL digestion and with the data obtained was able to develop a co-digestion mixture between dairy manure and chicken manure. With the ratio between dairy manure and chicken manure being 40:60, it was found that this ratio of co-digestion resulted in the larvae with the richest nutrient content, enhancing waste conversion efficiency of the BSFL. This study has shown that the use of organic waste in co-digestion must focus on implementing the process of mixing high fiber content with less-fibrous materials and explore the mechanisms as well as the magnitudes of the effect on the BSFL to ensure biodiesel production.

4. Microbial-Treated Substrates

A study has been conducted with the substrate mixture consisting of dairy manure, chicken manure and bacteria. The use of exogenous bacteria, *Bacillus* strains, assists the BSFL gut microbiome development in more efficiently reducing the waste capacity, utilizing the nutrients of the wastes and enhancing the production in the larval biomass. The ratio of dairy manure to chicken manure was 2:3 and this resulted in the lipid yield of 47.7% and protein yield of 53.9%. However, there was a significant increase in both of the yields when bacteria were added. The lipid yield was 67.8% while the protein yield was 71.2%. This shows that the usage of treatment with microbes utilizes a higher amount of lipid and protein compared to the controlled feed that contains only dairy manure and chicken manure. With the help of cellulose-degrading bacteria, a higher biomass promotes for a higher fat yield is promoted, as they enhance the digestion of the waste materials. Therefore, it is important for the selection of the bacteria in assisting the BSFL to ensure that the lignocellulose-rich waste is able to be managed successfully [16]. Soybean curd residue was also used as the feed of BSFL with the addition of a bacteria, *Lactobacillus buchneri*. The results shown by Somroo et al. [17] indicated that the lipid yields differed between when the feeds were only soybean curd residue (26.1%) or artificial feed (24.3%) and when bacteria were added to the feed, which resulted in an increase of the lipid yield (up to 30%). This gave a similar result for the protein yield as the insect–bacteria symbiosis increased the protein yield from 52.9% (soybean curd residue only) and 50.4% (artificial feed) to a much higher value of 55.3%. With this positive interaction, there is great benefit in the availability of the nutrients, playing a major role in the growth of BSFL, the development of the BSFL gut microbiota and the BSFL's production for digestive enzymes. This also shows that the use of symbiotic bacteria allows the success of the BSFL to adapt to new environments and new food sources while still being able to obtain positive growth and reproduction. When the treated rice straw with 39.7 g of glucose and 25.9 g of xylose underwent a fermentation process with *Saccharomyces cerevisiae*, the residues were mixed with enzymes containing hydrolyzed residues, such as lignocellulose, proteins and reducing sugars. The residue was then fed to the BSFL which underwent lipid extraction to yield a total of 5.2 g of lipid from 200 6-day-old BSFL. Additionally, 4.3 g of biodiesel was able to be produced from 200 BSFL. This shows that the nutritional source from BSFL diets consisting of lignocellulosic biomass can be another potential in lipid as well as biodiesel production. Having similar qualities as plant-based biodiesel, BSFL-based biodiesel is proven to be another alternative source of renewable energy [13]. Restaurant waste is heavily concentrated with lipids and protein. However, this substrate lacks the lignocellulose that the rice straw does not lack. If rice straws are used alone as the feed for the BSFL, the growth will be stilled because of the absence of nutrition. Therefore, using the ratio of restaurant waste to rice straw of 7:3 [18], a mixture was made. Rid-X contains natural bacteria that has the main function of breaking down the cellulase, lipase, protease and amylase of the rice straw as well as the solid waste of the restaurant waste. This helps and increases the efficiency of the conversion for BSFL, degrading the cellulose and hemicellulose much faster. More nutrition from the both of the substrates is available for consumption, and the digestion of the food is aided by the microbes. A total of 43.8 g of biodiesel was able to be produced from 2000 BSFL. The properties of the biodiesel were also investigated, and it was found that the fatty acids of the biodiesel were similar to rapeseed-based biodiesel. Thus, it is shown that the quality of the biodiesel, despite originating from different sources, can still hold a high quality in terms of performances.

From these results, although BSFL contains the microbes that can hydrolyze the cellulose content of the feed, the amount of the microbes in the gut may not be sufficient to digest a much larger amount of feed. Research must continue to test various types of microorganisms in undergoing treatment with the feed of BSFL that contains high amount of fiber. This is to observe the conversion efficiency of the bacteria to obtain a high quality fuel for biodiesel production.

5. Substrate Moisture Contents

According to Barry et al. [19], the conversion of waste to biomass of BSFL can be achieved if the consumption of food waste is given a particular care and attention in ensuring its efficiency. Therefore, different parameters need to be investigated, preponderantly the moisture content of the larval feeding substrate, in pushing towards a successful bioconversion. It was found that when study was carried out using almond hull as the main medium for BSFL, alteration of the moisture content in the hull could directly impact the growth of BSFL [20]. The results showed positive effects on the dry weight (0.013 to 0.46 mg/larvae) as well as the yield of harvested larvae (3.7×10^{-4} to 0.11) as the moisture content was increased from 480 to 680 g kg^{-1}. However, it was found that the larval consumption of hull decreased (from 15% to 13%) with increasing of the moisture content. Other studies that reported the effects of manipulating the moisture content of substrates on larval development presented opposite results. The BSFL had been found to grow bigger in terms of weight and needed less medium for consumption as the moisture content was increased [21]. The larval growth rate was also greatly affected when moisture content was reduced, as reported by Cheng et al. [22]. Using almond hull, as reported by Palma et al. [20], showed much different results, perhaps due to the decrease of diffusion of oxygen into the medium. Oxygen diffusion was limited when the pores of the hull were not air-filled and were blocked by the moisture. This directly impacted the growth of bacterial activity and would disrupt the synergistic potential of microorganisms that contributed in the conversion of hull to larval biomass. Therefore, more study needs to be conducted to observe the trends that affect the larval growth and substrate consumption resulting from manipulation of the moisture content, as the role of microorganisms plays a significant role in bioconversion of insect biomass.

6. Substrate Aeration

Managing the substrate aeration while growing the BSFL can improve the overall larval growth through engineering to acquire the right and suitable environment for medium digestion by larvae. Significant effects when manipulating the aeration content towards the development of larvae had been determined by experiment by Palma [20], using waste from almond as the BSFL feeding medium. It was found that increasing the aeration rate gave rise to a positive fit to the nonlinear regression of the BSFL weight and yield. Accordingly, the maximum larval weight and yield could be achieved at 95% at aeration of 0.57 and 0.05 mL/min g dry weight, respectively; this also contributed greatly towards the consumption of almond hull. It could be deduced that the aeration had a direct impact towards the growth of larvae, and the bed depth of the substrate may play a major role as well. This was because the anaerobic condition occurred from the oxygen utilization by both BSFL and microorganisms surpassing the oxygen being supplied from diffusion from the bed surface. This caused the larvae to migrate elsewhere to obtain nutrients that otherwise could be obtained when the larvae migrated to a deeper depth of substrate. When aeration rate was dropped, the larval consumption of hull was negatively affected as well, showing that the presence of microorganisms could heavily impact the environment for rearing BSFL. Therefore, the oxygen content supplied by aeration should be considered of the great importance in insuring that the growth of BSFL does not negatively affect the rate of waste conversion. Moreover, the calcium content was also investigated, and it was found that increasing oxygen content during larval growth would generally increase the larval calcium content. The consumption of hull by BSFL impacted the uptake of calcium from the almond hull and later affected the larval biomass compositions. According to Liu et al. [23], calcium as the mineral element of BSFL was needed for the cuticle formation. Another study, conducted by Wong et al. [24], showed that the harvesting of BSFL at different calcium or chitin levels could directly affect the lipid content since the accumulated body fat tissues were needed during metamorphosis.

7. Inactivation Methods

Various ways that the BSFL could be inactivated were reported by Larouche et al. [25]. Grinding was the first method of mechanical disruption in which the larvae were homogenized at 15,000 rpm in their study. The larvae were packed under 95% vacuum on high hydrostatic pressure with 600 MPa of pressure treatment. The next method involved heating the BSFL, i.e., via blanching, where the larvae were immersed in boiling water for 40 s. Desiccation was another method of heating the BSFL. This approach required the larvae to be located in the air oven with the temperature set to 60 °C for 30 min. The other type of larval inactivation method was freezing, where the larvae were either frozen at the temperature of −20 °C or −40 °C for one hour. Freezing the larvae also could be completed by using liquid nitrogen in a vacuum package for 40 s. Finally, asphyxiation was the last inactivation method reported by Larouche et al. [25], in which the larvae were initially vacuum packaged and subsequently stored at the temperature of 27 °C for 120 h. The atmosphere was modified either to contain 100% carbon dioxide or nitrogen gas. The larvae were then stored at the temperature of 27 °C for 120 or 144 h. The ether extract for each of the larval samples was conducted using petroleum ether as extraction solvent. It was found that the larval lipid contents were higher when the inactivation method of asphyxiation (CO_2 = 15.9%; N_2 = 16.6%; vacuum = 15.9%) was used as opposed to heating (desiccation = 13.4%; blanching = 14.5%), freezing (−20 °C = 12.8%; −40 °C = 12.4%; liquid nitrogen = 12.6%) or mechanical disruption (grinding = 11.9%; high hydrostatic pressure = 12.0%).

The inactivation method of BSFL was lately studied even further, as it was important to be explored to ensure the investigation of the composition of larval lipid could be exploited for biodiesel production and also support a higher value of uses. The characteristics of larval lipid distributions during the processing and storage of BSFL biomass must be unveiled since there is currently a lack of this information. Caligiani et al. [26] directly related the inactivation method of BSFL with extracted lipid characteristics. The inactivation methods reported were blanching and freezing. In their study, half of the first sample was provided in a frozen condition and stored at the temperature of −20 °C, while the other half was ground and freeze-dried until it reached residual moisture of 10% before being kept at the temperature of −20 °C. The next set of larval samples were obtained alive. The larvae were killed by blanching the prepupae in hot water at the temperature of 100 °C for 40 s before storing at the temperature of −20 °C prior to the lipid analyses. Another live larval sample was stored directly at the temperature of −20 °C until use. During the extraction, the Soxhlet lipid extractor that either used diethyl ether or petroleum ether as the solvent was compared with the use of chloromethane solvent. The results of the first sample inactivated by using freezing before the arrival showed that there was no significant difference of using the different extraction solvents via the Soxhlet method. The next set of samples were obtained alive, and diethyl ether was used, resulting in the lipid yield was 13.0 ± 1% when inactivated by freezing; while using blanching, the lipid yield was 13.3 ± 0.8% from BSFL biomass. This also proved that the BSFL was a good source of lipid, unlike other type of insects. However, as compared with the method employing chloromethane, the larval lipid extracted was slightly lower (9.11%). In the live larval sample that was also inactivated via freezing, it was found that both of these samples had a high free fatty acids content. This also explained that the low lipid yield when using the chloromethane for extraction was due to the loss of fatty acid salts in the aqueous phase. However, when the live BSFL were blanched before being frozen, the loss of fatty acid was negligible. When the freezing method was applied towards the BSFL, the amount of acyl glycerols was drastically reduced, most likely due to the activation of the lipase, releasing the free fatty acids. However, the free fatty acids were not used for biodiesel production as a reaction with acyl glycerols was needed during the transmethylation process. When the BSFL was inactivated by blanching, a thermal pretreatment method, the lipid fraction was observed to be stable as it was mainly composed of triacylglycerols. This may be due to the thermal environment deactivating the lipase activity in the BSFL, as it did not damage or influence the lipid fraction conspicuously, preserving it for transmethylation process in producing biodiesel.

8. BSFL-Based Biodiesel

Black soldier fly larval biomass has become an attractive candidate as a renewable source of energy due to its high lipid content. Transesterification is a process for biodiesel production from larval biomass in which the extracted lipid will react with alcohol. It has become essential to ensure that the lipid conversion during biodiesel production is at its highest efficiency. Surendra et al. [27] carried out an investigation to determine the fatty acid compositions of BSFL fats or lipids for biodiesel production. It was found that the BSFL had a very high amount of lauric acid (44.9% ± 1.5%) as compared with the crop-based biodiesel such as soybean oil (negligible) and palm oil (0.1%), a trait that was significant in terms of biodiesel production. The saturated fat was found to be 67% of total fatty acid while soybean was known to only have 11% and palm oil to have only 37% of total fatty acids. On the other hand, the BSFL had a proportion of 28% fatty acids being of unsaturated fat, lesser than that of soybean (85%) and palm oil (55%). The quality of the biodiesel was known to be greatly affected by the composition of fatty acids in a substrate. In this case, the BSFL-based biodiesel was shown to have a significant amount of saturated fatty acids and a low concentration of unsaturated fatty acids, making it an ideal substrate for a high quality of biodiesel production. Thus, the biodiesel would have a viscosity with much lower value and a more stabilized property in terms of its oxidative state. Additionally, the process of transesterification of larval oil that has been extracted must be efficiently processed in ensuring the biodiesel production is of the highest quality.

9. Transesterification of Larval Lipids

An optimum condition for executing the transesterification of BSFL lipids was investigated by Li et al. [28]. The harvested larval biomasses were initially fed with three different substrates individually, namely, cattle manure, pig manure and chicken manure. There is a two-step process during the conversion of larval lipids into fuel. The first step was the acid-catalyzed esterification of fatty acids. This step was to decrease the amount of acids in the BSFL lipids that were extracted and that acted as the pretreatment for the conversion process. The next step was the typical alkaline-catalyzed transesterification. One of the reaction conditions that was optimized was the esterification temperature for 1 h of reaction time using the methanol to larval lipid ratio of 8:1. It was found that as the temperature was increased from 55 to 85 °C, the conversion of fatty acids in the crude lipids to biodiesel increased from 73% to 92%. This demonstrated a positive relationship between temperature and conversion of fatty acids, as it could be directly related to the efficiency of mass transferred with increasing temperature that caused the crude lipid to be more soluble. Accordingly, the temperature of 75 °C was found to be optimal temperature for the esterification process. Another reaction condition investigated was the molar ratio of methanol to larval crude lipid. It was found that the maximum conversion of fatty acids was achieved at 90% when the optimum molar ratio used was 8:1. When a much lower ratio was used instead, the conversion was found to be incomplete. The reaction time was another reaction condition investigated, and the converted fatty acids were found to be 73% to 90% at the reaction times of 30 and 60 min, respectively. The biodiesels produced from the crude lipids of BSFL fed with chicken manure, pig manure and cattle manure were 91.4%, 57.8% and 35.6%, respectively, through the optimum transesterification condition of 30 min at 65 °C with molar ratio of methanol to lipid at 6:1 while using 0.8% NaOH as the catalyst of the reaction. The biodiesel was tested, and it contained a high percentage of saturated fatty acids at 67.6%. This meant the biodiesel produced would have a high oxidative stability value, an excellent trait for biodiesel storage. The optimal transesterification conditions were further tested and used during the experimentation with BSFL-based biodiesel derived from waste grease of restaurants [29]. The two-step process which consisted of the acid-catalyzed esterification and alkaline-catalyzed transesterification was carried out using 1% H_2SO_4 as the reaction catalyst with reaction temperature set at 75 °C, molar ratio of methanol to lipid at 8:1 and 1 h of reaction time. For the alkaline-catalyzed transesterification, the methanol-to-lipid ratio was kept at 6:1 with 0.8% NaOH as the reaction catalyst. The biodiesel produced was 23.6 g from 1000 g of solid residual fraction of restaurant waste fed to 1000 BSFL. The conversion rate of free fatty acids attained was 91.9%, with the total yield of biodiesel of 2.4%. Li et al. [10] also investigated the conversion of BSFL fat to biodiesel using the dairy manure as the main larval substrate. After conducting the two-step transesterification process, 15.8 g of biodiesel was able to be produced from 1.2 kg of dairy manure. The larval biodiesel also contained 58.2% saturated fatty acids and 39.8% of unsaturated fatty acids with overall quality satisfying the EN 14214 standard.

10. Co-Solvent for Transesterification of Larval Lipids

The conventional way of biodiesel production has generally presented some problems, such as consuming a high amount of energy, that make the process costly. Therefore, a direct transesterification involving fewer steps was suggested and investigated by Nguyen et al. [30]. Methanol was used in prior studies as both the solvent for lipid extraction and reactant for lipid transesterification. However, an excess amount of methanol could weaken the function of catalyst, reducing the yield of biodiesel. In the study by Nguyen et al. [30], co-solvents such as n-hexane, n-pentane, chloroform, acetone and petroleum ether were individually mixed with methanol during the direct transesterification process. With the capability to dissolve long chain triglycerides, these co-solvents showed high potential efficiency in extracting the larval lipid, yielding higher amounts of biodiesel. This would prevent lipid loss during the process as less solvent and energy were consumed. When the solvents were mixed with the methanol at the volume ratio of 1.17:1, a high yield of biodiesel was observed. The use of the mentioned solvent should bring positive effects, as the co-solvents are generally capable in dissolving the lipid and short-chain alcohol used as the homogenous catalyst. It was found that by using n-hexane as the co-solvent, the highest yield of biodiesel (63.37%) could be obtained as compared with acetone (54.83%), chloroform (48.50%) and petroleum ether (35.67%). The effects of volume ratios between the n-hexane and methanol were also investigated and it was observed that biodiesel could be yielded at a higher amount when a much lower volume ratio was used. This was probably because the high methanol content in the reaction would lead to a higher molar ratio between the methanol and lipid, leading to the higher reaction yield. Thus, the optimum volume ratio of n-hexane to methanol was determined to be 1:2. The methanol to biomass ratio of the reaction would increase, promoting the conversion yield. However, the frequency of collision between the lipid and methanol would decrease if excess solvent was used in the reaction. This would, overall, result in the increase of heat and mass transfer resistance, decreasing the conversion yield. Other than that, in a similar study with the presence of free fatty acids from the BSFL fat, the catalyst selected was sulfuric acid. As the catalyst loading was increased from 0.4 to 1.2 mL, the yield of biodiesel had also increased from 48.93% to 65.87%, respectively. The polymerization of unsaturated fatty acids was activated with the presence of excess sulfuric acid at high reaction temperature and long reaction time. The increase in reaction temperature would result in the increase of biodiesel yield (87.67% at temperature of 130 °C). The optimal temperature was found to be 120 °C as, although the extraction efficiency increased with enhancement of reaction rate, there was no significant difference between 120 and 130 °C. The biodiesel yields also increased with the reaction time and managed to reach the equilibrium state at reaction time longer than 90 min. Therefore, the highest biodiesel yield was able to be obtained at 94.14% with the reaction temperature of 120 °C using n-hexane:methanol volume ratio of 1:2, the solvent at 12 mL, the catalyst loaded at 1.2 mL and reaction time of 90 min. The biodiesel yield was also observed to be increased from 4.73% with no co-solvent to 63.37% when n-hexane was used. Therefore, n-hexane was proven to be the suitable co-solvent for the reaction involving larval lipid transesterification.

11. Biological Lipase Catalyst for Transesterification of Larval Lipids

The two-step process of transesterification involves acid-catalyzed esterification and alkali-catalyzed transesterification. However, some complications have been presented when sulfuric acid and sodium hydroxide were used in the process, such as damaging the equipment and complications in removing the dissolved catalysts. Therefore, in the study by Nguyen et al. [31] for biodiesel production, methanol catalyzed with lipase was used during the transesterification of BSFL lipid. In the biodiesel production, several lipases were tested such as Novozym 435, Lipozyme TL-IM, Lipozyme RM-IM and lipase PS. Transesterification was then carried out using 10% lipase mixed with methanol and lipid at the molar ratio of 3:1 and reaction temperature and time of 20 °C and 4 h, respectively. According to the results, the Novozym 435 presented a biodiesel yield of 56.78% as opposed to Lipozyme RM-IM (42.18%), porcine pancreas lipase (23.46%), *Candida rugosa* lipase (22.82%), *Rhizopus oryzae* lipase (21.56%), amano lipase G (13.85%) and amano lipase PS (12.56%). This showed that Novozym 435 had the highest catalytic conversion property and also could be used repeatedly. With the enzyme loading at 20%, the maximum yield for biodiesel could be achieved with the methanol:lipid molar ratio of 6.33. Lipase deactivation, however, may occur when the methanol level exceeded the required amount and would cause a reversal in the aforementioned trend. The yield of the reaction would then decrease. In terms of the reaction temperature, the highest yield could be achieved at 40 °C, as higher temperatures would deactivate the activities of enzymes. Enzyme loading played another part in the reaction, as low loading would affect the temperature and later the biodiesel yield. On the other hand, at high level of enzyme loading the temperature did not bring any significant effect towards the yield. Therefore, any problem rising from the temperature could be overcome by increasing the concentration of enzyme in the reaction. In conclusion, to obtain a maximum yield of biodiesel of 97.65%, the molar ratio of methanol to lipid was set at 6.33:1, with 20% enzyme loading and 26 °C as the reaction temperature for 9.48 h. This experiment had shown positive results, and this should be an encouragement in using the green enzyme-catalyzed process to produce biodiesel.

However, a high level of methanol and ethanol would still cause the deactivation of lipase functions. This was because the absorption of alcohol on the surface immobilized the lipase. Therefore, to overcome this problem, Nguyen et al. [32] investigated the effects of methyl acetate, which is an acyl acceptor that would increase the rate of reaction during transesterification. It was observed that using a high amount of methyl acetate had no effects on the stability as well as the activity of enzymes in replacing the alcohol. To obtain a high amount of biodiesel yield using lipase-catalyzed transesterification (Novozym 435 chosen as catalyst) using methyl acetate as the acyl acceptor for BSFL in the production of biodiesel, several reaction conditions were tested and observed for their effects. The reaction conditions observed were the molar ratio of methyl acetate to lipid and enzyme loading. Biodiesel yield decreased with low ratio, as this increased the loading of enzyme in the reaction. This was because it caused the polymer beads to aggregate with the immobilized enzymes, and this disrupted the mass transfer, enabling the enzyme to react with the oil–water interface flexibly. The conversion yield would be lowered as a result. However, at a higher molar ratio, the enzyme loading would increase, thus increasing the biodiesel yield. The temperature of the reaction was also observed, and it was found that that there was no significant increase of biodiesel yield with the increase in temperature. The optimal ratio of methyl acetate to lipid on the other hand was found to be 12:1. In between the enzyme loading and temperature, the yield would increase as the temperature was increased with any amount of loaded catalyst. However, the deactivation of enzyme occurred at high temperature. Therefore, for Novozym 435, biodiesel yield was produced at the highest amount at 39.5 °C reaction temperature. Concisely, in order to obtain a maximum yield of biodiesel, 12 h of reaction time, 14.64:1 molar ratio of methyl acetate to lipid, 17.58% enzyme loaded and 39.50 °C reaction temperature should be used in a lipase-catalyzed transesterification reaction using Novozym 435 as the catalyst. With the proven high biodiesel yield using the optimized conditions mentioned, BSFL lipid has become a reliable source of energy and can be further developed in future [33,34].

12. Conclusions

In short, this review has demonstrated that the BSFL biomass can be the source of protein and lipid for energy. The rearing conditions of BSFL can be systematically optimized to allow the accumulation of more larval protein and lipid in the fat body. Upon the harvesting, the lipid in the form of larval fat can be extracted and transesterified for producing a mixture of fatty acid methyl esters of biodiesel. In this regard, the extraction and transesterification processes can be optimized as well to maximize the BSFL-based biodiesel production. The residual BSFL biomass after the lipid extraction is a protein-rich larval biomass and can be served as the feedstock for animal feed production. In determining the optimum various larval processing conditions, the preparation of substrates, rearing of BSFL and eventually biodiesel production are vital in ensuring the maximum yield from BSFL biomass. The mixture of substrate added with 6% of xylose prompted 34.60% lipid content from the BSFL biomass. Microbial-treated substrate such as dairy manure and chicken manure mixture inoculated with *Bacillus* strains would yield 67.8% lipid and 71.2% protein. Addition of Rid-X to a mixture of restaurant wastes and rice straw could get 43.8 g of biodiesel from 2000 BSFL. To ensure the optimum conditions during rearing period, the moisture content of the substrate should be in the range of 480 to 680 g/kg. The increase in moisture content would result in the decrease in feed consumption. The aeration of the substrate should also be at 95% to achieve maximum larval weight and yield of dry weight as well as a positive growth of the larvae. To inactivate the mature BSFL for maximum yield, it has been shown that if petroleum ether was used as the extraction solvent during lipid extraction, asphyxiation would result in higher lipid content. If diethyl ether was used in the Soxhlet method, both blanching and freezing inactivation methods could be employed. The optimum transesterification conditions were determined to be at 30 min at 65 °C with molar ratio of methanol to lipid at 6:1 while using 0.8% NaOH as the catalyst in the reaction. During direct transesterification, hexane was recommended as the co-solvent, and sulfuric acid as the catalyst. The reaction temperature should be at 120 °C using hexane:methanol volume ratio of 1:2, the solvent at 12 mL, the catalyst loaded at 1.2 mL and reaction time of 90 min. Novozym 435 can be added in a lipase-catalyzed transesterification reaction with methyl acetate added to replace the methanol and ethanol. Studies have shown that with 12 h of reaction time, 14.64:1 of molar ratio between methyl acetate to lipid, 17.58% enzyme loaded and 39.50 °C of reaction temperature could ensure obtaining the maximum yield of biodiesel. Therefore, since the detailed laboratory-proven information with regard to the upstream and downstream of BSFL biomass production is currently accessible, the mass production of this feedstock at industrial scale should be assessed to unveil its feasibility concerning the cost and long-term environmental sustainability. Lastly, the authors of this review believe that the BSFL biomass could potentially arise as the new and unconventional feedstock for protein in replacing the traditional fishmeal if it is not used for biodiesel.

Author Contributions: Conceptualization, S.H. and J.W.L.; resources, K.K. and W.K.; writing—original draft preparation, S.H. and C.Y.W.; writing—review and editing, M.K.L. and P.L.S.; visualization, W.D.O.; supervision, T.L.C. and C.S.K.; funding acquisition, J.W.L., C.S.K. and K.K. All authors have read and agreed to the published version of the manuscript.

Funding: Financial supports from Yayasan Universiti Teknologi PETRONAS via YUTP-FRG with the cost center 015LC0-282 and Ministry of Education Malaysia under HICoE with the cost center of 015MA0-052 are gratefully acknowledged.

Acknowledgments: One of the authors, Kunlanan Kiatkittipong, wishes to thank the financial support received from the King Mongkut's Institute of Technology Ladkrabang, KMITL with the Grant no. KREF046209.

Conflicts of Interest: The authors declare no conflict of interest.

References

1. Barnwal, B.K.; Sharma, M.P. Prospects of biodiesel production from vegetable oils in India. *Renew. Sustain. Energy Rev.* **2005**, *9*, 363–378. [CrossRef]
2. Boocock, D.G.B.; Konar, S.K.; Mao, V.; Sidi, H. Fast one phase oil-rich processes for preparation of vegetable oil methyl esters. *Biomass Bioenergy* **1996**, *11*, 43–50. [CrossRef]
3. Atabani, A.E.; Silitonga, A.S.; Badruddin, I.A.; Mahlia, T.M.I.; Masjuki, H.H.; Mekhilef, S. A comprehensive review on biodiesel as an alternative energy resource and its characteristics. *Renew. Sustain. Energy Rev.* **2012**, *16*, 2070–2093. [CrossRef]
4. Sitepu, I.R.; Sectric, R.; Ignatia, L.; Levin, D.; German, J.B.; Gillies, L.A.; Almada, L.A.G.; Boundy-Mills, K.L. Manipulation of culture conditions alters lipid content and fatty acid profiles of a wide variety of known and new oleaginous yeast species. *Bioresour. Technol.* **2013**, *144*, 360–369. [CrossRef] [PubMed]
5. Xin, L.; Hong-ying, H.; Ke, G.; Ying-xue, S. Effects of different nitrogen and phosphorus concentrations on the growth, nutrient uptake, and lipid accumulation of a freshwater microalga *Scenedesmus* sp. *Bioresour. Technol.* **2010**, *101*, 5494–5500. [CrossRef] [PubMed]
6. Pinzi, S.; Leiva, D.; López-García, I.; Redel-Macías, M.D.; Dorado, M.P. Latest trends in feedstocks for biodiesel production. *Biofuels Bioprod. Biorefining* **2014**, *8*, 126–143. [CrossRef]
7. Manzano-Agugliaro, F.; Sanchez-Muros, M.J.; Barroso, F.G.; Martínez-Sánchez, A.; Rojo, S.; Pérez-Bañón, C. Insects for biodiesel production. *Renew. Sustain. Energy Rev.* **2012**, *16*, 3744–3753. [CrossRef]
8. Leong, S.Y.; Kutty, S.R.M.; Malakahmad, A.; Tan, C.K. Feasibility study of biodiesel production using lipids of *Hermetia illucens* larva fed with organic waste. *Waste Manag.* **2016**, *47 Pt A*, 84–90. [CrossRef]
9. Lalander, C.; Diener, S.; Zurbrügg, C.; Vinnerås, B. Effects of feedstock on larval development and process efficiency in waste treatment with black soldier fly (*Hermetia illucens*). *J. Clean. Prod.* **2019**, *208*, 211–219. [CrossRef]
10. Li, Q.; Zheng, L.; Qiu, N.; Cai, H.; Tomberlin, J.K.; Yu, Z. Bioconversion of dairy manure by black soldier fly (*Diptera: Stratiomyidae*) for biodiesel and sugar production. *Waste Manag.* **2011**, *31*, 1316–1320. [CrossRef]
11. Newton, L.; Sheppard, C.; Watson, D.W.; Burtle, G.; Dove, R. *Using the Black Soldier Fly, Hermetia Illucens, as a Value-Added Tool for the Management of Swine Manure*; North Carolina State University: Raleigh, NC, USA, 2005.
12. Arango Gutiérrez, G.P.; Vergara Ruiz, R.A.; Me-jía Vélez, H. Analisis composicional, microbiológico y digestibilidad de la proteína de la harina de larvas de Hermetia illuscens L (*Diptera: Stratiomyiidae*) en Angelópolis-An-tioquia, Colombia. *Revista Facultad Nacional de Agronomía-Medellín* **2004**, *57*, 2491–2500.
13. Li, W.; Li, M.; Zheng, L.; Liu, Y.; Zhang, Y.; Yu, Z.; Ma, Z.; Li, Q. Simultaneous utilization of glucose and xylose for lipid accumulation in black soldier fly. *Biotechnol. Biofuels* **2015**, *8*, 117. [CrossRef]
14. Li, W.; Li, Q.; Wang, Y.; Zheng, L.; Zhang, Y.; Yu, Z.; Chen, H.; Zhang, J. Efficient bioconversion of organic wastes to value-added chemicals by soaking, black soldier fly (*Hermetia illucens L.*) and anaerobic fermentation. *J. Environ. Manag.* **2018**, *227*, 267–276. [CrossRef] [PubMed]
15. Rehman, K.U.; Cai, M.; Xiao, X.; Zheng, L.; Wang, H.; Soomro, A.A.; Zhou, Y.; Li, W.; Yu, Z.; Zhang, J. Cellulose decomposition and larval biomass production from the co-digestion of dairy manure and chicken manure by mini-livestock (*Hermetia illucens L.*). *J. Environ. Manag.* **2017**, *196*, 458–465. [CrossRef] [PubMed]
16. Rehman, K.U.; Rehman, R.U.; Somroo, A.A.; Cai, M.; Zheng, L.; Xiao, X.; Rehman, A.U.; Rehman, A.; Tomberlin, J.K.; Yu, Z.; et al. Enhanced bioconversion of dairy and chicken manure by the interaction of exogenous bacteria and black soldier fly larvae. *J. Environ. Manag.* **2019**, *237*, 75–83. [CrossRef] [PubMed]
17. Somroo, A.A.; Rehman, K.U.; Zheng, L.; Cai, M.; Xiao, X.; Hu, S.; Mathys, A.; Gold, M.; Yu, Z.; Zhang, J. Influence of *Lactobacillus buchneri* on soybean curd residue co-conversion by black soldier fly larvae (*Hermetia illucens*) for food and feedstock production. *Waste Manag.* **2019**, *86*, 114–122. [CrossRef] [PubMed]
18. Zheng, L.; Hou, Y.; Li, W.; Yang, S.; Li, Q.; Yu, Z. Biodiesel production from rice straw and restaurant waste employing black soldier fly assisted by microbes. *Energy* **2012**, *47*, 225–229. [CrossRef]
19. Barry, T. Evaluation of the economic, social, and biological feasibility of bioconverting food wastes with the black soldier fly (*hermetia illucens*). Ph.D. Thesis, University of North Texas, Denton, TX, USA, 2014.
20. Palma, L.; Ceballos, S.J.; Johnson, P.C.; Niemeier, D.; Pitesky, M.; VanderGheynst, J.S. Cultivation of black soldier fly larvae on almond byproducts: Impacts of aeration and moisture on larvae growth and composition. *J. Sci. Food Agric.* **2018**, *98*, 5893–5900. [CrossRef] [PubMed]

21. Cammack, J.A.; Tomberlin, J.K. The impact of diet protein and carbohydrate on select life-history traits of the black soldier fly *Hermetia illucens* (L.) (Diptera: Stratiomyidae). *Insects* **2017**, *8*, 56. [CrossRef] [PubMed]
22. Cheng, J.Y.K.; Chiu, S.L.H.; Lo, I.M.C. Effects of moisture content of food waste on residue separation, larval growth and larval survival in black soldier fly bioconversion. *Waste Manag.* **2017**, *67*, 315–323. [CrossRef] [PubMed]
23. Liu, X.; Chen, X.; Wang, H.; Yang, Q.; Ur Rehman, K.; Li, W.; Cai, M.; Li, Q.; Mazza, L.; Zhang, J.; et al. Dynamic changes of nutrient composition throughout the entire life cycle of black soldier fly. *PLoS ONE* **2017**, *12*, e0182601. [CrossRef] [PubMed]
24. Wong, C.-Y.; Rosli, S.-S.; Uemura, Y.; Ho, Y.C.; Leejeerajumnean, A.; Kiatkittipong, W.; Cheng, C.-K.; Lam, M.-K.; Lim, J.-W. Potential protein and biodiesel sources from black soldier fly larvae: Insights of larval harvesting instar and fermented feeding medium. *Energies* **2019**, *12*, 1570. [CrossRef]
25. Larouche, J.; Deschamps, M.-H.; Saucier, L.; Lebeuf, Y.; Doyen, A.; Vandenberg, G.W. Effects of killing methods on lipid oxidation, colour and microbial load of black soldier fly (*hermetia illucens*) larvae. *Animals* **2019**, *9*, 182. [CrossRef] [PubMed]
26. Caligiani, A.; Marseglia, A.; Sorci, A.; Bonzanini, F.; Lolli, V.; Maistrello, L.; Sforza, S. Influence of the killing method of the black soldier fly on its lipid composition. *Food Res. Int.* **2019**, *116*, 276–282. [CrossRef] [PubMed]
27. Surendra, K.C.; Olivier, R.; Tomberlin, J.K.; Jha, R.; Khanal, S.K. Bioconversion of organic wastes into biodiesel and animal feed via insect farming. *Renew. Energy* **2016**, *98*, 197–202. [CrossRef]
28. Li, Q.; Zheng, L.; Cai, H.; Garza, E.; Yu, Z.; Zhou, S. From organic waste to biodiesel: Black soldier fly, *Hermetia illucens*, makes it feasible. *Fuel* **2011**, *90*, 1545–1548. [CrossRef]
29. Zheng, L.; Li, Q.; Zhang, J.; Yu, Z. Double the biodiesel yield: Rearing black soldier fly larvae, *Hermetia illucens*, on solid residual fraction of restaurant waste after grease extraction for biodiesel production. *Renew. Energy* **2012**, *41*, 75–79. [CrossRef]
30. Nguyen, H.C.; Liang, S.-H.; Li, S.-Y.; Su, C.-H.; Chien, C.-C.; Chen, Y.-J.; Huong, D.T.M. Direct transesterification of black soldier fly larvae (*Hermetia illucens*) for biodiesel production. *J. Taiwan Inst. Chem. Eng.* **2018**, *85*, 165–169. [CrossRef]
31. Nguyen, H.C.; Liang, S.-H.; Doan, T.T.; Su, C.-H.; Yang, P.-C. Lipase-catalyzed synthesis of biodiesel from black soldier fly (*Hermetica illucens*): Optimization by using response surface methodology. *Energy Convers. Manag.* **2017**, *145*, 335–342. [CrossRef]
32. Nguyen, H.C.; Liang, S.-H.; Chen, S.-S.; Su, C.-H.; Lin, J.-H.; Chien, C.-C. Enzymatic production of biodiesel from insect fat using methyl acetate as an acyl acceptor: Optimization by using response surface methodology. *Energy Convers. Manag.* **2018**, *158*, 168–175. [CrossRef]
33. Lim, J.W.; Mohd-Noor, S.N.; Wong, C.Y.; Lam, M.K.; Goh, P.S.; Beniers, J.J.A.; Oh, W.D.; Jumbri, K.; Ghani, N.A. Palatability of black soldier fly larvae in valorizing mixed waste coconut endosperm and soybean curd residue into larval lipid and protein sources. *J. Environ. Manag.* **2019**, *231*, 129–136. [CrossRef] [PubMed]
34. Mohd-Noor, S.N.; Wong, C.Y.; Lim, J.W.; Mah-Hussin, M.I.A.; Uemura, Y.; Lam, M.K.; Ramli, A.; Bashir, M.J.K.; Tham, L. Optimization of self-fermented period of waste coconut endosperm destined to feed black soldier fly larvae in enhancing the lipid and protein yields. *Renew. Energy* **2017**, *111*, 646–654. [CrossRef]

Review

Liquid Biphasic System: A Recent Bioseparation Technology

Kuan Shiong Khoo [1], Hui Yi Leong [2], Kit Wayne Chew [3], Jun-Wei Lim [4], Tau Chuan Ling [5], Pau Loke Show [1,*] and Hong-Wei Yen [6,7,*]

1 Department of Chemical and Environmental Engineering, Faculty of Science and Engineering, University of Nottingham Malaysia, Jalan Broga, Semenyih 43500, Malaysia; kuanshiong.khoo@hotmail.com
2 College of Chemical and Biological Engineering, Zhejiang University, Hangzhou 310027, China; leong.hy@zju.edu.cn or isabellelhy@gmail.com
3 School of Mathematical Sciences, Faculty of Science and Engineering, University of Nottingham Malaysia, Jalan Broga, Semenyih 43500, Malaysia; KitWayne.Chew@nottingham.edu.my
4 Department of Fundamental and Applied Sciences, Centre for Biofuel and Biochemical Research, Institute of Self-Sustainable Building, Universiti Teknologi PETRONAS, 32610 Seri Iskandar, Malaysia; junwei.lim@utp.edu.my
5 Institute of Biological Sciences, Faculty of Science, University of Malaya, Kuala Lumpur 50603, Malaysia; tcling@um.edu.my
6 Taipei Economic and Cultural Office, Prague, Czech Republic, Evropská 33C, 160 00 Praha 6, Czechia
7 Department of Chemical and Material Engineering, Tunghai University, Taiwan 407
* Correspondence: PauLoke.Show@nottingham.edu.my (P.L.S.); hwyen@most.gov.tw (H.-W.Y.)

Received: 16 December 2019; Accepted: 17 January 2020; Published: 23 January 2020

Abstract: A well-known bioseparation technique namely liquid biphasic system (LBS) has attracted many researchers' interest for being an alternative bioseparation technology for various kinds of biomolecules. The present review begins with an in-depth discussion on the fundamental principle of LBS and this is followed by the discussion on further development of various phase-forming components in LBS. Additionally, the implementation of various advance technologies to the LBS that is beneficial towards the efficiency of LBS for the extraction, separation, and purification of biomolecules was discussed. The key parameters affecting the LBS were presented and evaluated. Moreover, future prospect and challenges were highlighted to be a useful guide for future development of LBS. The efforts presented in this review will provide an insight for future researches in liquid-liquid separation techniques.

Keywords: liquid biphasic system; aqueous two-phase system; aqueous biphasic system; purification; separation; recovery; biomolecules

---

1. Introduction

The most trending research in the downstream biotechnology industries focuses on the production of various bio-based products from renewable sources. Examples of these sources are microalgae, fruit, lignocellulose biomass, secondary product, and crop waste. Separation and purification techniques for the recovery of biomolecules (e.g., proteins, carotenoids, and lipids) requires a precise operating condition to ensure high value end-products can be obtained [1,2]. Established extraction techniques such as membrane separation, chromatography-based method, ultrafiltration, and precipitation usually involve multiple step operations, complex pathways, time consuming operations, high energy inputs, and high cost for the recovery and extraction processes [3–5]. With that said, researchers are putting tremendous efforts in developing a new separation and purification techniques which can be performed in a one-step extraction process within a shorter period of time. On the other hand, the extraction solvents in a process that can be reused and recycled will lower the overall processing cost [6]. As for food and pharmaceutical applications, this requires an alternative non-toxic and environmentally friendly extraction solvents [7].

A well-established bioseparation technology namely liquid biphasic system (LBS) has attracted numerous researchers' attentions in the separation and purification of biomolecules [8]. It is also known as liquid-liquid extraction technology in the downstream processing. The concerns associated from the conventional extraction method has been overcome by using liquid biphasic separation techniques. The liquid biphasic extraction technology is comprised of two liquids which is separated by an interfacial layer when the mixture of two incompatible liquids is beyond the critical condition. Generally, the characteristics of the phase-forming components creates a physico-chemical interaction which can easily acclimatize the target biomolecules to be partitioned to either the top or bottom phase depending on the selectivity of the components. Furthermore, various assisted technologies such as bubbling, ultrasound, and electrolysis have been incorporated into the LBS to enhance the effectiveness of biomolecules separation [9–12]. The application of the LBS has been applied for the extraction, separation, and purification of proteins, lipids, and carotenoids from microalgae [2,13].

This review article strives to summarize the cognitive knowledge and previous experimental research dealing with LBS for extraction and purification of various biomolecules. This review begins with the principles and fundamentals of LBS, followed by the various type of biphasic systems were presented. Recent works related with advance technologies such as bubble-, ultrasound, and electricity-assisted LBS were evaluated and assessed. Additional information regards to the quantification (i.e., partition coefficient, selectivity, separation efficiency, and recovery yield) and composition of LBS were tabulated with provided references in Sections 2 and 3. Each section in this review would allow the readers to understand the development of LBS technologies. Moreover, the key parameters affecting the extraction efficiency in LBS, advantages and drawbacks of LBS were comprehensively discussed. In addition, future prospect and challenges associated with LBS were also discussed. This review article has a significant impact on the liquid-liquid extraction and purification for various biotechnological products, which serve as a resourceful tool for researchers dealing with extraction of biomolecules using LBS.

2. Liquid Biphasic System

Liquid biphasic system (LBS) or commonly known as aqueous two-phase system (ATPS) has been long introduced for the separation, recovery, and purification of biomolecules, and it is the current research trend adopted in the separation and purification technology. It was started back in 1896 when Martinus Willem Beijerinck accidentally mixed an aqueous starch solution with gelatin and found that an immersible layer was formed between both the aqueous solutions [14,15]. This idea of LBS as an analytical separation technique was sparked by Per-Åke Albertsson in the 1960s who discovered the phenomenon by mixing two different polymers (e.g., polyethylene glycol and dextran) resulting in an aqueous medium containing two separable phases [14,16,17]. This application was then extended to several generations of scientists and engineers who have been working in the industrial biotechnology field. Figure 1 shows a schematic diagram of the principles of LBS.

The LBS is well-known for the extraction of different biotechnological materials such as proteins, lipids, and carotenoids [1,18,19]. The specialty of LBS compared to traditional organic solvent extraction techniques is the composition of the phase-forming components which contains large amount of water while maintaining a low interfacial layer that separates both phases. It can be either used to separate proteins from cellular debris or to purify targeted proteins from contaminated proteins. Likewise, LBS has the capability of directing the target biomolecules by partitioning them to the top phase for extraction [20]. Conventional polymer-based LBS which possess a low ionic system is generally used for the separation and purification of biomolecules which are sensitive toward ionic condition [16]. Nevertheless, polymer-based LBS was neglected due to lack of compatibility between high ionic strength biomolecules, expensive phase-forming components, and its high viscosity system. Further development in LBS using different phase-forming components such as alcohol-, ionic liquids-, deep-eutectic solvent- and surfactant-based was utilized to replace the conventional polymer-based LBS.

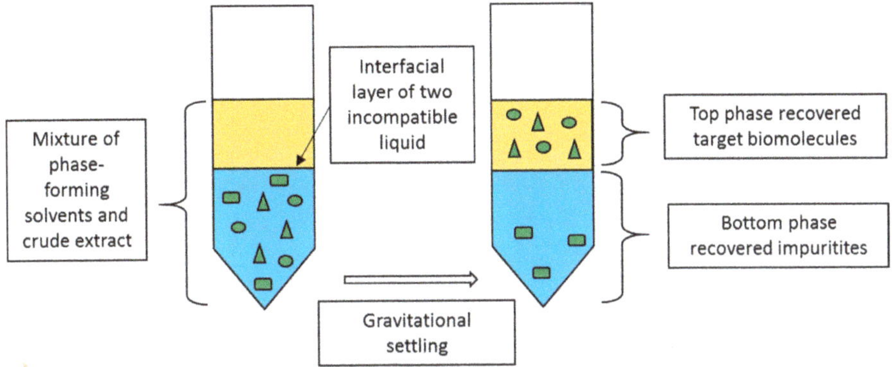

Figure 1. Schematic diagram of the principle in the liquid biphasic system (LBS).

The selective partitioning of the LBS allows the extraction of biomolecules to be operated in a single-step process compared to traditional extraction techniques which require multiple operation steps. LBS possess an environmental-friendly, inexpensive, ease of scaling-up, rapid, and efficient techniques for recovery and purification of biomolecules. During the planning stage, it is crucial to understand the complexity of the physical and chemical interaction reaction throughout the partitioning process in the LBS [21]. The selection of various parameters which are compatible to the system properties are important to achieve an optimal extraction, recovery, and purification condition. It is also important to evaluate the interactions during the selection of various parameters (e.g., salt precipitation, crystallization, and absence of biphasic system) as it may affect the findings. Lastly, is to assess the effect of each process parameters on the product recovery and purity [21].

Fundamental principles for the formation of LBS requires a phase diagram or also called the binodal curve where these provide a set of information regarding the two-phase formation and their required concentration in the top and bottom phases [22]. A detailed study has been evaluated previously by Iqbal et al., (2016) on the tie line length (TLL) and slope tie line (STL) for the construction of phase diagrams [23]. Binodal curves can be constructed using three methods namely, turbidometric titration, cloud point, and node determination method for predetermined phase diagram [22,24,25]. Moreover, the partition coefficient (K) LBS is to evaluate the equilibrium relationship between the top and bottom phase in the LBS. However, there is still lack of studies reporting on the theory and chemistry of these phase forming mixtures in the LBS which is a gap to-be-filled. Apart from that, factors that affect the partition coefficient can be manipulated using electrical, hydrophobicity-phase forming components, bio-specific affinity, molecular size, and surface area to understand the physico-chemical properties of the partitioning mechanism in the LBS.

2.1. Polymer-Based LBS

The conventional polymer-based LBS is typically made up of two polymers (e.g., polyethylene glycol (PEG) and dextran) and PEG-salt combinations (e.g., phosphate-, sulphate-, and citrate-based) as the phase-forming components. The purpose of using polymer-based LBS is that the chemical composition of a non-ionic characteristics toward an ionic environment is compatible towards biomolecules having low ionic strength [16]. Aside from that, the phase forming component from polymer-based has the ability to be recycled and reused for subsequent extraction process and this reduces the cost of polymers phase-forming component [26]. Polymer-based LBS are commonly used for protein extraction due to its poor hydrophilic and hydrophobic interaction in polymer/salt-based

LBS. However, it is important to maintain concentration of salt solution as high salt concentration may denature and damage the fragile protein in the system.

In most work, conventional polymer-based LBS has been replaced by using thermo-separating polymers as the phase-forming component to overwhelm the limitation of polymer-based LBS such as high viscosity and difficulties in recycling process [27,28]. Thermo-separating polymers are random, di-block, and tri-block co-polymers of ethylene oxide (EO) and propylene oxide (PO) [29]. Thermo-separating polymers have a low cloud point temperature (≤47 °C) which is suitable to achieve temperature-induced phase separation where a target protein can be recovered from the polymer [30]. Generally, a back-extraction process such as ultrafiltration, diafiltration, and crystallization is needed to separate the target protein from the polymer. However, an in-depth understanding on the mechanism by the polymer phase-forming component for the recovery of biomolecules is still poorly understood. This shows a gap for future researchers to further explore the fundamental principles of this LBS extraction technique.

Several studies have been conducted involving cyclodextrin glycosyltransferase (CGTase) from *Bacillus cereus*. Ng et al., (2012) reported that the TLL of 41.2% (*w*/*w*), volume ratio (V_R) of 1.25, pH 7, and crude loading (*w*/*w*) of 20% were the optimal conditions to recover cyclodextrins using polymer-based LBS with ethylene oxide–propylene oxide (EOPO) 3900 and two phosphate salts [31]. This experiment showed that the highest CGTase was purified up to 13.1-fold with a yield of 87% recovered in the EOPO-rich top phase. However, this experiment did not discuss the time period in cyclodextrins recovery. Another research carried out by Lin et al. [32] with modified method using flotation technique and the combination of PEG 8000 and potassium phosphate salt. The optimum conditions in cyclodextrins (CDs) recovery was optimized at 18% (*w*/*w*) PEG 8000 and 7.0% (*w*/*w*) potassium phosphate with TLL of 27.2% (*w*/*w*), V_R of 3.0, pH 7, and crude loading (*w*/*w*) of 20%. The experiment showed that the recovery of CDs was affected by alternating each of the parameters such as TLL, V_R, and pH where the purification factor (P_{FT}), which corresponded to the highest CGTase purity up to 21.8 with a yield of 97.1%, was recovered in the PEG-rich top phase within a short period [32].

A similar approach utilizing polymer-based LBS was employed for the recovery of lignin peroxidase from *Amauroderma rugosum* (Blume and T. Nees) [33]. However, this experiment used a lower molecular weight (PEG 600) for a high purification of lignin peroxidase. Generally, this approach showed that a higher molecular weight polymer reduces the purification factor of lignin peroxidase due to the interaction of PEG and hydrophobic enzyme. An optimal condition in lignin peroxidase recovery was optimized at 15% (*w*/*w*) PEG 600 and 16% (*w*/*w*) dipotassium phosphate with highest purification factor of 1.33 ± 0.62 and recovery yield of 72.18 ± 8.50%.

2.2. Organic Solvent-Based LBS

Organic solvent-based LBS consists of various water-miscible alcohols (e.g., methanol, ethanol, 1-propanol, and 2-propanol) and inorganic salts. This form of LBS has been utilized to overcome the limitation of polymer-based LBS to improve the recovery of biomolecules from the phase-forming component [7]. The use of alcohol as the phase-forming components can easily recover the biomolecules by evaporating the alcohol from the top phase. A recent study also showed a greener approach using food grade alcohol such as ethanol and 2-propanol compared to the conventional polymer-based LBS for the extraction and recovery of carotenoids from microalgae [7]. Additionally, the phase-forming component can reduce the cost of the process by recycling and reusing the alcohol using rotary evaporator for the next extraction process. Despite its advantages, the drawbacks of using alcohol, especially methanol, as the phase-forming component is the toxicity and hazardous effects towards the environment.

Ooi et al. (2009) reported a study on purification of lipase from *Burkholderia pseudomallei* using alcohol/salt-based LBS [19]. The best lipase recovery was achieved in LBS composed of 16% (*w*/*w*) of 2-propanol, 16% (*w*/*w*) of potassium phosphate and 4.5% (*w*/*v*) sodium chloride with a purification factor of 13.5 along with the yield of 99%. The presence of alcohol component in LBS also did not inhibit the enzymatic activity of purified lipase. The effect of NaCl on lipase partitioning was found to generate an electrical potential difference in the LBS [34]. An increase in the salt concentration could generate an electrostatic potential that strongly expelled the negatively charge biomolecules toward the water-miscible alcohol in top phase, thus resulting in a high recovery yield.

Lin et al., (2013) conducted a study using alcohol/salt-based LBS to recover the intracellular human recombinant interferon-α2b (IFN-α2b) from *Escherichia coli* [34]. A different variety of combinations between alcohol-based top phase (ethanol, 1-propanol and 2-propanol) and salt phase (ammonium sulfate, dipotassium hydrogen phosphate, and monosodium citrate) were conducted. LBS composed of 18% (*w/w*) of propanol and 22% (*w/w*) ammonium sulfate in 1% (*w/w*) sodium chloride was reported to be the optimal conditions for the purification of IFN-α2b achieving a purification factor of 16.2 with the yield of 74.6%. Ammonium sulfate salt was selected due to its high level of pH in the system which provided a high purification factor of IFN recovery. As the pH environment in LBS increased, the contaminant protein and IFN protein were partitioned toward water-miscible alcohol top phase. This is mainly due to the negatively charge protein which tends to partition to the top phase and repels from the salt-rich bottom phase [34].

A recent study conducted on a recyclability test utilizing 1-propanol and ammonium sulfate system for the phlorotannin recovery from *Padina australis* and *Sargassum binderi* [35]. The highest recovery of phlorotannin were 76.1% and 91.67% with purification factor of 2.49 and 1.59 from *Padina australis* and *Sargassum binderi*, respectively. A consistent recovery of phlorotannin was obtained after conducting two cycles of the system. This showed a feasible and eco-friendly approach of utilizing the alcohol-based LBS for biomolecules extraction.

2.3. *Ionic Liquid-Based LBS*

A new trend of research by using ionic liquids (ILs) have been an alternative organic compound and non-volatile green solvent in the downstream processes. Their remarkable properties such as negligible vapor pressure, low melting point and high thermal stability have received numerous attention from researchers [36,37]. ILs are composed with tuneable physico-chemical properties of cationic and anionic ions [38]. The cationic part of ILs usually consists of choline cation, ammonium cation, quaternary ammonium or phosphonium, and guanidium cation. As for the anionic part, it consists of environmentally friendly sources such as carboxylic acid, amino acid and biological buffers. Thus, replacing ILs as the phase-forming component in LBS would be beneficial for the extraction and purification of specific target biomolecules from complex crude extract [39]. Additionally, ILs have also been employed for various applications such as electrolytes (e.g., fuel cells, batteries and sensors), CO_2 capture, lubricants, and fuel additives. The cost of reactant for the synthesis of ILs are expensive. Therefore, it is important for ILs to be recycle- and reuse-able to ensure that ILs-based LBS are more feasible and applicable in the bioprocessing industries for the next extraction processes. A review by Ostadjoo et al., (2017) revealed the green and environmentally friendly, 1-ethyl-3-methylimidazolium acetate ([C2mim][OAc]) for its potential features in the field of lignocellulose biomass dissolution and biopolymer processing [40–42]. Yet, there is still insufficient studies related to their toxicity and eco-friendliness on scaling up these ILs, especially imidazole- and pyridinium-based ILs. Here we recommended that these ILs need to be further fabricated by replacing environmentally-friendly anionic part such as carboxylic acid, amino acid and biological buffers in order to minimize their toxicity in various application.

Gutowski et al. (2003) reported that by mixing imidazole-based ILs and a kosmotropic salt (i.e., K_3PO_4) would lead to the formation of a biphasic system [43]. This research had gained interest investigating the phase separation behavior of IL-based LBS. The study on protein extraction using IL-based LBS in a single step was conducted by Du et al. (2007). The researchers had successfully extracted the protein from human urine into the IL-rich top phase with a distribution of 10 and enrichment factor of 5 [44]. Apart from that, Ng et al. (2014) investigated the purification of CGTase from *Bacillus cereus* fermentation broth in IL/salt LBS, composing of 35% (*w/w*) of (Emim)BF_4 and

18% (*w*/*w*) of sodium carbonate with the addition of 3% (*w*/*w*) of NaCl [45]. The optimized operating conditions showed that the IL-based LBS was a promising approach for the purification and recovery of CGTase in a single step operation attaining a high purification factor of 13.86 and yield of 96.23%. Ng et al. (2014) also reported that it was crucial in the selection of salt such as citrate and carbonate ions as they played an important role in LBS formation and was able to attract water molecules toward them by forming strong intermolecular interaction [45].

Chang et al., (2018) used a series of alkyl bromide imidazole for the extraction of C-phycocyanin (CPC) from *Spirulina platensis* and found that the longer the alkyl chain, C_8MIM-Br enhanced the extraction efficiency of CPC [46]. The results indicated that by using C_8MIM-Br/salt LBS the maximum extraction efficiency, partition coefficient, and separation factor of CPC were 99.0%, 36.6, and 5.8 respectively. ILs-based LBS demonstrated an efficient and feasible separation technique for the extraction of various biomolecules from complex crude extract. This was supported by a recent study that evaluated the protein partitioning in ILs-based LBS composed of Iolilyte 221 PG and citrate salts was found to be feasible but complex depending on various factors such as concentration of phase-forming component, pH, temperature, ionic strength, and chemical nature of the target biomolecules [47]. Proteins are negatively charged particles therefore it favours a system pH (≥6.50) higher than the isoelectric point of protein. Moreover, the partition coefficient for tie-line length within 38–76% were reference points for specific protein (e.g., bovine serum albumin and rubisco) to be partitioned at the top phase.

2.4. Deep-Eutectic-Solvent-Based LBS

Deep-eutectic-solvents (DESs) are defined as a subclass from ILs because of their similarity in physical and chemical properties of ILs [48]. The behavior exhibited from DESs are contributed from hydrogen bonding, whereas ILs are dominated by ionic interactions [49]. DESs are more environmentally friendly as compared to ILs (e.g., imidazole- and pyridinium-based ILs) which are toxic and non-biodegradable. The synthesis of DESs is by combining hydrogen bond acceptors (e.g., quaternary ammonium and phosphonium salts) and hydrogen bond donors (e.g., alcohols, carboxylic acid, and amide). A major advantage from DESs are their charge delocalization properties which are responsible for the decrease in melting point of mixture relative to the raw material [50]. The bottleneck from using ILs such as high cost and complex synthesis route have been solved by these DESs. By having the similar characteristic as ILs and exhibiting some distinguishing features, including ease of synthesis, low cost, and valuable for industrial application, DESs have gained interest in many fields especially in LBS [51].

Choline chloride (ChCl) is a convention quaternary salt used to synthesize DESs. ChCl-based DESs have the same advantages with ILs besides showing excellent biodegradability and low toxicity [52]. Zeng et al. (2014) had performed the extraction of bovine serum albumin (BSA) using four different kind of DESs, namely, choline chloride (ChCl)-urea, tetramethylammonium chloride (TMACl)-urea, tetrapropylammonium bromide (TPMBr)-urea, and ChCl-methylurea [53]. The extraction efficiency of BSA under the optimum LBS conditions composed of 0.7 g mL^{-1} ChCl-urea and 2.0 mL dipotassium phosphate, K_2HPO_4 could reach up to 100.5% that collectively highlighted the advantages of the DES-based LBS for the extraction of protein. Unfortunately, this work was unable to back-extract the target protein free from the DES-LBS because of the hydrophilicity characteristic of DES in the aqueous solution.

A similar work with different DESs was investigated by Pang et al. (2017) using DES-based LBS which composed of choline chloride-polyethylene glycol (ChCl-PEG or DES) and sodium carbonate were applied for the extraction of specific protein (i.e., BSA and papain) [52]. ChCl-based DES was prepared by mixing two compounds, 0.68 g mL^{-1} ChCl and 0.1 g mL^{-1} PEG 2000 at the molar ratio of 20:1, stirring up to 100 °C until a homogenous colorless liquid was formed. The result showed that the DES-Na_sCO_3 LBS under the optimum condition had successfully obtained a high extraction efficiency of BSA (95.16%) and papain (90.95%). Moreover, the back-extraction of target protein was performed by extracting 1 mL DES top phase followed by the addition of ammonium sulfate $(NH_4)_2SO_4$ and 0.45 mL ethanol to form a new LBS. However, it was found that by increasing the concentration in the salt-rich bottom concentration would lower the efficiency of the back extraction.

A modified DES-based LBS using ultrasonic-assisted were employed for the extraction of ursolic acid from Cynomorium songaricum Rupr [54]. This approach was compared to the convention ultrasonic-extraction method. The recovery yield of ursolic acid was comparable. However, the presence of LBS promotes a higher purification of ursolic acid. The recovery yield of ursolic acid was 22.10 ± 0.44 mg/g with purification factor of 42.41 ± 0.84% as compared to conventional ultrasonic-extraction method where the recovery yield was only 20.9 ± 0.79 mg/g with a low purification factor of 20.17 ± 0.77%.

2.5. Surfactant/Detergent-Based LBS

Surfactant-based LBS is the transformation of phase-forming component from conventional polymer-based LBS. The surfactant-based LBS is formed when both cationic and anionic surfactants are separated into two immiscible liquid phases which consist of a high concentration than critical micelle concentration (CMC) and at certain molar ratio of cationic and anionic surfactant composition. This novel approach of surfactant-LBS has gained interest mainly due to the combination phase which exist in many different forms (i.e., spherical micelles, rod-like micelles, or vesicles) by simply alternating different composition and concentration of surfactants [55]. The principle of surfactant-based LBS used the cloud point extraction (CPE) system in which the non-ionic surfactant is heated above the cloud point temperature, causing dehydration of detergent for the phenomenon of phase separation to occur [56]. The surfactant-LBS consists of one surfactant-rich phase and the other is the surfactant-dilute phase. The organic contaminant will partition into the surfactant-rich phase and will then aggregate and concentrate at that phase. The presence of small amount of remediated water in the contaminant will remain in the surfactant-dilute phase. Surfactant-based LBS is commonly used to separate hydrophobic and amphiphilic molecules by solubilization and partitioning of membrane-bound substances.

Surfactant-based LBS composed of 24% (*w*/*w*) Triton X-100 and 20% (*w*/*w*) xylitol was used for the purification of lipase from pumpkin seeds [57]. The results showed that the surfactant-based LBS had the ability to partition the lipase into the top surfactant-rich phase and leave the impurities at the bottom xylitol-rich phase. The proposed optimized method had successfully recovered the enzyme with purification factor of 16.4 and yield of 97%. This study also demonstrated that the recovery phase component could be recycled up to five runs with a high percentage of recovery of 97%. However, it was noted that there was a significant decrease in recovery of the phase component after the fifth cycle in which could be mainly due to the accumulation of impurities present in the phase component.

An example of surfactant-based LBS extraction was conducted by Sankaran et al. (2018) using surfactant and xylitol under the optimum operation condition of 25% *w*/*w* of xylitol concentration, 15% (*w*/*w*) Triton X-100, 80% *w*/*w* of crude lipase, 4 mL of top phase, 35 mL of bottom phase, pH 7, and 15 min of flotation time showed the maximum lipase extraction and efficiency of 3.63 and 86.46% [58]. In addition, the recyclability of both components in surfactant-LBS extraction makes this an excellent process, as this innovative method was practical and feasible to be applied in the biotechnology industry for extraction of other biomolecules. Table 1 summarizes the extraction of biomolecules using various types of phase-forming component in LBS.

Table 1. Extraction of biomolecules using various types of phase-forming components in LBS.

Type of LBS	Composition of LBS	Type of Feedstock	Biomolecule	Selectivity	Partition Coefficient, K	Purification Factor, P_{FT}	Recovery Yield (%)	Ref.
Polymer/salt-based	EOPO 3900 and two phosphate salts	*Bacillus cereus cyclodextrin glycosyltransferase*	Cyclodextringlycosyltransferase (CGTase)	3.19	17.54	5.30	87.0	[31]
	18% (*w*/*w*) PEG 8000 and 7.0% (*w*/*w*) potassium phosphate salts	*Bacillus cereus cyclodextrin glycosyltransferase*	Cyclodextringlycosyltransferase (CGTase)	-	-	21.8	97.1	[32]
	15% (*w*/*w*) PEG 600 and 16% (*w*/*w*) dipotassium phosphate	*Amauroderma rugosum*	Lignin peroxidase	-	-	1.33 ± 0.62	2.18 ± 8.50	[33]
Alcohol/salt-based	18% (*w*/*w*) 2-propanol and 22% (*w*/*w*) ammonium sulfate, $(NH_4)_2SO_4$	*Escherichia coli*	Interferon (IFN)/ Glycoproteins	-	0.82	16.24	74.64	[34]
	16% (*w*/*w*) 2-propanol and 16% (*w*/*w*) potassium phosphate	*Burkholderia pseudomallei*	Lipase	287.5	-	13.5	99.3	[19]
	33.5% (*w*/*w*) of 2-propanol and 10% (*w*/*w*) ammonium sulfate	*Padina australis*	Phlorotannin	-	-	2.49	76.1	[35]
	25% (*w*/*w*) of 2-propanol and 12.5% (*w*/*w*) ammonium sulfate	*Sargassum binderi*	Phlorotannin	-	-	1.59	91.67	[35]
Ionic-liquid based	35% (*w*/*w*) of (Emim)BF_4 and 18% (*w*/*w*) of sodium carbonate Na_2CO_3	Fermentation broth	*Bacillus cereus* cyclodextrin glycosyltransferase (CGTase)	9.66	-	51.0	96.00	[45]
	C_8MIM-Br and tri-potassium phosphate	*Spirulina platensis*	C-phycocyanin (CPC)	5.8	36.6	-	99.00	[46]
Deep-eutectic solvent based	0.7 g mL^{-1} ChCl-urea and 2.0 mL dipotassium phosphate, K_2HPO_4	Protein	Bovine serum albumin (BSA)	-	-	-	99.6 99.7 and 100.0 BSA	[53]
	Choline chloride and PEG 2000, molar ratio of 20:1	Protein	Bovine serum albumin and papain	-	-	-	Bovine serum albumin (95.16), papain (90.95)	[52]
	36% (*w*/*w*) ChCl-glucose and 25% (*w*/*w*) dipotassium phosphate, K_2HPO_4	Ursolic acid	*Cynomorium songaricum* Rupr.	-	-	42.41 ± 0.84	22.10 ± 0.44 mg/g	[54]
Surfactant/detergent based	24% (*w*/*w*) Triton X-100 and 20% (*w*/*w*) xylitol	*Cucurbita moschata*	Lipase	-	-	16.4	97.0	[57]
	25% (*w*/*w*) of xylitol concentration, 15% (*w*/*w*) Triton X-100	*Burkholderia cepacia*	Lipase	2.62	-	2.56	86.46	[58]

3. Advance Technologies Integrated with LBS

3.1. Bubble-Assisted LBS

Bubble-assisted LBS or known as liquid biphasic flotation (LBF) is the combination of LBS and solvent sublation (SS), in which the biphasic medium composed of organic solvent and aqueous salt solution is aerated by air bubbles (e.g., nitrogen and oxygen) in promoting the adsorption of target biomolecules during the separation process [8] (refer to Figure 2a). SS is an adsorptive bubble separation technique introduced by Sebba who suggested that the use of an immiscible thin organic solvent layer overlaid on top of the liquid bulk as a modification of ion flotation [59]. LBF has accommodated the ease for extraction of high value biomolecules such as protein, lipase, astaxanthin, and betacyanin [8,10,60,61]. The theory of LBF system is the phenomenon of surface-active biomolecules having a sorption mechanism between the air bubbles surfaces. The bubbles then arise and dissolve in an organic solvent phase on top of the aqueous solution in the system [11]. With the presence of bubble-assistance in LBS, this could intensively strengthen the adsorption mechanism produced by the bubble transportation; thus, this system is feasible for separation and extraction of biomolecules. Figure 2a illustrates the set-up of bubble-assisted LBS.

A pilot-scale LBF consisting of 0.9 L of 50% (*w*/*w*) of 1-propanol and 1.5 L of 250 g/L ammonium sulfate salt, $(NH_4)_2SO_4$ had been developed for direct recovery of lipase derived from *Burkholderia cepacian* [62]. The purpose of this study was to conduct a comparison between the recovery of lipase on pilot-scale and small-scale LBF processes. Preshna et al., (2016) had reported that the pilot-scale alcohol/salt LBF system acquired a purification factor of 12.2, efficiency of 88%, and a recovery yield of 93.27% which was feasible for purification of lipase to be implemented into the industrial scale processes [62].

Leong et al. (2018) utilized LBF which composed of 10 mL of 100% ethanol, 20 mL of 200 g/L K_2HPO_4 salt solution, 1 g FE (peel or flesh of red-purple pitaya), and 15 min flotation time for betacyanins extraction [8]. The results under the optimum conditions of LBF revealed that the betacyanins extractions from 1 g FE of peel in alcohol-rich top phase was 95.989 ± 1.708% with separation efficiency and partition coefficient of 88.361 ± 1.708% and 24.168 ± 2.949%, respectively. The recovery from 1 g FE of flesh was 95.488 ± 0.213% with separation efficiency and partition coefficient of 94.886 ± 0.060% and 21.195 ± 1.030%, respectively [8]. The objective of this work showed that the LBF has a great potential in bioseparation technology as compared with other extraction techniques such as diffusion extraction, ultrafiltration, and reverse osmosis in which only able to recover 70–75% of betacyanins [63].

Rather than using alcohol-based LBS, a recent study had showed the extraction of α-Lactalbumin from whey used a different phase forming component (i.e., PEG 1000 and citrate salts) along with bubble-assisted technologies showed a separation efficiency and purification fold of 87.54% and 5.33 [64]. The advantages of this study had showed the feasibility of bubble-assisted technology compared to conventional liquid-liquid extraction providing a low processing cost, rapid, and good separation yield. However, a further study is required to fulfil the gaps in the bubble-assisted technology. This is to ensure a better understanding regarding the mass transfer and the development of kinetics model of LBF in the separation of biomolecules.

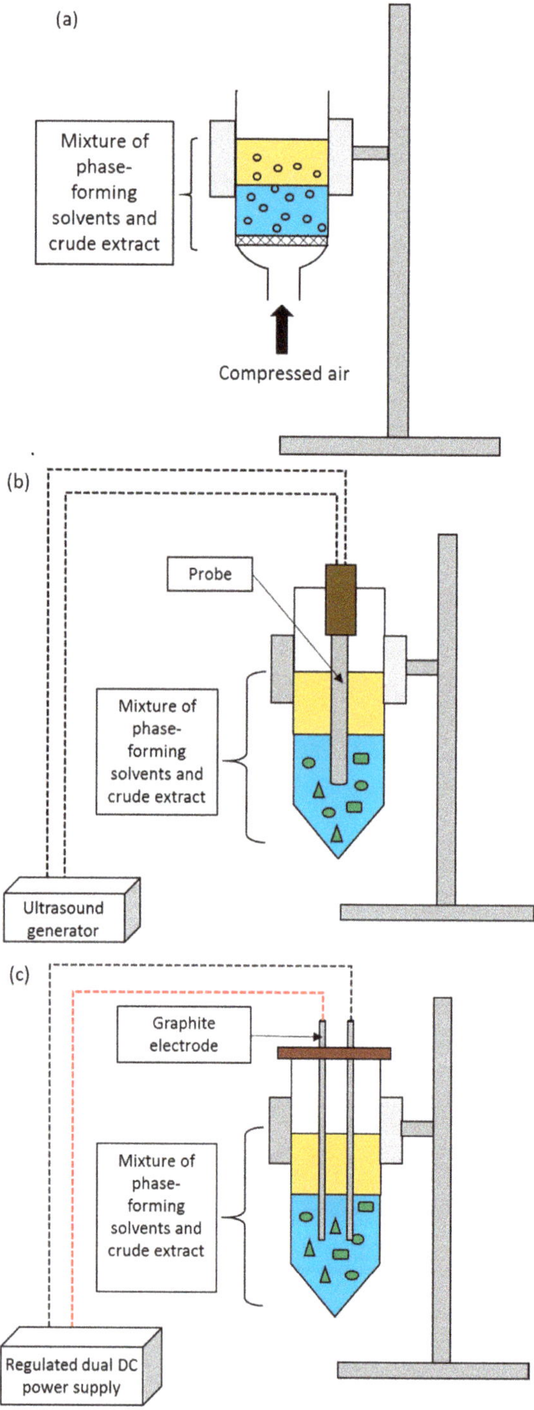

Figure 2. Schematic diagram of (**a**) bubble-assisted LBS, (**b**) ultrasound-assisted LBS, and (**c**) electricity-assisted LBS.

3.2. Ultrasound-Assisted LBS

In the biotechnology processes, cell disruption is considered as the most important process for higher extraction and recovery yield. Ultrasound-assisted LBS is an integrated technique which has been extensively acknowledged by researchers due to its effective properties of cell disruption [65,66]. The advantages of ultrasound-assisted LBS includes low operating cost, less energy consumption and short period of time requirement [67]. The fundamental of ultrasound irradiation is the high shear forces produced from cavitation bubbles of ultrasonic waves and mechanical shears which enhanced the cell disruption for effective biomolecules extraction [68]. Figure 2b shows a schematic set-up of ultrasound-assisted LBS.

A recent study conducted by Sankaran et al. (2018) utilized the application of ultrasound-assisted LBS for extraction of protein from *Chlorella vulgaris* FSP-E microalgae [12]. The authors found that the ultrasound-assisted LBS had the ability to break down the rigid cell wall, followed by the release of protein for extraction. The maximum efficiency and yield of protein were 75% and 65.4%, respectively [6]. An integrated system of ultrasound and LBF was used to compare the effectiveness recovery of the release protein into the solution for extraction [69]. It was reported that the ultrasound-assisted LBF had better advantages over the ultrasound-assisted LBS, driven by its higher concentration coefficient and a better separation efficiency. This was mainly due to the presence of air bubbles which enabled the adsorption of surface-active proteins from the bottom phase to the top phase. As a result, this led to a higher separation efficiency and recovery yield. This integrated sugaring-out ultrasound-assisted LBF under the optimum conditions composed of 100% (*w*/*w*) acetonitrile, 200 g/L glucose concentration, biomass concentration of 0.6% with 5 min of 5 s ON/10 s OFF pulse mode, and at a flow rate of 100 cc/min had given rise to the protein separation efficiency and recovery yield of 86.38% and 93.33%, respectively.

Aside from that, ultrasound-assisted extraction has also been widely employed for the cell disruption of lignocellulose biomass from plants [70]. The extraction of phenylethanoid glycosides (e.g., echinacoside and acteoside) from *Cistanche deserticola* stems using ultrasonic-assisted LBS successfully recovered 27.56 and 30.23 mg/g, respectively [71]. This approach showed that ultrasonic-assisted LBS were efficient, eco-friendly and cheap method for extracting and enriching biomolecules from lignocellulose biomass. However, it is crucial to monitor the process temperature when dealing with ultrasonic irradiation. The high shear forces produced from the cavitation bubbles of sonic wave would generate a high temperature process which will degrade or deform the target biomolecules resulting in an unfavorable low extraction yield. Another supporting research of using the application of ultrasound-assisted LBS was the extraction and separation of antioxidants such as xylooligosaccharides (sugar) and phenolic compound from wheat. In ultrasound-assisted LBS composed of 23.8% (*w*/*w*) ammonium sulfate, 24.3% (*w*/*w*) ethanol, 1.2% (*w*/*w*) biomass loading with ultrasound wave (30 Hz, 500 W, 10 min), extraction yielded the highest recovery of sugar and phenols were 16 mg/g and 2.67 mg/g dry material [72]. This showed that implementation of ultrasound improved the efficiency of extraction of wheat chaff in LBS yielding 1.3–2 times higher, respectively than those without ultrasound.

3.3. Electricity-Assisted LBS

Electricity-assisted LBS (see Figure 2c) is a promising mild cell disintegration extraction technique for recovery of biomolecules. For instance, the electricity treatment such as pulsed electric field (PEF) demonstrates the conceptualization of the initiation of short electrical pulses in the order of magnitude of ms or μs subjecting the charge in the cell membrane which is sufficient to perform a rearrangement or disruption of membrane and lead to the pore formation. This process is also known as electroporation. However, an optimum condition is required as PEF is dependent on the intensity of the treatment and cell characteristics in which pore formation is reversible or irreversible [73–75]. PEF treatment also increased the mass transfer energy of the system. By combining both PEF and LBS would be an advantage for an efficient extraction of treated sample. This combination is known as an electropermeabilization where the presence of electric and extractive solvent improves the release of intracellular compound from treated sample [76]. Moreover, electricity treatment not only provides higher extraction efficiency of biomolecules but also a greener approach in the biotechnology industries.

Lam et al., (2017) investigated the operating condition required to release selective proteins from the cell wall of *Chlamydomonas reinhardtii* (cc-124) strain and the cell wall deficient mutant strain (cc-400) using PEF treatment without the presence of LBS [77]. The results showed that after PEF treatment, with operating condition of 5–7.5 kV/cm, 1–10 pulses, and a pulse length of 0.05–0.2 ms on the cell wall, deficient mutant (cc-400) was on average three times higher than cell wall strain (cc-124) with average protein yield of 31 ± 6% protein and 11 ± 3% protein. Additional experiments utilizing PEF treatment with low energy input (range between 0.01 and 0.5 kWh/kg_{DW}) were also conducted on cell wall deficient mutant strain (cc-400) with a maximum recovery of 30% at 0.04 kWh/kg_{DW}. Furthermore, the results obtained from PEF treatment with low energy input was compared with bead beating which only obtain an average of 34 ± 4.2% proteins.

A recent work conducted by Leong et al. (2019) on betacyanins extraction from peel and flesh of red-purple pitaya using the liquid biphasic electric flotation (LBEF) [76] had reported that this new integration process of electricity supplied in LBF system could cause an electropermeabilization of red-purple pitaya membrane structure and improve the betacyanins extraction from red-purple pitaya. An optimum system composed of 100% (*w*/*w*) ethanol, 200 g/L of dipotassium hydrogen phosphate (K_2HPO_4) with 15 min floatation time (flow rate of 20–30 cc/min), and applied up to 3 V of voltage using graphitic electrodes showed the highest separation efficiency of betacyanins concentration (98.383 ± 0.215% for peel and 96.576 ± 0.0083% for flesh, respectively) [76]. Table 2 summarizes the advance technologies integrated with LBS for the extraction of biomolecules.

Table 2. Extraction of biomolecules using various types of advance technologies integrated in LBS.

Assisted Technology	Composition of LBS	Type of Assisted Employed	Type of Feedstock	Biomolecule	Time	Extraction Efficiency, E (%)	Partition Coefficient, K	Recovery Yield (%)	Ref
Bubble-assisted LBS or Liquid biphasic flotation (LBF)	50% (*w/w*) of 1-propanol and 250 g/L ammonium sulfate salt, $(NH_4)_2SO_4$	Flotation system (compressed air 0.5 bar)	*Burkholderia cepacian*	Lipase	30 min	88.0	-	93.27	[62]
	100% ethanol, 20 mL of 200 g/L dipotassium phosphate K_2HPO_4	Flotation system (compressed air 0.5 bar)	*Hylocereous polyrhizus*	Lipase	15 min	E for peel and flesh were 88.361 ± 1.708%, 94.886 ± 0.060%.	K value of peel and flesh were 24.168 ± 2.949, 21.195 ± 1.030.	Recovery for peel and flesh were 95.488 ± 0.213, 94.886 ± 0.060.	[8]
	0.5 g/mL PEG 1000, 35 mL of 0.40 g/mL trisodium citrate $Na_3C_6H_5O_7$	Flotation system (30 mL/min flow velocity)	Whey	a-lactalbumin	42 min	87.54	-	-	[64]
Ultrasound-assisted LBS	100% (*w/w*) acetonitrile and 200 g/L glucose solution.	Ultrasound irradiated for 5 min of 5 s ON/10 s OFF pulse mode and flotation system	*Chlorella vulgaris FSP-E*	Protein	5 min	86.38	-	93.33 of protein recovered	[6]
	20% (*w/w*) ethanol and 23.5% ammonium sulfate	Ultrasound irradiated (300 W, 37 min)	*Cistanche deserticola* Y. C. Ma stems	Phenylethanoid glycosides	37 min	Echinacoside and acteoside were 5.35 and 6.22 mg/g dry weight	-	Echinacoside and acteoside were 27.56 and 30.23 mg/g dry weight	[71]
	24.3% (*w/w*) ethanol and 23.8% (*w/w*) ammonium sulfate	Ultrasound irradiated (30 Hz, 500 W, 10 min),	Wheat chaff	Xylooligosaccharides (sugar) and phenolic compound	10 min	72.79 ± 3.98	3.91	Recovery of sugar and phenols were 16 mg/g and 2.67 mg/g	[72]
Electricity-assisted LBS	Without LBS	PEF treatment (5–7.5 kV/cm, 1–10 pulses and a pulse length of 0.05–0.2 ms)	Cell wall *C. reinhardtii* strain (cc-124) and cell wall deficient mutant strain (cc-400)	Protein	10 min/pulse	-	-	Cell wall strain (cc-124) and cell deficient (cc-400) with average protein yield of 31 ± 6 protein and 11 ± 3 protein.	[77]
	100% (*w/w*) ethanol, 200 g/L of dipotassium hydrogen phosphate (K_2HPO_4)	PEF treatment (3 V of voltage using graphitic electrodes) and 15 min flotation system	Peel and flesh of *Hylocereus polyrhizus*	Betacyanins	15 min	E for peel and flesh were 98.383 ± 0.215 and 96.576 ± 0.083	K for peel and flesh were 100.814 ± 7.324 and 24.883 ± 1.052	Betacyanins concentration (98.383 ± 0.215 for peel and 96.576 ± 0.0083 for flesh	[76]

4. Key Parameters Affecting LBS

4.1. Type and Molecular Weight of Polymer

In polymer-salt based LBS, the polymer phase component is crucial as it exhibits different degrees of hydrophobicity on target biomolecules partitioning. As the molecular weight of polymer increases, the hydrophobicity also increases due to the long hydrocarbon chain of monomers. This effect causes a reduction in free volume of the polymer-rich top phase, forcing the target biomolecules to be partitioned to the bottom phase. On the other hand, low molecular weight polymer will decrease the purification factor for target biomolecules as it will be partitioned together with contaminant proteins at the polymer-rich top phase [31,32]. Therefore, it is important in selecting an optimum condition for the hydrophobicity of polymers to obtain the maximum recovery of target compounds.

The effect of molecular weight has been discussed with the used of polymers such as PEG and potassium phosphate salt for the recovery of cyclodextringlycosyltransferase (CGTase) from *Bacillus cereus* [32]. In this work, the different molecular weights of PEG (e.g., PEG 4000, 6000, 8000, 10,000, and 20,000) were used in the LBF system for the CGTase extraction at a constant crude extract to volume ratio of 1.0:3.0. It was found that the maximum purification factor of 7.26 and 97.1% recovery of CGTase were achieved composed of 18.0% (*w*/*w*) PEG 8000 and 7.0% (*w*/*w*) potassium phosphates LBS. Well, as for the lowest molecular weight, PEG 4000 and highest molecular weight, PEG 20,000 showed a purification factor of 2.25 and 3.23, respectively. This indicated that the low molecular weight polymer (PEG 4000) withdraw contaminant biomolecules to the polymer-rich top phase and the high molecular weight (PEG 20,000) would engender a more viscous phase, resulting in the decrease of free volume of polymer-rich top phase caused by volume exclusion effect. In most cases, it is recommended to start with a low molecular weight, depending on the product compatibility while optimizing the partitioning condition.

In addition, one of the limitations of using PEG and salt as the phase-forming component in LBS is that most of them cannot be recycled for the next process. The non-recyclable phase-forming component makes the overall LBS in downstream processes to be unfavorable as it causes environmental pollution and increases cost operation [78]. To improve the recyclability of phase-forming component in the LBS process, another similar research replaced using thermo-separating polymer (EOPO) as the phase component for the purification and recovery of CGTase [31]. The recovery of EOPO after recyclability was more than 80% verifying the viability of recyclable characteristics. This simple, rapid and recyclable feature show that the LBS process is a promising and attractive approach for the recovery and purification of target biomolecules.

4.2. Type and Concentration of Alcohol

The use of different alcohols (e.g., methanol, ethanol, 1-propanol, and 2-propanol) with different concentrations in the LBS will affect the overall recovery yield of target biomolecules. The exposure of active site from the implementation of organic solvent helps to maintain the enzyme's open conformation and bind the target compounds to the alcohol-rich top phase. A larger amount of alcohol is favorable as it will enhance the target biomolecules buoyancy and stability towards the interface layer.

Santos et al. (2016) conducted an experiment on extraction of caffeine from coffee bean and guaraná seed and reported the possibilities to manipulate the partitioning of caffeine to either the alcohol-rich top phase and salt-rich bottom phase [79]. For caffeine to be partitioned at alcohol-rich top phase, an increase in the concentration of 2-propanol caused the increment in the "caffeine-water" interaction. This effect will promote the biomolecules to be partitioned at the alcohol-rich top phase. Meanwhile, methanol was selected for caffeine to be partitioned at the salt-rich bottom phase. The purpose of selecting methanol was due to its low partition coefficient; therefore, increasing the tendency of caffeine to be partitioned at the salt-rich bottom phase.

A recent study on recovery of glycyrrhizic acid (GA) and liquiritin (LQ) from Chinese licorice root (*Glycyrrhiza uralensis Fisch*) reported that 87% GA and 94% LQ were successful obtained at alcohol-rich top phase under the optimum condition of 25% (*w*/*w*) ethanol and 30% (*w*/*w*) K_2HPO_4 in the LBS [80]. The effect of alcohol concentrations from 14 to 34% (*w*/*w*) and the extraction efficiency and partition coefficient were studied. By increasing the alcohol concentration to 26% in the system, the extraction efficiency and partition coefficient increased for both GA and LQ biomolecules. However, the extraction efficiency and partition coefficient decreased when the alcohol concentration was increased to 34%. This was due to the large amount of water-soluble alcohol in the alcohol-rich top phase interacting with the water molecules and causing the biomolecules to be partitioned to the salt-rich bottom phase [81]. This term was also referred as "volume exclusion" effect. In general, the selection of alcohol is mainly dependent on the target biomolecules from the complex crude extract. Each target biomolecule has their respective physico-chemical properties and therefore, it is difficult to govern a specific optimum condition for extraction and separation in LBS.

4.3. Type and Concentration of Salt

In the LBS, it is critical in selecting the type of salts as the phase-forming component since it can significantly affect the solubility and interaction of the target biomolecules. When the salt is added into a solution, the surface tension of water will increase which then leads to the increase of hydrophobic interaction between protein and water [82]. Few studies had shown that a high saturation level of salt concentration will cause a reduction in solubility of target biomolecules due to the higher salting-out ability of salt [20,36]. Lu et al., (2016) reported that the ability of salt solution and hydrophilic alcohol solution to form a biphasic system was mainly dependent on the Gibbs free energy of salt hydration [83]. The alteration in environmental phase system and behavior of biomolecules partitioning is utilized by the different salt components [84]. Different salts used for the LBS were based on their capability to support hydrophobic interaction between biomolecules [85]. According to the Hofmeister series, the salting-out ability of anions are arranged in the following order: $SO_4^{2-} > HPO_4^{2-} >$ citrate$^{3-} > F^- > Cl^- > Br^- > I^- > NO^{-3} > ClO^{4-}$ [25]. However, an optimum condition is required in order to obtain the maximum recovery of target biomolecules. It is also important to select a biodegradable and eco-friendly salt to ensure a more sustainable green approach in utilizing the LBS.

The effect of various salts used has been studied with the use of potassium dihydrogen phosphate (KH_2PO_4), magnesium sulfate ($MgSO_4$) and ammonium sulfate ($(NH_4)_2SO_4$) for the extraction of protein from *Chlorella sorokiniana* microalgae [10]. In this study, the salt concentration of 250 g/L were selected for each salt (KH_2PO_4, $MgSO_4$, and $(NH_4)_2SO_4$) as an optimum condition in the LBF. It was found that the KH_2PO_4, $MgSO_4$, and $(NH_4)_2SO_4$ exhibited high separation efficiency of 97.85%, 97.74%, and 97.74%, respectively. However, an observation was found using KH_2PO_4 solution where a white solid was formed and deposited around the interface at flotation time of 1.5 min, showing its incapability for the separation process. This formation happened when the properties of salt having a low solubility. Thus, an addition process is required to melt the solid salt solution. Another observation found using $MgSO_4$ solution was the absence of interface in the LBF after a flotation time of 4 min. In contrast, it was observed that only $(NH_4)_2SO_4$ solution could clearly render the highest recovery yield and purification values of 56.06% and 68.99%, respectively. The possible explanation was $(NH_4)_2SO_4$ has a lower molecular weight as compared with KH_2PO_4 and $MgSO_4$. As a conclusion, the extraction of protein is more favorable in the alcohol-rich top phase with increasing partitioning coefficient (K) when a low molecular weight salts is used [18,34]. However, the selection of various salts is still dependent on the compatibility of LBS and interaction among biomolecules.

The study of salt concentration was continued by using ammonium sulfate at the concentration range of 100 to 300 g/L. The effect of increasing salt concentration tends to increase the protein recovery yield. As supported by Phong et al. (2016), it was stated that the salting-out effect would occur at a higher salt concentration, the presence of ions tended to decrease the solubility of protein in the salt-rich bottom phase [9]. A further increase in salt concentration would decrease the protein recovery percentage. It was recommended to start with a minimum salt concentration of 20% (*w*/*w*) until the optimum condition was obtained rather introducing a high salt concentration abruptly.

4.4. pH System

The partitioning of target biomolecules can be affected by the pH system in LBS, due to a change in charges and solute properties of solute. The net charge of the target biomolecule becomes negative when the pH value is greater than the isoelectric point (pI) and positive when pH value is lower than the pI. If the net charge is equal to zero, both pH and pI values are equal [86]. Generally, it is found that in higher pH system would induce a positive dipole moment causing the partition coefficient to increase; therefore, favor the partitioning of negatively charge target biomolecules towards the polymer-rich top phase [87,88].

The partitioning of polyhydroxyalkanoate (PHA) from *Cupriavidus necator* H-16 in the thermoseparating-based LBS showed a good setup in altering the pH system as compared with conventional PEG-based LBS [89]. PHA showed a purification factor and recovery yield of 3.67% and 63.5%, respectively, at the pH 6 which was better than the conventional PEG-based LBS that had zero recovery of PHA in the top phase when pH was less than 7. However, there was a sudden drop in PHA recovery yield of 46.4% when the pH was adjusted to 8.0 to 8.8 in the system. Another study of extraction of BSA had shown that the different pH values could alter the net charge of targeted compound [90]. It was reported that the pH value increased from 6.0 to 9.0 which was larger than the isoelectric point of BSA (pI = 4.8) resulted in a maximum recovery yield of 84.32%. However, the high pH value is not favorable in the LBS since it can induce the protein denaturation.

Another experiment of antioxidants (i.e., xylooligosaccharides and phenol) extraction from wheat chaff explored the effect of pH on LBS [72]. The influence of pH value ranging from 2.5 to 7.0 was studied in the case of partitioning parameters of antioxidant such as recovery and partition coefficient. A maximum recovery of sugar ranging from 96% to 99% was obtained at pH 7.0 but the recovery of phenol decreased which could be explained by the phenol compound having a low pKa value of 4.5. In extend, at pH values near pKa such as pH 4.0 was reported that the partitioning of xylooligosaccharides was more towards the ethanol-rich top phase and phenol was more toward salt-rich bottom phase at the highest recovery of 75% and 77%, respectively. Hence, it is important to examine the effect of pH at the optimum condition to enhance the purification factor and recovery yield of the target biomolecules as it could be damaged or denatured by varying the sensitivity of pH conditions.

4.5. Temperature

The effect of temperature is dependent on the type of phase-forming components used in the LBS and stability of target biomolecules from denaturation. A change in temperature also affects the viscosity and density of the interface in the LBS. In most cases, the optimum temperature within the range of 20 to 40 °C was utilized for maximum recovery and partitioning of target biomolecules. The effect of temperature on the extraction efficiency of CPC from *Spirulina platensis* microalgae was studied and the maximum extraction efficiency up to 99.0% was achieved near the temperature range of 308 K [46]. It was found that lowering the temperature to 298 K caused the rate of CPC recovery to decrease, resulting in a low extraction efficiency. The influence of temperature on extraction efficiency study of BSA and papain was evaluated [52]. However, the studies showed that the extraction efficiency of both BSA and papain decreased when the temperature was increased. This phenomenon was due to the increasing temperature which could inhibit the interaction of amino acid and surface water of protein, resulting in less efficiency of protein extraction [89]. Hence, the effect of temperature should be taken into consideration as the extraction efficiency of the biomolecule is dependent on the range of temperature in the LBS.

5. Future Prospect and Challenges of LBS Application

The use of LBS is a promising separation technique for extraction of valuable biomolecules. The LBS can serve as an analytical tool to understand the chemical properties and behavior of target biomolecules. However, developing LBS as an alternative way for separation and purification for large-scale industrial application does encounter some key challenges that have to be re-addressed. One of the major concerns regarding the LBS is the partitioning coefficient (K) of the biomolecules into the top phase which is mainly dependent on the key parameters. It is time-consuming and crucial to investigate each of the key parameters in order to determine the optimum condition for maximum purification and recovery of biomolecules. The selection of phase-forming components should also be made concerning to their biocompatibility, hazards and biodegradability. Therefore, this favors an alternative phase-forming component which is more environmentally friendly and highly biodegradable in the aquatic environment.

Another challenge that needs to be addressed in the LBS is the extraction of biomolecules from natural sources. Regardless of various studies reporting the efficiency of LBS in the extraction of biomolecules from natural sources and microbial fermentation broths, it is still difficult to understand the biomolecules partition behavior, particularly when a complex crude feedstock is added into the LBS. Moreover, the contaminants might have the similar characteristics with the target biomolecules in the crude feedstock. This will cause lower extraction efficiency during the interaction with the extraction medium in the LBS. The lack of understanding on the partition behavior remains a challenge in utilizing the LBS for the recovery of biomolecules.

Furthermore, there is still some unexplored technologies such as magnetic and microwave approach which can be integrated with LBS for enhancing biomolecules extraction efficiency. The implementation of these new advanced technologies would be beneficial to enhance the knowledge in LBS. Moreover, it has been proven that the present assisted technologies using bubble-, ultrasound-, and electricity-assisted technologies showed a promising prospect in the recovery and purification of biomolecules. However, these assisted technologies required an in-depth study due to the lack of knowledge between its physico-chemical mechanism aligned with the LBS.

To maximize the large-scale use of LBS requires an ideal optimization technique at where the system can perform at its best desired performance for various application. A recent review by Torres-Acosta et al., (2019) has comprehensively evaluated the strategies to incorporate the LBS technologies in the industry [91]. One of the most frequent optimization techniques used is univariate optimization or known as one-factor-at-a-time (OFAT) analysis is where a single parameter at the time after the other is selected based on its best performance. Aside from that, response surface methodology (RSM) is another optimization technique found in most literature studies. This optimization technique composed of a statistical design that allows a simultaneous variation of several parameters compared to OFAT which depends on a single parameter at a time. Lastly, genetic algorithms (GA) is less frequently used compared to RSM however can deliver excellent results. The fundamental of GA involves the natural genetic inheritance (genotype) which relate specifically to the raw information of LBS components such as concentration of alcohol, salt, pH system, and temperature and then interprets the results (i.e., recovery yield, partition coefficient, and separation efficient) based on the characteristic of the LBS. The advantages of this GA compared to RSM is that GA does not require a regression or model tool as the optimizing approach, as the LBS is based on the previous results. In general, these optimization strategies are one of the best-selling points to make LBS to be implemented at industrial scales.

Aside from that, another strategy which could beneficial to the LBS is to study the recyclability and reusability of the phase-forming components in LBS. This is to ensure that the LBS not only can be employed as a separation and purification technique, but also promoting a sustainable low-cost process in the downstream processes. On top of that, the implementation of extractive technologies such as fermentation, cell disruption, bioconversion, crystallization, distillation, and metallurgy can be proposed along with the LBS to allow the production and purification tasks to occur in one-step process.

The advantages of this extractive technologies prevent the inhibition of the product and enhance the stability of biomolecules in the production stage. These benefits from extraction technologies should be further explored for the future development of LBS.

6. Conclusions

LBS is a simple, selective, scalable, and efficient tool to be utilized in downstream processing for the purification and recovery of biomolecules. However, it is still yet favorable to be used at the commercial scale as the complexity of the partitioning mechanism is difficult to predict. The challenges associated with the LBS techniques such as economic feasibility and the understanding of partition behavior need to be addressed to ensure the applicability in biotechnology industries. It is believed that more development along with various kind of technologies integrated in LBS will be discovered in the future. Hence, promoting the LBS to be used in commercial applications in recovering various high value bio-based products.

Author Contributions: Writing—original draft, K.S.K.; Writing—review & editing, K.S.K., H.Y.L., K.W.C. and J.-W.L.; Conceptualization, P.L.S.; Supervision & Funding acquisition, P.L.S., T.C.L. and H.-W.Y. All authors have read and agreed to the published version of the manuscript.

Funding: This work was supported by the Fundamental Research Grant Scheme, Malaysia [FRGS/1/2019/STG05/UNIM/02/2].

Conflicts of Interest: The authors declare that they have no conflicts of interests.

References

1. Khoo, K.S.; Lee, S.Y.; Ooi, C.W.; Fu, X.; Miao, X.; Ling, T.C.; Show, P.L. Recent Advances in Biorefinery of Astaxanthin from Haematococcus pluvialis. *Bioresour. Technol.* **2019**, *288*, 121606. [CrossRef] [PubMed]
2. Tham, P.E.; Ng, Y.J.; Sankaran, R.; Khoo, K.S.; Chew, K.W.; Yap, Y.J.; Malahubban, M.; Aziz Zakry, F.A.; Show, P.L. Recovery of Protein from Dairy Milk Waste Product Using Alcohol-Salt Liquid Biphasic Flotation. *Processes* **2019**, *7*, 875. [CrossRef]
3. Azevedo, A.M.; Rosa, P.A.; Ferreira, I.F.; Aires-Barros, M.R. Chromatography-free recovery of biopharmaceuticals through aqueous two-phase processing. *Trends Biotechnol.* **2009**, *27*, 240–247. [CrossRef] [PubMed]
4. Drexler, I.L.; Yeh, D.H. Membrane applications for microalgae cultivation and harvesting: A review. *Rev. Environ. Sci. Bio* **2014**, *13*, 487–504. [CrossRef]
5. Grossmann, L.; Ebert, S.; Hinrichs, J.; Weiss, J. Effect of precipitation, lyophilization, and organic solvent extraction on preparation of protein-rich powders from the microalgae Chlorella protothecoides. *Algal. Res.* **2018**, *29*, 266–276. [CrossRef]
6. Sankaran, R.; Manickam, S.; Yap, Y.J.; Ling, T.C.; Chang, J.-S.; Show, P.L. Extraction of proteins from microalgae using integrated method of sugaring-out assisted liquid biphasic flotation (LBF) and ultrasound. *Ultrason. Sonochem.* **2018**, *48*, 231–239. [CrossRef]
7. Khoo, K.S.; Chew, K.W.; Ooi, C.W.; Ong, H.C.; Ling, T.C.; Show, P.L. Extraction of natural astaxanthin from Haematococcus pluvialis using liquid biphasic flotation system. *Bioresour. Technol.* **2019**, *290*, 121794. [CrossRef]
8. Leong, H.Y.; Ooi, C.W.; Law, C.L.; Julkifle, A.L.; Ling, T.C.; Show, P.L. Application of liquid biphasic flotation for betacyanins extraction from peel and flesh of Hylocereus polyrhizus and antioxidant activity evaluation. *Sep. Purif. Technol.* **2018**, *201*, 156–166. [CrossRef]
9. Phong, W.N.; Le, C.F.; Show, P.L.; Chang, J.S.; Ling, T.C. Extractive disruption process integration using ultrasonication and an aqueous two-phase system for protein recovery from *Chlorella sorokiniana*. *Eng. Life Sci.* **2017**, *17*, 357–369. [CrossRef]
10. Phong, W.N.; Show, P.L.; Teh, W.H.; Teh, T.X.; Lim, H.M.Y.; binti Nazri, N.S.; Tan, C.H.; Chang, J.-S.; Ling, T.C. Proteins recovery from wet microalgae using liquid biphasic flotation (LBF). *Bioresour. Technol.* **2017**, *244*, 1329–1336. [CrossRef]
11. Lee, S.Y.; Khoiroh, I.; Ling, T.C.; Show, P.L. Aqueous Two-Phase Flotation for the Recovery of Biomolecules. *Sep. Purif. Rev.* **2016**, *45*, 81–92. [CrossRef]

12. Sankaran, R.; Show, P.L.; Cheng, Y.-S.; Tao, Y.; Ao, X.; Nguyen, T.D.P.; Van Quyen, D. Integration Process for Protein Extraction from Microalgae Using Liquid Biphasic Electric Flotation (LBEF) System. *Mol. Biotechnol.* **2018**, *60*, 749–761. [CrossRef] [PubMed]
13. Koyande, A.K.; Chew, K.W.; Rambabu, K.; Tao, Y.; Chu, D.T.; Show, P.L. Microalgae: A potential alternative to health supplementation for humans. *Food Science and Human Wellness.* **2019**, *8*, 16–24. [CrossRef]
14. Chew, K.W.; Chia, S.R.; Krishnamoorthy, R.; Tao, Y.; Chu, D.-T.; Show, P.L. Liquid biphasic flotation for the purification of C-phycocyanin from *Spirulina platensis* microalga. *Bioresour. Technol.* **2019**, *288*, 121519. [CrossRef] [PubMed]
15. Van Berlo, M.; Luyben, K.C.A.; van der Wielen, L.A. Poly (ethylene glycol)–salt aqueous two-phase systems with easily recyclable volatile salts. *J. Chromatogr. B Biomed. Sci. Appl.* **1998**, *711*, 61–68. [CrossRef]
16. Grilo, A.L.; Raquel Aires-Barros, M.; Azevedo, A.M. Partitioning in aqueous two-phase systems: Fundamentals, applications and trends. *Sep. Purif. Rev.* **2016**, *45*, 68–80. [CrossRef]
17. Albertsson, P. Fractionation of particles and macromolecules in aqueous two-phase systems. *Biochem. Pharmcol.* **1961**, *5*, 351–358. [CrossRef]
18. Albertsson, P.A. *Partitioning of Cell Particles and Macromolecules*; Wiley: New York, NY, USA, 1986.
19. Asenjo, J.A.; Andrews, B.A. Aqueous two-phase systems for protein separation: Phase separation and applications. *J. Chromatogr. A* **2012**, *1238*, 1–10. [CrossRef]
20. Ooi, C.W.; Tey, B.T.; Hii, S.L.; Kamal, S.M.M.; Lan, J.C.W.; Ariff, A.; Ling, T.C. Purification of lipase derived from Burkholderia pseudomallei with alcohol/salt-based aqueous two-phase systems. *Process Biochem.* **2009**, *44*, 1083–1087. [CrossRef]
21. Zhao, L.; Peng, Y.-L.; Gao, J.-M.; Cai, W.-M. Bioprocess intensification: An aqueous two-phase process for the purification of C-phycocyanin from dry Spirulina platensis. *Eur. Food Res. Technol.* **2014**, *238*, 451–457. [CrossRef]
22. Rosa, P.A.J.; Ferreira, I.F.; Azevedo, A.M.; Aires-Barros, M.R. Aqueous two-phase systems: A viable platform in the manufacturing of biopharmaceuticals. *J. Chromatogr. A* **2010**, 1217. [CrossRef] [PubMed]
23. Raja, S.; Murty, V.R.; Thivaharan, V.; Rajasekar, V.; Ramesh, V. Aqueous two phase systems for the recovery of biomolecules–a review. *Sci. Technol.* **2011**, *1*, 7–16. [CrossRef]
24. Iqbal, M.; Tao, Y.; Xie, S.; Zhu, Y.; Chen, D.; Wang, X.; Huang, L.; Peng, D.; Sattar, A.; Shabbir, M.A.B. Aqueous two-phase system (ATPS): An overview and advances in its applications. *Biol. Proced. Online* **2016**, *18*, 18. [CrossRef] [PubMed]
25. Hatti-Kaul, R. *Aqueous Two-Phase Systems: Methods and Protocols*; Springer Science & Business Media: Berlin/Heidelberg, Germany, 2000; Volume 11.
26. Raja, S.; Murty, V.R. Development and evaluation of environmentally benign aqueous two phase systems for the recovery of proteins from tannery waste water. *ISRN Chem. Eng.* **2012**. [CrossRef]
27. Johansson, H.-O.; Feitosa, E.; Junior, A.P. Phase diagrams of the aqueous two-phase systems of poly (ethylene glycol)/sodium polyacrylate/salts. *Polymers* **2011**, *3*, 587–601. [CrossRef]
28. Hou, D.; Li, Y.; Cao, X. Synthesis of two thermo-sensitive copolymers forming aqueous two-phase systems. *Sep. Purif. Technol.* **2014**, *122*, 217–224. [CrossRef]
29. Tan, Z.; Li, F.; Zhao, C.; Teng, Y.; Liu, Y. Chiral separation of mandelic acid enantiomers using an aqueous two-phase system based on a thermo-sensitive polymer and dextran. *Sep. Purif. Technol.* **2017**, *172*, 382–387. [CrossRef]
30. Leong, Y.K.; Lan, J.C.W.; Loh, H.S.; Ling, T.C.; Ooi, C.W.; Show, P.L. Thermoseparating aqueous two-phase systems: Recent trends and mechanisms. *J. Sep. Sci.* **2016**, *39*, 640–647. [CrossRef]
31. Show, P.L.; Tan, C.P.; Shamsul Anuar, M.; Ariff, A.; Yusof, Y.A.; Chen, S.K.; Ling, T.C. Extractive fermentation for improved production and recovery of lipase derived from Burkholderia cepacia using a thermoseparating polymer in aqueous two-phase systems. *Bioresour. Technol.* **2012**, 116. [CrossRef]
32. Ng, H.S.; Tan, C.P.; Mokhtar, M.N.; Ibrahim, S.; Ariff, A.; Ooi, C.W.; Ling, T.C. Recovery of Bacillus cereus cyclodextrin glycosyltransferase and recycling of phase components in an aqueous two-phase system using thermo-separating polymer. *Sep. Purif. Technol.* **2012**, *89*, 9–15. [CrossRef]
33. Lin, Y.K.; Show, P.L.; Yap, Y.J.; Tan, C.P.; Ng, E.-P.; Ariff, A.B.; Annuar, M.S.B.M.; Ling, T.C. Direct recovery of cyclodextringlycosyltransferase from Bacillus cereus using aqueous two-phase flotation. *J. Biosci. Bioeng.* **2015**, *120*, 684–689. [CrossRef] [PubMed]

34. Jong, W.Y.L.; Show, P.L.; Ling, T.C.; Tan, Y.S. Recovery of lignin peroxidase from submerged liquid fermentation of Amauroderma rugosum (Blume & T. Nees) Torrend using polyethylene glycol/salt aqueous two-phase system. *J. Biosci. Bioeng.* **2017**, *124*, 91–98. [PubMed]
35. Lin, Y.K.; Ooi, C.W.; Tan, J.S.; Show, P.L.; Ariff, A.; Ling, T.C. Recovery of human interferon alpha-2b from recombinant Escherichia coli using alcohol/salt-based aqueous two-phase systems. *Sep. Purif. Technol.* **2013**, *120*, 362–366. [CrossRef]
36. Chia, S.R.; Show, P.L.; Phang, S.-M.; Ling, T.C.; Ong, H.C. Sustainable approach in phlorotannin recovery from macroalgae. *J. Biosci. Bioeng.* **2018**, *126*, 220–225. [CrossRef]
37. Smiglak, M.; Metlen, A.; Rogers, R.D. The Second Evolution of Ionic Liquids: From Solvents and Separations to Advanced Materials Energetic Examples from the Ionic Liquid Cookbook. *Acc. Chem. Res.* **2007**, *40*, 1182–1192. [CrossRef]
38. Lee, S.Y.; Vicente, F.A.; e Silva, F.A.; Sintra, T.E.; Taha, M.; Khoiroh, I.; Coutinho, J.O.A.; Show, P.L.; Ventura, S.P. Evaluating self-buffering ionic liquids for biotechnological applications. *ACS Sustain. Chem. Eng.* **2015**, *3*, 3420–3428. [CrossRef]
39. Zhang, S.; Sun, N.; He, X.; Lu, X.; Zhang, X. Physical properties of ionic liquids: Database and evaluation. *J. Phys. Chem. Ref. Data* **2006**, *35*, 1475–1517. [CrossRef]
40. Lee, S.Y.; Khoiroh, I.; Ooi, C.W.; Ling, T.C.; Show, P.L. Recent advances in protein extraction using ionic liquid-based aqueous two-phase systems. *Sep. Purif. Rev.* **2017**, *46*, 291–304. [CrossRef]
41. Ostadjoo, S.; Berton, P.; Shamshina, J.L.; Rogers, R.D. Scaling-up ionic liquid-based technologies: How much do we care about their toxicity? Prima facie information on 1-ethyl-3-methylimidazolium acetate. *Toxicol. Sci.* **2017**, *161*, 249–265. [CrossRef]
42. Sun, N.; Rahman, M.; Qin, Y.; Maxim, M.L.; Rodríguez, H.; Rogers, R.D. Complete dissolution and partial delignification of wood in the ionic liquid 1-ethyl-3-methylimidazolium acetate. *Green Chem.* **2009**, *11*, 646–655. [CrossRef]
43. Swatloski, R.P.; Spear, S.K.; Holbrey, J.D.; Rogers, R.D. Dissolution of cellose with ionic liquids. *J. Am. Chem. Soc.* **2002**, *124*, 4974–4975. [CrossRef] [PubMed]
44. Gutowski, K.E.; Broker, G.A.; Willauer, H.D.; Huddleston, J.G.; Swatloski, R.P.; Holbrey, J.D.; Rogers, R.D. Controlling the aqueous miscibility of ionic liquids: Aqueous biphasic systems of water-miscible ionic liquids and water-structuring salts for recycle, metathesis, and separations. *J. Am. Chem. Soc.* **2003**, *125*, 6632–6633. [CrossRef] [PubMed]
45. Du, Z.; Yu, Y.L.; Wang, J.H. Extraction of proteins from biological fluids by use of an ionic liquid/aqueous two-phase system. *Chem. A Eur. J.* **2007**, *13*, 2130–2137. [CrossRef] [PubMed]
46. Ng, H.S.; Ooi, C.W.; Show, P.L.; Tan, C.P.; Ariff, A.; Moktar, M.N.; Ng, E.-P.; Ling, T.C. Recovery of Bacillus cereus cyclodextrin glycosyltransferase using ionic liquid-based aqueous two-phase system. *Sep. Purif. Technol.* **2014**, *138*, 28–33. [CrossRef]
47. Chang, Y.-K.; Show, P.-L.; Lan, J.C.-W.; Tsai, J.-C.; Huang, C.-R. Isolation of C-phycocyanin from Spirulina platensis microalga using Ionic liquid based aqueous two-phase system. *Bioresour. Technol.* **2018**, *270*, 320–327. [CrossRef]
48. Garcia, E.S.; Ruiz, C.A.S.; Tilaye, T.; Eppink, M.H.; Wijffels, R.H.; van den Berg, C. Fractionation of proteins and carbohydrates from crude microalgae extracts using an ionic liquid based-aqueous two phase system. *Sep. Purif. Technol.* **2018**, *204*, 56–65. [CrossRef]
49. Abbott, A.P.; Boothby, D.; Capper, G.; Davies, D.L.; Rasheed, R.K. Deep eutectic solvents formed between choline chloride and carboxylic acids: Versatile alternatives to ionic liquids. *J. Am. Chem. Soc.* **2004**, *126*, 9142–9147. [CrossRef]
50. Shishov, A.; Bulatov, A.; Locatelli, M.; Carradori, S.; Andruch, V. Application of deep eutectic solvents in analytical chemistry. A review. *Microchem. J.* **2017**, *135*, 33–38. [CrossRef]
51. Paiva, A.; Craveiro, R.; Aroso, I.; Martins, M.; Reis, R.L.; Duarte, A.R.C. Natural deep eutectic solvents–solvents for the 21st century. *ACS Sustain. Chem. Eng.* **2014**, *2*, 1063–1071. [CrossRef]
52. Dai, Y.; van Spronsen, J.; Witkamp, G.-J.; Verpoorte, R.; Choi, Y.H. Natural deep eutectic solvents as new potential media for green technology. *Anal. Chim. Acta* **2013**, *766*, 61–68. [CrossRef]
53. Pang, J.; Sha, X.; Chao, Y.; Chen, G.; Han, C.; Zhu, W.; Li, H.; Zhang, Q. Green aqueous biphasic systems containing deep eutectic solvents and sodium salts for the extraction of protein. *RSC Adv.* **2017**, *7*, 49361–49367. [CrossRef]

54. Zeng, Q.; Wang, Y.; Huang, Y.; Ding, X.; Chen, J.; Xu, K. Deep eutectic solvents as novel extraction media for protein partitioning. *Analyst* **2014**, *139*, 2565–2573. [CrossRef]
55. Zhang, X.; Teng, G.; Zhang, J. Deep eutectic solvents aqueous two-phase system based ultrasonically assisted extraction of ursolic acid (UA) from *Cynomorium songaricum* Rupr. *Chem. Eng. Commun.* **2019**, *206*, 419–431. [CrossRef]
56. Weschayanwiwat, P.; Kunanupap, O.; Scamehorn, J.F. Benzene removal from waste water using aqueous surfactant two-phase extraction with cationic and anionic surfactant mixtures. *Chemosphere* **2008**, *72*, 1043–1048. [CrossRef] [PubMed]
57. Selber, K.; Tjerneld, F.; Collén, A.; Hyytiä, T.; Nakari-Setälä, T.; Bailey, M.; Fagerström, R.; Kan, J.; Van Der Laan, J.; Penttilä, M. Large-scale separation and production of engineered proteins, designed for facilitated recovery in detergent-based aqueous two-phase extraction systems. *Process Biochem.* **2004**, *39*, 889–896. [CrossRef]
58. Amid, M.; Manap, M.; Hussin, M.; Mustafa, S. A novel aqueous two phase system composed of surfactant and xylitol for the purification of lipase from pumpkin (Cucurbita moschata) seeds and recycling of phase components. *Molecules* **2015**, *20*, 11184–11201. [CrossRef] [PubMed]
59. Sankaran, R.; Show, P.L.; Yap, Y.J.; Tao, Y.; Ling, T.C.; Tomohisa, K. Green technology of liquid biphasic flotation for enzyme recovery utilizing recycling surfactant and sorbitol. *Clean Technol. Environ. Policy* **2018**, *20*, 2001–2012. [CrossRef]
60. Sebba, F. *Ion Flotation*; Elsevier: Amsterdam, The Netherlands, 1962; Volume 30.
61. Sankaran, R.; Show, P.L.; Lee, S.Y.; Yap, Y.J.; Ling, T.C. Integration process of fermentation and liquid biphasic flotation for lipase separation from *Burkholderia cepacia*. *Bioresour. Technol.* **2018**, *250*, 306–316. [CrossRef]
62. Mathiazakan, P.; Shing, S.Y.; Ying, S.S.; Kek, H.K.; Tang, M.S.; Show, P.L.; Ooi, C.-W.; Ling, T.C. Pilot-scale aqueous two-phase floatation for direct recovery of lipase derived from *Burkholderia cepacia* strain ST8. *Sep. Purif. Technol.* **2016**, *171*, 206–213. [CrossRef]
63. Bi, P.-y.; Dong, H.-r.; Yuan, Y.-c. Application of aqueous two-phase flotation in the separation and concentration of puerarin from Puerariae extract. *Sep. Purif. Technol.* **2010**, *75*, 402–406. [CrossRef]
64. Jiang, B.; Wang, L.; Na, J.; Zhang, X.; Yuan, Y.; Liu, C.; Feng, Z. Environmentally-friendly strategy for separation of α-lactalbumin from whey by aqueous two phase flotation. *Arab. J. Chem.* **2018**, *13*, 3391–3402. [CrossRef]
65. Wu, X.; Joyce, E.M.; Mason, T.J. The effects of ultrasound on cyanobacteria. *Harmful Algae* **2011**, *10*, 738–743. [CrossRef]
66. Wang, M.; Yuan, W.; Jiang, X.; Jing, Y.; Wang, Z. Disruption of microalgal cells using high-frequency focused ultrasound. *Bioresour. Technol.* **2014**, *153*, 315–321. [CrossRef]
67. Gerde, J.A.; Montalbo-Lomboy, M.; Yao, L.; Grewell, D.; Wang, T. Evaluation of microalgae cell disruption by ultrasonic treatment. *Bioresour. Technol.* **2012**, *125*, 175–181. [CrossRef] [PubMed]
68. Chemat, F.; Rombaut, N.; Sicaire, A.; Meullemiestre, A.; Abert-vian, M.; Fabiano-Tixier, A.; Abert-vian, M. Ultrasonics Sonochemistry Ultrasound assisted extraction of food and natural products. Mechanisms, techniques, combinations, protocols and applications. A review. *Ultrason. Sonochem.* **2017**, *34*, 540–560.
69. Pakhale, S.V.; Vetal, M.D.; Rathod, V.K. Separation of bromelain by aqueous two phase flotation. *Sep. Sci. Technol.* **2013**, *48*, 984–989. [CrossRef]
70. Rahim, A.H.A.; Khoo, K.S.; Yunus, N.M.; Hamzah, W.S.W. Ether-Functionalized Ionic Liquids as Solvent for Gigantochloa Scortechini Dissolution. Proceedings of AIP Conference Proceedings, Cesme-Izmir, Turkey, 26–30 May 2019; p. 020025.
71. Dong, B.; Yuan, X.; Zhao, Q.; Feng, Q.; Liu, B.; Guo, Y.; Zhao, B. Ultrasound-assisted aqueous two-phase extraction of phenylethanoid glycosides from *Cistanche deserticola* YC Ma stems. *J. Sep. Sci.* **2015**, *38*, 1194–1203. [CrossRef] [PubMed]
72. Đorđević, T.; Antov, M. Ultrasound assisted extraction in aqueous two-phase system for the integrated extraction and separation of antioxidants from wheat chaff. *Sep. Purif. Technol.* **2017**, *182*, 52–58. [CrossRef]
73. Eing, C.J.; Bonnet, S.; Pacher, M.; Puchta, H.; Frey, W. Effects of nanosecond pulsed electric field exposure on Arabidopsis thaliana. *IEEE Trans. Dielectr. Electr. Insul.* **2009**, *16*, 1322–1328. [CrossRef]
74. Kotnik, T.; Frey, W.; Sack, M.; Meglič, S.H.; Peterka, M.; Miklavčič, D. Electroporation-based applications in biotechnology. *Trends Biotechnol.* **2015**, *33*, 480–488. [CrossRef]

75. Luengo, E.; Condón-Abanto, S.; Álvarez, I.; Raso, J. Effect of pulsed electric field treatments on permeabilization and extraction of pigments from Chlorella vulgaris. *J. Membr. Biol.* **2014**, *247*, 1269–1277. [CrossRef]
76. Leong, H.Y.; Ooi, C.W.; Law, C.L.; Julkifle, A.L.; Katsuda, T.; Show, P.L. Integration process for betacyanins extraction from peel and flesh of Hylocereus polyrhizus using liquid biphasic electric flotation system and antioxidant activity evaluation. *Sep. Purif. Technol.* **2019**, *209*, 193–201. [CrossRef]
77. 't Lam, G.P.; van der Kolk, J.A.; Chordia, A.; Vermuë, M.H.; Olivieri, G.; Eppink, M.H.; Wijffels, R.H. Mild and selective protein release of cell wall deficient microalgae with pulsed electric field. *ACS Sustain. Chem. Eng.* **2017**, *5*, 6046–6053. [CrossRef]
78. Phong, W.N.; Show, P.L.; Chow, Y.H.; Ling, T.C. Recovery of biotechnological products using aqueous two phase systems. *J. Biosci. Bioeng.* **2018**, *126*, 273–281. [CrossRef] [PubMed]
79. Santos, S.B.; Reis, I.A.; Silva, C.P.; Campos, A.F.; Ventura, S.P.; Soares, C.M.; Lima, Á.S. Selective partition of caffeine from coffee bean and guaraná seed extracts using alcohol–salt aqueous two-phase systems. *Sep. Sci. Technol.* **2016**, *51*, 2008–2019. [CrossRef]
80. Huang, L.; Li, W.; Feng, Y.; Fang, X.; Li, J.; Gao, Z.; Li, H. Simultaneous recovery of glycyrrhizic acid and liquiritin from Chinese licorice root (Glycyrrhiza uralensis Fisch) by aqueous two-phase system and evaluation biological activities of extracts. *Sep. Sci. Technol.* **2018**, *53*, 1342–1350. [CrossRef]
81. Wang, Y.; Liu, Y.; Han, J.; Hu, S. Application of water-miscible alcohol-based aqueous two-phase systems for extraction of dyes. *Sep. Sci. Technol.* **2011**, *46*, 1283–1288. [CrossRef]
82. Wingfield, P. Protein precipitation using ammonium sulfate. *Curr. Protoc. Protein Sci.* **1998**, *13*, A. 3F. 1–A. 3F. 8.
83. Lu, Y.; Yu, M.; Tan, Z.; Yan, Y. Phase equilibria and salt effect on the aqueous two-phase system of polyoxyethylene cetyl ether and sulfate salt at three temperatures. *J. Chem. Eng. Data* **2016**, *61*, 2135–2143. [CrossRef]
84. Yang, L.; Huo, D.; Hou, C.; He, K.; Lv, F.; Fa, H.; Luo, X. Purification of plant-esterase in PEG1000/NaH2PO4 aqueous two-phase system by a two-step extraction. *Process Biochem.* **2010**, *45*, 1664–1671. [CrossRef]
85. Goja, A.M.; Yang, H.; Cui, M.; Li, C. Aqueous two-phase extraction advances for bioseparation. *J. Bioprocess. Biotechnol.* **2013**, *4*, 1–8.
86. Karr, L.J.; Shafer, S.G.; Harris, J.M.; Van Alstine, J.M.; Snyder, R.S. Immuno-affinity partition of cells in aqueous polymer two-phase systems. *J. Chromatogr. A* **1986**, *354*, 269–282. [CrossRef]
87. Andrews, B.; Schmidt, A.; Asenjo, J. Correlation for the partition behavior of proteins in aqueous two-phase systems: Effect of surface hydrophobicity and charge. *Biotechnol. Bioeng.* **2005**, *90*, 380–390. [CrossRef] [PubMed]
88. Olivera-Nappa, A.; Lagomarsino, G.; Andrews, B.A.; Asenjo, J.A. Effect of electrostatic energy on partitioning of proteins in aqueous two-phase systems. *J. Chromatogr. B* **2004**, *807*, 81–86. [CrossRef] [PubMed]
89. Chakraborty, A.; Sen, K. Impact of pH and temperature on phase diagrams of different aqueous biphasic systems. *J. Chromatogr. A* **2016**, *1433*, 41–55. [CrossRef] [PubMed]
90. Chow, Y.H.; Yap, Y.J.; Tan, C.P.; Anuar, M.S.; Tejo, B.A.; Show, P.L.; Ariff, A.B.; Ng, E.-P.; Ling, T.C. Characterization of bovine serum albumin partitioning behaviors in polymer-salt aqueous two-phase systems. *J. Biosci. Bioeng.* **2015**, *120*, 85–90. [CrossRef] [PubMed]
91. Torres-Acosta, M.A.; Mayolo-Deloisa, K.; González-Valdez, J.; Rito-Palomares, M. Aqueous Two-Phase Systems at Large Scale: Challenges and Opportunities. *Biotechnol. J.* **2019**, *14*, 1800117. [CrossRef] [PubMed]

Review

Anaerobic Co-Digestion of Wastewater Sludge: A Review of Potential Co-Substrates and Operating Factors for Improved Methane Yield

Wei Ling Chow [1], Siewhui Chong [1], Jun Wei Lim [2], Yi Jing Chan [1,*], Mei Fong Chong [3], Timm Joyce Tiong [1], Jit Kai Chin [4] and Guan-Ting Pan [5,*]

1 Department of Chemical and Environmental Engineering, University of Nottingham Malaysia, Broga Road, Semenyih 43500, Selangor Darul Ehsan, Malaysia; Weiling.chow@hotmail.co.uk (W.L.C.); Faye.chong@nottingham.edu.my (S.C.); Joyce.tiong@nottingham.edu.my (T.J.T.)

2 Department of Fundamental and Applied Sciences, Centre for Biofuel and Biochemical Research, Institute of Self-Sustainable Building, Universiti Teknologi PETRONAS, Seri Iskandar 32610, Perak Darul Ridzuan, Malaysia; Junwei.lim@utp.edu.my

3 28, Jalan Pulau Tioman U10/94, Taman Greenhill, Shah Alam 40170, Selangor Darul Ehsan, Malaysia; Chong_mei_fong@yahoo.com

4 Department of Chemical Sciences, School of Applied Sciences, University of Huddersfield, Queensgate, Huddersfield HD1 3DH, UK; J.chin@hud.ac.uk

5 Department of Chemical Engineering and Biotechnology, National Taipei University of Technology, No. 1, Section 3, Zhongxiao E Rd, Da'an District, Taipei City 106, Taiwan

* Correspondence: Yi-jing.chan@nottingham.edu.my (Y.J.C.); Gtpan@ntut.edu.tw (G.-T.P.); Tel.: +60-3-8924-8773 (Y.J.C.); +886-2-2771-2171 (G.-T.P.)

Received: 20 August 2019; Accepted: 20 September 2019; Published: 1 January 2020

Abstract: Anaerobic digestion has been widely employed in waste treatment for its ability to capture methane gas released as a product during the digestion. Certain wastes, however, cannot be easily digested due to their low nutrient level insufficient for anaerobic digestion, thus co-digestion is a viable option. Numerous studies have shown that using co-substrates in anaerobic digestion systems improve methane yields as positive synergisms are established in the digestion medium, and the supply of missing nutrients are introduced by the co-substrates. Nevertheless, large-scale implementation of co-digestion technology is limited by inherent process limitations and operational concerns. This review summarizes the results from numerous laboratory, pilot, and full-scale anaerobic co-digestion (ACD) studies of wastewater sludge with the co-substrates of organic fraction of municipal solid waste, food waste, crude glycerol, agricultural waste, and fat, oil and grease. The critical factors that influence the ACD operation are also discussed. The ultimate aim of this review is to identify the best potential co-substrate for wastewater sludge anaerobic co-digestion and provide a recommendation for future reference. By adding co-substrates, a gain ranging from 13 to 176% in the methane yield was accomplished compared to the mono-digestions.

Keywords: anaerobic digestion; co-digestion; wastewater; biogas production; methane yield, sludge

---

1. Introduction

Many industries face difficulties in treating their high-strength wastewaters. These wastewaters usually consist of a high chemical oxygen demand (COD) of more than 10,000 mg/L, or biochemical oxygen demand (BOD) of more than 5000 mg/L. For instance, pineapple industrial wastewater and palm oil mill effluent, generated at a rate of 0.5 to 0.75 tonnes per tonne of fresh fruit bunch, have a COD of 50,000–80,000 mg/L [1,2]. These high-strength wastewaters need to undergo a series of treatment processes to meet the local allowable discharge limits. Anaerobic co-digestion (ACD) has been recognized as an effective effluent treatment for industrial wastes, due to its ability to counter-balance nutrient insufficiency and economic feasibility. ACD is a process of adding energy rich organic waste materials to wastewater digesters in the absence of oxygen, as bacteria break down organic materials and produce biogas. The most common processes involve a major amount of a main basic substrate, for example, manure or sewage sludge, which is mixed and digested together with minor amounts of a single, or a variety of additional substrate [3].

ACD has strong potential to contribute to both pollution control and energy recovery. One of the major advantages of ACD is that it increases the efficiency of organic waste degradation and thus biogas production. The capture of methane, which can be used as an energy source, is contributed by diverting the wastes from landfills to wastewater treatment facilities. Co-digestion also forms an ideal nutrient balance resulting in an increase of digestion performance and biogas yields. Because of the use of co-substrates, which establish positive synergisms in the digestion medium, and act as a supplier of missing nutrients, biogas yields are found to be improved from mono-digestion [4,5]. In addition, co-digestion gives a diversion opportunity to reduce landfill space, besides aiding most municipalities to achieve their recovery goals. Therefore, wastewater treatment facilities can expect to see a cost savings from incorporating wastes into anaerobic digesters with mono-digestion systems. These include reduced-energy costs resulting from production of on-site power from biogas yields.

While there have been many studies on the advantages of ACD over mono-digestion [6], and its effective outcome in treating waste, little to no research has reviewed the feasibility of various co-substrates associated with their influential factors on co-digestion, and their optimum operating conditions. Although co-digestion carries numerous benefits, many researchers have still encountered difficulties in performing co-digestion, which have sometimes led to system upset, mainly due to inappropriate substrate ratios and operating conditions. This challenge was demonstrated by Wan, et al. [7], who reported a digestion failure when adding 75% or more of fat, oil and grease feed, which may have caused acidification of the digester. This is also supported by Astals, et al. [8], who reported system upset resulting in low methane yield when incorporating more than 4% of crude glycerol into the digester. Wan, et al. [7] further added that a relatively short hydraulic retention time (HRT) of ten days employed might also contribute to the digestion failure, due to washout of microorganisms during removal of treated wastewater.

Despite the possible disadvantages of ACD, which likely have occurred through lack of knowledge of ACD practice, this review discusses how results from numerous laboratory, pilot, and full-scale ACD studies can be used to address the feasibility and potential of different co-substrates used for ACD using domestic wastewater sludge as the main substrate. The review starts with the summaries and discussions of the anaerobic co-digestion results from recent literature, and ends with a comparative analysis to identify the best co-substrate for the ACD of wastewater sludge. At the end, a recommendation on how the anaerobic co-digestion can be integrated into a wastewater treatment plant (WWTP) is presented. This review aims to serve as a guideline for other researchers to achieve optimum operating conditions for co-digestion leading to minimal operational problems and maximum biogas production.

2. Potential Co-Substrates for Anaerobic Co-Digestion of Wastewater Sludge

In a wastewater treatment plant, wastewater sludge such as sewage sludge, waste activated sludge, and thickened waste activated sludge, are usually sent to an anaerobic digester to remove solids and generate biogas. As shown in Table 1, the potential co-substrates used in anaerobic co-digestion of wastewater sludge are identified to be the organic fraction of municipal solid wastes (OFMSW), food waste (FW), agricultural wastes (AW), crude glycerol (CG), and fat, oil and grease (FOG). In order to achieve the optimal carbon-to-nitrogen (C/N) ratio for anaerobic digestion, which is around 20–30 [9], wastewater sludge with a low C/N ratio of 6–10 can be co-digested with the co-substrate with a higher C/N ratio to counterbalance the nutrients and avoid inhibition that leads to system instability.

Table 1. Carbon-to-nitrogen (C/N) ratio of the primary and co-substrates reviewed in this study.

Common Primary Wastewater Substrates		C/N Ratio	Ref.
Sewage sludge (SS)		6–10	[10]
Waste activated sludge (WAS)		10	[11]
Thickened waste activated sludge (TWAS)		6–9	[12]
Common Co-Substrates	**Example**	**C/N Ratio**	**Ref.**
Organic fraction of municipal solid wastes (OFMSW)		11–21	[13]
Food waste (FW)		11–15	[14]
Agricultural wastes (AW):	Rice straw	50–53	[12]
	Potatoes	35–60	[12]
	Corn stalks/straw	50–56	[12]
	Sugar cane/bagasse	140–150	[12]
	Sugar beet	35–40	[12]
	Grass/trimmings	12–16	[12]
	Fallen leaves	50–53	[12]
	Horse manure	20–25	[12]
	Pig manure	6–14	[12]
	Cow dung	16–25	[12]
Crude glycerol (CG)		68	[15]
Fat, oil and grease (FOG)		22	[7]

2.1. *Organic Fraction of Municipal Solid Waste (OFMSW) and Food Waste (FW)*

Municipal solid waste consists of diverse, discarded items such as product packaging, grass clippings, furniture, clothing, bottles, food scraps, newspapers, appliances, paint, and batteries, originating from residential, commercial and institutional locations [16]. The U.S. Environmental Protection Agency's (EPA) definition of municipal solid waste, however, does not include industrial, hazardous, or construction and demolition waste [16]. The organic fraction of municipal solid waste on the other hand, can be defined as the biodegradable fraction consisting of food waste, kitchen waste, leaf, grass clippings, flower trimmings and yard waste, of which food waste occupies the highest proportion [17].

The C/N ratio of OFMSW is around 11–21, and moisture content around 70–82% [13]; whereas for food waste, the C/N ratio is around 11–15 [14] and moisture content of 82–86% [18]. Because of its high biodegradability and suitable C/N ratio, OFMSW and food waste are potential co-substrates for anaerobic digestion. Fruit and vegetable waste, on the other hand, is a source-selected food waste consisting of only waste fruits and vegetables.

Table 2 shows the summary of studies utilizing OFMSW, FW or FVW as co-substrates for the anaerobic co-digestion of wastewater sludge. Cabbai, et al. [19] demonstrated an improvement of 16% and 48% in methane yield when co-digesting sewage sludge with FVW and FW in a volatile solids (VS) ratio of 1:0.23 and 1:2.09, respectively. However, the VS removal reduced by 25–40%. Dai, et al. [14] showed an increase in methane yield from 25% to 67%, and VS removal from 51% to 70%, when the VS ratio of dewatered sludge and FW decreased from 2.4:1 to 0.4:1. These two studies indicated that a higher portion of the FW/FVW co-substrate brought about enhancement in methane yield compared to mono-digestion of wastewater sludge alone. This meets expectations, because FW contains higher biodegradable substrates for anaerobic digestion compared to sewage sludge, which is quite refractory to hydrolysis [20]. The limitation of using FW as a co-substrate is in the monitoring of the Na^+ and free-ammonia nitrogen (FAN) contents in the substrate, which needs to be less than 4 g L^{-1} for Na^+ [14] and 600 mg/L for FAN [21] to avoid system instability. Excess Na^+ could interfere with the metabolic process of the microbial activity [22], and excess FAN could lead to volatile fatty acid (VFA) accumulations [21].

Co-digesting wastewater sludge with OFMSW under mesophilic conditions was shown to achieve a methane yield enhancement of 59% and 89% in Silvestre, et al. [23] and Cavinato, et al. [24], respectively. The methane yield enhancement reached 233% when the co-digestion occurred under thermophilic conditions [24].

Their results contradict Al bkoor Alrawashdeh, et al. [25] who showed a 44% reduction of methane yield when the sewage sludge was co-digested with OFMSW. The same study [25] showed that anaerobic digestion of waste untreated sludge alone gave the highest methane yield and VS removal compared to treating waste activated sludge or co-digestion with OFMSW. Despite similar VS and C/N ratios of the wastewater sludge and OFMSW, the differences in the methane yield enhancement and VS removal could be due to the heterogeneous OFMSW sample used in the experiments. In Silvestre, et al. [23] and Cavinato, et al. [24], the OFMSW came from a municipal solid waste facility, whereas in Al bkoor Alrawashdeh, et al. [25], the OFMSW originated from pre-selected household waste. Nevertheless, co-digestion with OFMSW generally aids in methane yield, as the higher C/N ratio, VS and COD concentration of OFMSW (in the ranges of 17–36, 20–30%, and 207–641 g/kg, respectively) could counterbalance the organic matter and nutrient deficiencies in sewage sludge during anaerobic co-digestion, thus improving the metabolic activity of the biomass, and, as a consequence, the enhancement in methane yield [26–28]. However, in order to incorporate OFMSW as a co-substrate for sewage sludge anaerobic co-digestion, it is strongly recommended that the impurities in the OFMSW be removed prior to the anaerobic co-digestion process to avoid hydrodynamic problems, and blockage of the pump system [28].

Table 2. Summary of studies utilizing FW, FVW or OFMSW as co-substrate in anaerobic co-digestion of wastewater sludge.

	Primary Substrate	Co-Substrate [1]	Mixing Ratio [2]	Mode [3] (Volume)	T [4]	HRT (d)	OLR (gVS $L^{-1}d^{-1}$)	VS Removal (%)			Methane Yield (L CH_4 g^{-1} VS_{added})			Ref.
								Mono	Co	Improved %	Mono	Co	Improved (%)	
1	Sewage sludge	FW	60:40 VS	C (4 L)	M T-M	8 7	3.5 6.1	- -	42 45	- -	- -	0.18 0.20	- -	[29]
2	Dewatered sludge	FW	2.4:1 VS 0.9:1 VS 0.4:1 VS	C (6 L)	M	30	4–6	38	51 62 70	34 63 84	0.24	0.30 0.35 0.40	25 46 67	[14]
3	Sewage sludge	FVW + FW	1:0.23 VS 1:2.09 VS	B (1.2 L)	M	17 26	-	50	30 40	−40 −25	0.25	0.29 0.37	18 47	[19]
4	Sewage sludge	FVW	100:20 V	C (100 L)	M	10–14	2.1	22	24	9	-	0.38	-	[30]
5	Waste activated sludge	OFMSW	50:50 V	C (380 L)	M T	23.5 22.3	1.6 1.66		- -	- -	0.09	0.17 0.30	89 233	[24]
6	Sewage sludge	OFMSW	46:54 VS	C (5.5 L)	M	20	1.9	36	70	94	0.22	0.35	59	[28]
7	Waste activated sludge	OFMSW	70:30 W 50:50 W	B (0.2 L)	M	100	-	65	58 58	−11 −11	0.25	0.22 0.14	−12 −44	[25]
8	Waste untreated sludge	OFMSW	70:30 W 50:50 W	B (0.2 L)	M	100	-	72	66 60	−8 −17	0.44	0.29 0.28	−34 −36	[25]

[1] Co-substrate: FW, food waste; FVW, Fruit and vegetable waste; OFMSW, the organic fraction of municipal solid waste. [2] Mixing ratio: V, by volume, VS, by volatile solids; W, by weight. [3] Mode: B, batch; C, continuous. [4] T, Temperature: M, mesophilic (28–38 °C); T, thermophilic (40–56 °C).

2.2. Crude Glycerol

Crude glycerol, the main byproduct from biodiesel production, is a high-strength organic matter with an average COD of 1,200,000 mg/L and BOD of 97,080 mg/L [15]. It consists of glycerol, alcohol, salts, water, free fatty acids, heavy metals, methyl esters, and unreacted mono-, di-, and tri-glycerides [31,32]. The transesterification process produces biodiesel and crude glycerol at a volumetric ratio of 10:1 [33]. In 2011, with a total of 5.1 million tonnes of glycerol produced, only 40% were used, with the remaining 3 million tonnes a surplus. It is estimated that the global production of glycerol will reach 7.66 million tonnes in 2020 [34]. Therefore, glycerol management and its associated waste or value-added technologies play an important role.

The use of crude glycerol is limited because of the impurities, which affect its physical, chemical and biological properties [35]. There are currently two consumer markets for glycerol, the first being the already-existing market with the demand for high-purity glycerol, and the second being the use of lower-purity crude glycerol from biodiesel production. Due to limited market demand of the latter, biodiesel-producing industries are treating crude glycerol as industrial waste [35]. There are, however, some other industries not engaging in biodiesel production who are looking for alternative routes for crude glycerol as a raw material in other products [36].

One of the uses of crude glycerol is as a co-substrate for anaerobic digestion of low-strength wastes. Crude glycerol is difficult to treat biologically, due to its low nitrogen content and extreme pH; however, due to its high carbon content and anaerobic biodegradability, it can be used as a co-substrate in the anaerobic digestion of low-strength wastes/wastewater, such as sewage sludge. The high alkalinity level of crude glycerol with a pH of 10.3 [37], can also act as a buffer for acidic waste.

Table 3 shows the studies using crude glycerol as the co-substrate in anaerobic co-digestion of wastewater sludge. When 1 vol% of crude glycerol was added to the anaerobic co-digestion of sewage sludge, the methane yield was increased by 115% in Fountoulakis, et al. [38], and 176% in Rivero, et al. [39], resulting in 0.56 and 2.1 L CH_4 g^{-1} VS_{added}, respectively. The high methane yield in the latter was due to the use of optimized two-stage anaerobic digestion processes (acidogenic and methanogenic) and chemically pre-treated sewage sludge [39]. Despite the decrease in VS removal, dos Santos Ferreira, et al. [40] showed that the optimum crude glycerol addition was 0.5 vol% to achieve 73% enhancement in methane yield. The pilot-scale study by Razaviarani, et al. [41] showed that the optimum organic loading rate should be around 1.04 g VS L^{-1} d^{-1}, corresponding to 1.1 vol% of glycerol addition to avoid process instability. Most of the studies were limited to using 1 vol% of crude glycerol as the co-substrate, as a higher dosage would result in reduction in alkalinity and thus, VFA accumulation and methanogen inhibition [41].

Table 3. Summary of studies using crude glycerol as a co-substrate in anaerobic co-digestion of wastewater sludge.

	Primary Substrate	Co-Substrate	Mixing Ratio [1]	Mode [2] (Volume)	T [3]	HRT (d)	OLR (g VS $L^{-1}d^{-1}$)	VS Removal (%)			Methane Yield (L CH_4 g^{-1} VS_{added})			Ref.
								Mono	Co	Improved %	Mono	Co	Improved (%)	
1	Sewage sludge	Crude glycerol	99:1 V	C (3 L)	M	23–25	-	−79	−32	60	0.26 [4]	0.56 [4]	115	[38]
2	Sewage sludge	Crude glycerol	98.9:1.1 V	SC (1200 L)	M	20	1.04	42	52	24	0.62	0.87	40	[41]
						13	1.94					0.92	21	
3	Pre-treated sewage sludge	Crude glycerol	99:1 V	SC (9 L)	M	11	2.11	-	88–92	-	0.76	0.88	16	[39]
						9	2.26					0.98	29	
						9	1.00					2.10	176	
4	Primary sludge	Crude glycerol	124:2–6 W	C (4 L)	M	32	1.25–1.9	-	-	-	0.35	0.55–0.75	57–114	[42]
5	Sewage sludge	Crude glycerol	99.5:0.5 W	C (0.85 L)	M	17	1.81–3.68	53	62	17	0.40	0.36	−10	[43]
			98:2 W					55	75	36	0.38	0.34	−11	
6	Sewage sludge	Crude glycerol	98.8:1.2 V	C (5.5 L)	M	20	1.2	36	57	58	0.22	0.29	31	[23]
7	Sewage sludge	Crude glycerol	99:1 V	B (0.05 L)	M	15-30	-	32.5	8	−75	0.045	0.010	−78	[40]
			99.7:0.3 V						21	−35		0.062	38	
			99.5:0.5 V						19	−42		0.078	73	
			99.3:0.7 V						16	−51		0.064	42	

[1] Mixing ratio: V, by volume, VS, by volatile solids; W, by weight. [2] Mode: B, batch; C, continuous; SC, semi-continuous. [3] T, Temperature: M, mesophilic (28–38 °C); T, thermophilic (40–56 °C). [4] Taking VS of crude glycerol as 800 g/L (averaged from [37,42,43]) unless specified.

A proper amount of crude glycerol added (1–1.2 vol% [23]) may aid in the specific biomass activity, as the degradation of crude glycerol results in the formation of more acetate, hydrogen, and propionate and, as a consequence, enhance the methane yield [23].

2.3. Agricultural Wastes

Agricultural waste (also called agro-waste) includes animal waste, food processing waste, crop waste, and hazardous and toxic agricultural waste [44]. Rice husks, coconut husks and shells, wheat straw, sugarcane fiber, and groundnut shells are examples of the most common agricultural plant residues. Animal waste includes cattle, pig and chicken manures. The major characteristics of manures are often associated with high nitrogen content and the presence of easily-formed sulfur, ammonia, and hydrogen sulfide gasses. Hence, mono-digestion of manures bring about excessive nutrients and organic matter in the digester, thus inhibiting the production of methane. However, due to its high nutrient content, it is a potential co-substrate for the main substrate with low nutrient content such as sewage sludge. Although most agricultural residues or energy crops are lignocellulosic biomasses, with carefully designed operating conditions, the anaerobic digestion process of these biomasses may have the ability to promote methane from hemicellulose, which destabilizes the recalcitrant biomass structure, allowing for improved solubilization of cellulose by commercial enzymes in the downstream processes [45,46]. Mono-digestions of other agricultural waste, especially farm waste, have been demonstrated to be successful [6,47]. Therefore, due to its large availability and variety, agricultural waste could be a potential co-substrate for anaerobic co-digestion of wastewater sludge.

Table 4 summarizes the studies using agricultural waste as co-substrates for wastewater sludge anaerobic co-digestion. Sugar beet pulp, olive and grape pomaces, paper pulp reject, cheese whey, sheep manure, and brewery spent grain were utilized. The use of 50 vol% of paper pulp reject was able to boost up to 131% of the methane yield compared to sewage sludge-anaerobic digestion alone, due to its mostly cellulose nature in powder form, which resulted in faster hydrolysis [48]. The use of 5 vol% and 10 vol% cheese whey were also shown to improve methane yield by 44% [49] and 121% [50], respectively. This enhancement is likely due to the improved C/N ratio of the feedstock mixture [51]. Maragkaki, et al. [50] compared the use of different co-substrates with a ratio of 5–10 vol% using grape residue, sheep manure, cheese whey, with crude glycerol and food waste for sewage sludge-anaerobic co-digestion. They found that crude glycerol was the best co-substrate, followed by food waste, which was comparable with cheese whey, and lastly grape residue. Sheep manure was found to yield no improvement, and deteriorated methane yield and VS removal [50]. Co-digesting sewage sludge with acid cheese whey was also found to have negligible improvement. However, with both acid cheese whey and brewery spent grain at a sewage sludge: acid cheese whey ratio of 89: 11, and an acid cheese whey: brewery spent grain ratio of 10 g: 1 L, a methane yield enhancement of 56% was achieved [52]. Thus, selection of a suitable co-substrate and optimization of the operating conditions according to the properties of the main substrate remains to be the major challenge before these co-substrates can be utilized in a WWTP.

Table 4. Summary of studies using agricultural wastes as co-substrate in anaerobic co-digestion of wastewater sludge.

	Primary Substrate	Co-Substrate	Mixing Ratio [1]	Mode [2] (Volume)	T [3]	HRT (d)	OLR (g VS L^{-1} d^{-1})	VS Removal (%)			Methane Yield (L CH_4 g^{-1} VS_{added})			Ref.
								Mono	Co	Improved %	Mono	Co	Improved %	
1	Sewage sludge	Sugar beet pulp	-	SC (5 L)	M	30	1.1	-	-	-	-	0.17	-	[53]
						20	1.2					0.10		
						15	1.8					0.16		
						10	2.1					0.34		
						6	5					0.06		
2	Waste activated sludge	Olive pomace	50:50 W	B (1.6 L)	M	30	-	40	45	12.5	0.16	0.21	31	[54]
3	Sewage sludge	Cheese whey	95:5 V	B (180 L)	M	24	1.1	19	25	32	0.25	0.36	44	[49]
4	Primary sludge	Paper pulp reject	50:50 V	B (0.75 L)	M	13	-	-	-	-	0.16	0.37	131	[48]
5	Waste activated sludge	Olive and grape pomaces	-	B (1.6 L)	M	30	-	35	52–55	49–57	0.06	0.07	17	[51]
6	Sewage sludge	Grape residue	95:5 V	C (3 L)	M	24	1	46	41	−11	0.17	0.22	29	[50]
		Sheep manure	95:5 V	C (3 L)			1.3	38	25	−34	0.17	0.14	−21	
		Cheese whey	90:10 V	C (1 L)			0.8	43	27	−37	0.14	0.31	121	
7	Sewage sludge	Acid cheese whey + brewery spent grain	90:10 V *	SC (40 L)	M	20	1.75	34	40	18	0.27	0.26	−4	[52]
			91:9 V *			18	2.9	34	42	24	0.27	0.27	0	
			89:11 V *			16.7	2.35	46	46	0	0.18	0.28	56	

[1] Mixing ratio: V, by volume, VS, by volatile solids; W, by weight; * Sewage sludge: Acid cheese whey: 90:10 V at a brewery spent grain mass: feedstock volume ratio of 10 g: 1 L. [2] Mode: B, batch; C, continuous, SC, semi-continuous. [3] T, Temperature: M, mesophilic (28 to 38 °C); T, thermophilic (40 to 56 °C).

2.4. Fat, Oil and Grease (FOG)

Fat, oil and grease is a term commonly used to describe lipid-rich waste material primarily from edible oil producers, food processing industries, slaughterhouses, and food wastes [55]. It can be grouped as grease trap, waste cooking oil (yellow grease), and interceptor wastes (brown grease) consisting of yellow grease, water and food solids [56]. Since FOG has the tendency to accumulate on pipe walls through a physical or chemical aggregation process, the direct discharge of FOG into municipal wastewater collection systems is illegal in many countries, because it would cost the municipalities a large amount of money in cleaning and maintenance [57]. Therefore, the current best treatment method for waste FOG is separate collection for recycling (to be used for biodiesel), or proper disposal.

FOG constituents include biodegradable lipids which present as neutral fats (i.e., triglycerides), and saturated or unsaturated long chain fatty acids (LCFAs) [58,59]. Generally, lipids in FOG are first hydrolyzed to glycerol and LCFAs in the anaerobic digestion process. Subsequently, LCFAs can be degraded anaerobically to acetate via β-oxidation pathways to short-chain fatty acids, acetate, hydrogen, and methane; whereas glycerol is degraded to acetate. These acetates are then anaerobically degraded to methane [57,60]. Since the degradable fraction of lipids is higher than the mainstream carbohydrate and proteins present in sewage sludge, anaerobic digestion of FOG theoretically could generate higher biogas production [61]. Although FOG is regarded as an energy, value-added material, and recognized as an advantage for incineration and biodiesel generation processes, anaerobic co-digestion with FOG is considered a proper treatment method, because it requires insignificant pre-treatment, and produces higher amounts of biogas [62,63].

Studies utilizing FOG as co-substrate in wastewater sludge anaerobic co-digestion are summarized in Table 5. Co-digestion studies with FOG are relatively more complete and established than other co-substrates discussed earlier. Results show a notably-improved methane yield ranging from 13 to 198%. Most of these studies showed that about 50:50% of sewage sludge and FOG to grease trap waste on a VS basis offered comparatively higher methane yield, achieving close to 0.50 L CH_4/g VS_{added} [64,65]. Higher than that (e.g., more than 60 VS% of FOG/grease trap waste) would result in biomass aggregation and thus, create a mass transfer limitation due to LCFA accumulation on and in the biomass aggregates [65,66]. The high LCFA content in the FOG/grease trap waste due to degradation of lipid-rich materials, is a known inhibitor of methanogenic bacteria, which would usually cause a lag between LCFA degradation and methane production [67], and lead to operational problems such as clogging and scum formation. Co-digesting with sewage sludge helps in diluting the inhibitive LCFA [4,64]. Due to the carbohydrate and/or protein content in the sewage sludge, such inhibition could be alleviated [68].

3. Factors Influencing Co-Digestion Performance

Several factors affect the functioning of anaerobic digesters and feasibility of co-digestion where sufficient control is needed to prevent reactor failure. A few of the major influences that greatly affect digester performances in co-digestion are mixing, co-substrate mixing ratio and nutrient balance, operating temperature, organic loading rates (OLR), and hydraulic retention time (HRT) in the digester. Based on the results as seen in Tables 2–5, these factors are further analyzed, reviewed, and discussed in the following sections to gain deeper insight and consideration for future research. Summarized optimum operating conditions are also presented.

Table 5. Summary of studies utilizing FOG as a co-substrate in wastewater sludge co-digestion.

	Primary Substrate	Co-Substrate [1]	Mixing Ratio [2]	Mode [3] (Volume)	T [4]	HRT (d)	OLR (g VS L^{-1} d^{-1})	VS Removal (%)			Methane Production (L CH_4 g^{-1} VS_{added})			Ref.
								Mono	Co	Improved %	Mono	Co	Improved %	
1	Sewage sludge	GTS	90:10 VS	B (2 L)	M	37	-	-	-	-	0.29	0.38	31	[69]
			75:25 VS	B (2 L)		37	-	-	-	-	0.29	0.42	45	
			40:60 VS	B (2 L)		37	-	-	-	-	0.29	0.60	107	
			90:10 VS	C (35 L)		13	2.5	45	55	22	0.24	0.27	13	
			70:30 VS	C (35 L)		13	2.4	45	58	29	0.24	0.30	25	
2	Sewage sludge	FOG	52:48 VS	C (2 L)	M	12	2.45–4.35	25	45	79	0.21	0.64	198	[70]
					T	13		25	51	66	0.25	0.67	169	
3	Sewage sludge	GTS	95:5 VS	SC (4 L)	M	16	1.95	-	-	-	0.278	0.37	34	[64]
			80:20 VS				2.19					0.44	59	
			72:28 VS				2.8					0.44	60	
			62:38 VS				3.13					0.45	61	
			54:46 VS				3.46					0.46	67	
			45:55 VS				4.01					0.32	14	
			29:71 VS				4.41					0.32	13	
4	Sewage sludge	GTS	96:4 VS	SC (5.5 L)	M	20	1.2	36	46	28	0.22	0.25	12	[71]
			77:23 VS				1.6		52	44		0.33	48	
			63:37 VS				1.7		56	56		0.29	33	
5	Thickened WAS	FOG	36:64 VS	SC (2 L)	M	15	2.34	40	57	43	0.25	0.60	137	[7]
6	Thickened WAS	GW	93:7 VS	C (200 L)	M	30	1.8	29	43	48	0.23	0.16	−32	[65]
			83:13 VS	C (200 L)		24	2		36	24		0.29	26	
			85:15 VS	C (3.4 L)		25	1.7		42	45		0.29	23	
			78:22 VS	C (3.4 L)		26	1.6		42	45		0.26	11	
			66:34 VS	C (3.4 L)		25	1.4		39	35		0.35	51	
			48:52 VS	C (3.4 L)		24	1.2		44	52		0.48	107	
			26:74 VS	C (3.4 L)		24	1		33	14		0.40	72	
			13:87 VS	C (3.4 L)		25	0.8		17	−41		0.14	−40	
7	Sewage sludge	GW	88:12 COD	C (5 L)	T	20	1.2	50	46	−8	0.22	0.25	15	[72]
			73:27 COD				2.1		45	−10		0.23	6.5	
			63:37 COD				1.9		44	−12		0.20	−6.5	
8	Thickened WAS	FOG	80:20 VS	B (0.25 L)	T	60	-	71	74	4	0.316	0.43	35	[73]
			60:40 VS						74	4		0.45	43	
			40:60 VS						76	7		0.49	56	
			20:80 VS						47	−33		0.10	−68	

[1] Co-substrate: GTS, grease trap sludge; FOG, fat oil and grease; GW, grease waste. [2] Mixing ratio: VS, by volatile solids; COD, by COD. [3] Mode: C, continuous; B, batch; SC, semi-continuous. [4] T, temperature: M, mesophilic (28 to 38°C); T, thermophilic (40 to 56°C).

3.1. Mixing

Mixing provides a good contact between microbes and substrates, improves the ability of bacterial population to obtain nutrients, and promotes homogenous feed to the digester. Mixing also reduces the forming of inhibitory intermediates such as scum, and stabilizes digestion conditions by preventing the development of temperature gradients [74]. Pockets of material may weaken the overall rate of the process at different stages of pH and temperature when mixing is inadequate. Mixing can be done through mechanical mixing, biogas recirculation, or through slurry recirculation [75]. Mixing also improves gas production compared to unmixed digesters [76]. This has been proven in the study performed by Rizk, et al. [77], in which ACD of wastewater treatment plant sludge and the OFMSW with no mixing system in a 70 L reactor under mesophilic conditions showed that most biogas and methane generated only at the first month of operation. This suggests that a mixing system needs to be installed in the digester, particularly in co-digestion, to ensure a sufficient homogeneity besides improving digester performance.

The effects of varying mechanical mixing conditions on co-digestion operations studied by Stroot, et al. [78] affirmed that a low mixing state (80 rpm) is more effective than a high mixing speed (200 rpm). This is because excessive mixing can interrupt the microorganisms in their ability to digest. This indicates that setting an optimum mixing speed is essential, as it could substantially affect the digestion and biogas production rates. However, Karim, et al. [76] mentioned that mixing during the start-up is not beneficial, because it lowers the digester pH, resulting in performance instability, and leading to delays in the start-up period. Hence, pH should be considered to obtain a suitable condition in the digester during mixing for the best digester performance. Some studies have also affirmed that intermittent mixing, known as semi-continuous mixing, is better than continuous mixing. Unlike continuous mixing, which breaks the syntrophic relationship between acetogens and methanogens [79], semi-continuous mixing provides sufficient time for microbial growth, which then enhances the mass transfer from the liquid-to-gas phase, and ultimately increases the methane yield [80]. Apart from the common impeller driven mixing method to aid mass and heat transfer, some novel mixing innovations have also been developed, such as the Pneu-mechanical mixer [81] and biogas recirculation using a new 'O' shaped diffuser design [79], which contributes to the synergistic effect of carbon dioxide acidification and, thus, reduces digester pH, ammonia control and methane enrichment.

3.2. Co-Substrate Mixing Ratio and Nutrient Balance

The C/N ratio represents the correlation between the amount of carbon and nitrogen present in organic matter. This ratio is the balance of food that a microbe needs to grow in order to perform digestion of organic matter. The optimum C/N ratios in anaerobic digesters are usually between 20:1 and 30:1 [82]. A rapid utilization of nitrogen by the methanogens results in a lower gas production, and this is indicated with a high C/N ratio. Conversely, a lower C/N ratio causes the ammonia deposition and pH to surpass 8.5, which is considerably toxic to methanogenic bacteria. Therefore, to achieve an ideal ratio of C/N in the co-digestion, it is suggested to have a mixture of waste with a low and high C/N ratio.

Various mixture ratios often substantially affect the outcomes, as the ratio reflects the nutrient balance in it. An ideal mixing ratio for OFMSW is recommended within 20–50 vol% to achieve optimum nutrient balance in the digester (Table 2). This recommendation has been verified by Heo, et al. [83], who conducted ACD of food wastes with sludge at a 50:50 ratio. The mixture of a high and low C/N ratio resulted in an improved ratio from 6 to 16. Although the obtained C/N ratio of 16 is considered below optimum, this improved ratio tended to have an overall better digestion performance in terms of effluent volatile solids concentration, buffer capacity, and methane production.

An optimum substrate mixing ratio for crude glycerol is relatively low at 0.5–1.2 vol%, as shown in Table 3. This is due to a high C/N ratio of crude glycerol at 68, indicating that a rapid consumption of nitrogen by the methanogens leads to low biogas production if added in high concentration. To ensure stable operation, the addition of glycerol should not exceed 4% under mesophilic conditions, to avoid adding excess nitrogen, which will trigger the generation of high acetic acid, eventually causing a decrease in pH. This would then cause the methanogens to consume for a longer period to adjust and overcome the pH, hence reducing methane production.

The substrate mixing ratio for agricultural wastes, on the other hand, is suggested within 5–50 vol% based on Table 4. The range appeared to be large as it is dependent on the selection of pairing substrates. Conversely, it is advised to keep the digestion ratio for FOG within 5–60 VS% to avoid an overly low or high C/N ratio based on Table 5. This is because the C/N ratio for FOG is variable depending on the type of FOG existing in either grease traps, waste cooking oil, or interceptor wastes forms. The C/N ratio for FOG can be as low as 9 or as high as 15. Hence, it is safe to start the ratio at approximately 50% for stable methane improvement without excessive or insufficient nutrients.

3.3. Operating Temperature

Co-digestion treatment under both mesophilic and thermophilic conditions are feasible depending on the substrates. Numerous researchers have highlighted the significant effects of temperature on the microbial community, process kinetics, and stability and methane yield. Lower temperatures during the process are known to decrease microbial growth, substrate utilization rates, and biogas production. Lower temperatures may also result in an exhaustion of cell energy, a leakage of intracellular substances or complete lysis [84,85]. Although higher temperatures allow a reduced volume of waste mass in small reactors, making it more effective in minimizing lipids, and could accommodate higher loading capacity, it also could result in lower biogas yield. This is due to the production of volatile gases such as ammonia, which suppress methanogenic activities [86]. Therefore, it is suggested that different temperature conditions are required to match with different substrates, due to the characteristics of each substrate.

A thermophilic condition is recommended as favorable for the OFMSW. This finding has been proven by Li, et al. [87] and Kim, et al. [29], who performed ACD of food waste OFMSW under a thermophilic condition, and observed a better performance than conventional the mesophilic process for methane production and digestion capability. Similarly, the study conducted by Lee, et al. [88] showed that a thermophilic condition is more favorable compared to mesophilic and hyperthermophilic. This may be the result of the slow adaptation of microorganism, which are likely to be degraded, and release nutrients under a hyperthermophilic condition, leading to the overall inhibition of methane production.

Likewise, a mesophilic condition is ideal for the ACD with crude glycerol. As the crude glycerol has a high alkaline level at pH averaging 10.3, it is usually co-digested with acidic waste to form a neutral pH. However, when the temperature increases, acetate is the first acid to increase. Subsequently, the more acetate formed, the more the main part of VFA concentration constrains methanogenic bacteria [89]. Various researchers, including Fountoulakis and Manios [90], Fountoulakis, et al. [38] and Rivero, et al. [39], have also demonstrated a significantly higher increase in methane under mesophilic with an average increase ranging from 115% to 176%, as reflected in Table 3.

As for co-digesting with agricultural waste (Table 4), most of the studies which can be found were operating under mesophilic conditions [6]. It was demonstrated in some studies that a thermophilic condition was beneficial for degradation of some agricultural plant residue waste such as maize [91], rice straw and risk husk [92]. However, a mesophilic condition is a better option for ACD with manures to overcome ammonia inhibition. According to Zeeman, et al. [93], ammonia increases with the increase of pH and temperature. Hence, the thermophilic condition will result in a higher concentration of free ammonia, which is regarded as an active component causing ammonia inhibition. This would result in higher VFA concentration, which negatively affects the activity of methanogenic bacteria and, hence, achieves a low methane yield.

On the other hand, ACD of FOG is more suitably treated under a mesophilic condition. A major concern of FOG is primarily the harmful effect of long-chain fatty acids on methanogenic bacteria, which coats the bacteria in a layer and constrains cell access to substrates, resulting in poor release of biogas [94–96]. Although the thermophilic condition allows the accessibility of lipids to microorganisms, and their lipolytic enzymes caused by the increased diffusion coefficients and lipid solubility in aqueous media, this has not been the case in recent studies. As evidenced in Table 5, Kabouris, et al. [70] showed that a mesophilic condition could achieve a greater improvement at 198% of methane yield compared to thermophilic at 169%. These results indicate that a thermophilic condition may increase nutrient release, and thermophilic bacteria may be more sensitive to LCFA inhibition than mesophilic bacteria. In addition, a thermophilic operation does not always offer sufficiently higher volatile solids reduction to justify the cost associated with the increased energy for heating.

3.4. Organic Loading Rate (OLR) and Hydraulic Retention Time (HRT)

OLR, a crucial parameter, describes the amount of volatile solids to be fed into the digester each day. Volatile solids signify the amount of organic solids that can be digested, while the remaining solids are non-degradable. The optimum OLR and HRT often depend on the type of substrates fed into the digester, since the substrates determine the activity level of biodegradation that will occur in the digester. ACD with OFMSW or food waste is suitable for treatment at a low-range OLR < 2 gVS $L^{-1}d^{-1}$ with HRT of 25 days (Table 2). This is due to the considerably high COD concentration of the OFMSW of food waste averaging 252,000 mg/L. Therefore, bacteria would need a longer time to degrade the high-strength OFMSW, and not overfeed at the same time, which could form tension in ACD. This recommendation has been verified by Babaee and Shayegan [97], who proposed that stable performance was achieved under an OLR of 1.4 g VS/L.day with a HRT of 25 days. This finding is also cohesive with the findings in Mata-Alvarez, et al. [98].

Similarly, it is advised to keep ACD with crude glycerol below an OLR 2 gVS $L^{-1}d^{-1}$ for VS with similar HRT of 20–30 days as shown in Table 3. It is also important to add glycerin gradually to allow bacteria to acclimatize to gain better results at increasing an OLR. Unlike the OFMSW and crude glycerol, ACD of agricultural waste can accommodate a higher OLR of up to 2.35 gVS $L^{-1}d^{-1}$ with a HRT of 17 days [52] (Table 4). This is due to the lower COD concentration averaging 131,000 mg/L [99,100] compared to crude glycerol. ACD with FOG, on the other hand, was found to be working effectively under a high OLR up to 4 gVS $L^{-1}d^{-1}$ with a HRT of 16 days, as shown in Table 5. A higher OLR could increase the carbon source in the treatment. However, an OLR higher than recommended may cause damage to technical components such as mixers or pumps, or require earlier maintenance than scheduled. Therefore, a HRT of 20–30 days could provide satisfactory co-digestion and stable biogas production. This is to allow sufficient time for the introduction of digesting bacteria to break down organic matter and to generate methane.

4. Summary and Challenges of ACD of Wastewater Sludge

Table 6 summarizes the findings from the review of utilizing OFMSW/FW, CG, AW and FOG as the co-substrates for sewage sludge-anaerobic co-digestion. Note that the comparison shown uses the recommended or widely-used OLR and HRT of 1–2.5 gVS $L^{-1}d^{-1}$ and 15–30 days, respectively, as the criteria, which is the common operating condition of a WWTP. The mixing ratio, preferred temperature mode, VS removal efficiency, and methane yield for each co-substrate are summarized. The ease and effectiveness of integrating these co-substrates in the anaerobic co-digestion of sewage sludge are then rated with the challenges outlined for future reference.

Table 6. Summary of reviews on co-substrates (at OLR = 1–2.5 gVS $L^{-1}d^{-1}$ and HRT = 15–30 days).

		SS	OFMSW/FW	CG	AW	FOG
1	Mixing ratio (%) with respect to SS [1]	100	20–50 V	0.5–1.2 V	5–50 V	5–50 VS
2	Preferred temperature mode [2]	M	T	M	M	M/T
3	VS removal %	22–50	24–70	52–72	25–55	36–58
4	Methane yield (L CH_4 g^{-1} VS_{added})	0.22–0.25	0.17–0.38	0.29–0.65	0.10–0.37	0.16–0.50
	Evaluation of the effectiveness of implementing anaerobic SS-co-digestion with the co-substrate in a WWTP:					
5	Benefits and challenges [6,23,39,49,55]		• Highly abundant; greatly helps divert waste from landfills. • Impurity level affects the operational condition; not ideal for WWTP design. • Effective source collection and segregation system; increases cost of management.	• Best co-substrate to improve VS removal and methane yield. • High anaerobic bio-degradability; low impact on digester solids management. • Simpler feed implementation and better hydrodynamics; lower operational problems. • Not commonly available; low perception of biodiesel companies.	• Similar to OFMSW/FW in terms of abundancy and impurity level control. • Agri-residues contain high level of lignocellulose affecting digestibility. • High solids content tends to result in rapid acidification; increases solid management cost. • Mixing and new reactor configuration are important.	• Most researched. • Transport limitation due to LCFAs, scum formation and sludge flotation problems; limited efficiency. • May be more economical than ACD in biodiesel production • Pre-treatment of FOG and properly designed sludge bed are important.
6	Rating		★★	★★★★	★	★★★

[1] Mixing ratio (%) with respect to sewage sludge: V, by volume; VS, by volatile solids. [2] Preferred temperature mode: M, mesophilic; T, thermophilic.

Based on the above analysis (Table 6), though the highly abundant OFMSW/FW and AW are able to improve VS removal and methane yield compared to SS mono-digestion alone, the costs associated with solids management, and a separate source collection and segregation system to alleviate the operational problems of the anaerobic digester would be increased [28]. And, while FOG is a potential co-substrate, it requires pre-treatment and a sludge bed operational condition design to avoid operational problems such as scum formation and biomass washout. A lifecycle assessment study by Tu [101] pointed out, however, that waste FOG in the processing of biodiesel is a more effective option compared to using it as an anaerobic co-substrate. Another approach known as dual-fuel can, nevertheless, be carried out to separate the higher-grade layer of FOG for biodiesel production and the lower-grade layer of FOG for anaerobic co-digestion [102,103]. Crude glycerol appears to be the best co-substrate candidate for SS anaerobic co-digestion, as it has high anaerobic biodegradability, which is able to significantly increase the OLR, while having minimal impact on the HRT [43]. Because it is in liquid form, crude glycerol requires a lower investment, as it requires simple feed implementation, and has a low impact on digestate solid management [39]. A typical anaerobic digester utilizes only 40–50% of the influent organic matter to convert to methane. Based on the outcome of this review, a recommendation is provided in Figure 1 that outlines a possible way of integrating anaerobic co-digestion using waste crude glycerol or lower-grade FOG into a WWTP to boost the methane yield, and thus offset the energy consumption.

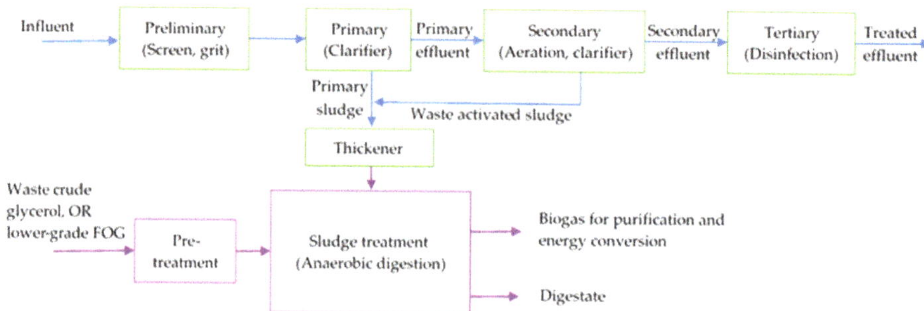

Figure 1. A suggested scheme for integrating anaerobic co-digestion of sewage sludge into a WWTP.

5. Conclusions

This paper summarizes the recent research on the ACD of wastewater sludge such as sewage sludge, waste activated sludge, and thickened waste activated sludge, with some potential co-substrates. The review shows that ACD of wastewater sludge involving co-substrates of OFMSW/FW, crude glycerol, agricultural waste, and FOG show potential in boosting biogas production and methane yield. ACD can also reduce costs from shared equipment, create easier handling of feedstock, and create a more stable process in general. The main challenges in ACD technology are process instability, which is mainly due to inappropriate substrate ratios, and operating conditions. Therefore, the anaerobic co-digestion system needs to be designed by considering the operating conditions such as a slow mixing system and a mixture of waste with low and high C/N ratio at the right mixing ratio. It can be concluded that out of these four potential co-substrates, crude glycerol at a mixing ratio of 0.5–1.2 vol%, and lower-grade FOG at 5-50 VS% are the most feasible for integration into a WWTP, in terms of VS removal, methane yield improvement, and operational and management costs. Future research is required focusing on the optimization of the operating parameters, kinetic and thermodynamic modeling, cost analysis, and lifecycle assessment to further evaluate its scale-up feasibility.

Author Contributions: Conceptualization and writing, W.L.C.; review and editing, S.C. and T.J.T.; supervision, Y.J.C. and M.F.C.; administration, G.-T.P.; funding acquisition, J.W.L. and J.K.C.

Funding: One of our authors, Jun Wei Lim, wishes to acknowledge the financial supports from the International Grant awarded by Universitas Muhammadiyah Purwokerto with grant number 015ME0-094 and Ministry of Education Malaysia under HICoE with grant number 015MA0-052.

Conflicts of Interest: The authors declare no conflicts of interest.

References

1. Chan, Y.J.; Chong, M.F.; Law, C.L. Performance and kinetic evaluation of an integrated anaerobic–aerobic bioreactor in the treatment of palm oil mill effluent. *Environ. Technol.* **2017**, *38*, 1005–1021. [CrossRef] [PubMed]
2. Musa, N.S.; Ahmad, W.A. Chemical oxygen demand reduction in industrial wastewater using locally isolated bacteria. *Malays. J. Fundam. Appl. Sci.* **2010**, *6*. [CrossRef]
3. Braun, R.; Wellinger, A. *Potential of Co-Digestion*; IEA: Paris, France, 2002.
4. Mata-Alvarez, J.; Mace, S.; Llabres, P. Anaerobic digestion of organic solid wastes. An overview of research achievements and perspectives. *Bioresour. Technol.* **2000**, *74*, 3–16. [CrossRef]
5. Long, J.H.; Aziz, T.N.; Francis III, L.; Ducoste, J.J. Anaerobic co-digestion of fat, oil, and grease (FOG): A review of gas production and process limitations. *Process Saf. Environ. Prot.* **2012**, *90*, 231–245. [CrossRef]
6. Sawatdeenarunat, C.; Surendra, K.; Takara, D.; Oechsner, H.; Khanal, S.K. Anaerobic digestion of lignocellulosic biomass: Challenges and opportunities. *Bioresour. Technol.* **2015**, *178*, 178–186. [CrossRef] [PubMed]
7. Wan, C.; Zhou, Q.; Fu, G.; Li, Y. Semi-continuous anaerobic co-digestion of thickened waste activated sludge and fat, oil and grease. *Waste Manag.* **2011**, *31*, 1752–1758. [CrossRef] [PubMed]
8. Astals, S.; Nolla-Ardèvol, V.; Mata-Alvarez, J. Anaerobic co-digestion of pig manure and crude glycerol at mesophilic conditions: Biogas and digestate. *Bioresour. Technol.* **2012**, *110*, 63–70. [CrossRef] [PubMed]
9. Mao, C.; Feng, Y.; Wang, X.; Ren, G. Review on research achievements of biogas from anaerobic digestion. *Renew. Sustain. Energy Rev.* **2015**, *45*, 540–555. [CrossRef]
10. Wu, C.; Li, Y.; Li, W.; Wang, K. Characterizing the distribution of organic matter during composting of sewage sludge using a chemical and spectroscopic approach. *RSC Adv.* **2015**, *5*, 95960–95966. [CrossRef]
11. Hallaji, S.M.; Kuroshkarim, M.; Moussavi, S.P. Enhancing methane production using anaerobic co-digestion of waste activated sludge with combined fruit waste and cheese whey. *BMC Biotechnol.* **2019**, *19*, 19. [CrossRef]
12. Rabii, A.; Aldin, S.; Dahman, Y.; Elbeshbishy, E. A review on anaerobic co-digestion with a focus on the microbial populations and the effect of multi-stage digester configuration. *Energies* **2019**, *12*, 1106. [CrossRef]
13. Campuzano, R.; González-Martínez, S. Characteristics of the organic fraction of municipal solid waste and methane production: A review. *Waste Manag.* **2016**, *54*, 3–12. [CrossRef] [PubMed]
14. Dai, X.; Duan, N.; Dong, B.; Dai, L. High-solids anaerobic co-digestion of sewage sludge and food waste in comparison with mono digestions: Stability and performance. *Waste Manag.* **2013**, *33*, 308–316. [CrossRef]
15. Chow, W.L.; Chan, Y.J.; Chong, M.F. A new energy source from the anaerobic co-digestion (acd) treatment of oleo chemical effluent with glycerin pitch. *Asia Pac. J. Chem. Eng.* **2015**, *10*, 556–564. [CrossRef]
16. EPA. *Municipal Solid Waste*; United States Environmental Protection Agency (EPA): Washington, DC, USA, 1996.
17. Paritosh, K.; Yadav, M.; Mathur, S.; Balan, V.; Liao, W.; Pareek, N.; Vivekanand, V. Organic fraction of municipal solid waste: Overview of treatment methodologies to enhance anaerobic biodegradability. *Front. Energy Res.* **2018**, *6*, 75. [CrossRef]
18. Cheng, J.Y.; Chiu, S.L.; Lo, I.M. Effects of moisture content of food waste on residue separation, larval growth and larval survival in black soldier fly bioconversion. *Waste Manag.* **2017**, *67*, 315–323. [CrossRef] [PubMed]
19. Cabbai, V.; Ballico, M.; Aneggi, E.; Goi, D. BMP tests of source selected OFMSW to evaluate anaerobic codigestion with sewage sludge. *Waste Manag.* **2013**, *33*, 1626–1632. [CrossRef] [PubMed]
20. Appels, L.; Baeyens, J.; Degrève, J.; Dewil, R. Principles and potential of the anaerobic digestion of waste-activated sludge. *Prog. Energy Combust. Sci.* **2008**, *34*, 755–781. [CrossRef]

21. Duan, N.; Dong, B.; Wu, B.; Dai, X. High-solid anaerobic digestion of sewage sludge under mesophilic conditions: Feasibility study. *Bioresour. Technol.* **2012**, *104*, 150–156. [CrossRef]
22. Mendez, R.; Lema, J.M.; Soto, M. Treatment of seafood-processing wastewaters in mesophilic and thermophilic anaerobic filters. *Water Environ. Res.* **1995**, *67*, 33–45. [CrossRef]
23. Silvestre, G.; Fernández, B.; Bonmatí, A. Addition of crude glycerine as strategy to balance the C/N ratio on sewage sludge thermophilic and mesophilic anaerobic co-digestion. *Bioresour. Technol.* **2015**, *193*, 377–385. [CrossRef] [PubMed]
24. Cavinato, C.; Bolzonella, D.; Pavan, P.; Fatone, F.; Cecchi, F. Mesophilic and thermophilic anaerobic co-digestion of waste activated sludge and source sorted biowaste in pilot-and full-scale reactors. *Renew. Energy* **2013**, *55*, 260–265. [CrossRef]
25. Al bkoor Alrawashdeh, K.; Pugliese, A.; Slopiecka, K.; Pistolesi, V.; Massoli, S.; Bartocci, P.; Bidini, G.; Fantozzi, F. Codigestion of untreated and treated sewage sludge with the organic fraction of municipal solid wastes. *Fermentation* **2017**, *3*, 35. [CrossRef]
26. Macias-Corral, M.; Samani, Z.; Hanson, A.; Smith, G.; Funk, P.; Yu, H.; Longworth, J. Anaerobic digestion of municipal solid waste and agricultural waste and the effect of co-digestion with dairy cow manure. *Bioresour. Technol.* **2008**, *99*, 8288–8293. [CrossRef] [PubMed]
27. Li, J.; Jha, A.K.; He, J.; Ban, Q.; Chang, S.; Wang, P. Assessment of the effects of dry anaerobic co-digestion of cow dung with waste water sludge on biogas yield and biodegradability. *Int. J. Phys. Sci.* **2011**, *6*, 3679–3688.
28. Silvestre, G.; Bonmatí, A.; Fernández, B. Optimisation of sewage sludge anaerobic digestion through co-digestion with OFMSW: Effect of collection system and particle size. *Waste Manag.* **2015**, *43*, 137–143. [CrossRef]
29. Kim, H.-W.; Nam, J.-Y.; Shin, H.-S. A comparison study on the high-rate co-digestion of sewage sludge and food waste using a temperature-phased anaerobic sequencing batch reactor system. *Bioresour. Technol.* **2011**, *102*, 7272–7279. [CrossRef]
30. Di Maria, F.; Micale, C.; Contini, S. Energetic and environmental sustainability of the co-digestion of sludge with bio-waste in a life cycle perspective. *Appl. Energy* **2016**, *171*, 67–76. [CrossRef]
31. Ma, J.; Van Wambeke, M.; Carballa, M.; Verstraete, W. Improvement of the anaerobic treatment of potato processing wastewater in a UASB reactor by co-digestion with glycerol. *Biotechnol. Lett.* **2008**, *30*, 861–867. [CrossRef]
32. Astals, S.; Nolla-Ardèvol, V.; Mata-Alvarez, J. Thermophilic co-digestion of pig manure and crude glycerol: Process performance and digestate stability. *J. Biotechnol.* **2013**, *166*, 97–104. [CrossRef]
33. Knothe, G.; Krahl, J.; Gerpen, J.V. *The Biodiesel Handbook*, 2nd ed.; AOCS Press: Urbana, IL, USA, 2010.
34. Anuar, M.R.; Abdullah, A.Z. Challenges in biodiesel industry with regards to feedstock, environmental, social and sustainability issues: A critical review. *Renew. Sustain. Energy Rev.* **2016**, *58*, 208–223. [CrossRef]
35. Monteiro, M.R.; Kugelmeier, C.L.; Pinheiro, R.S.; Batalha, M.O.; da Silva César, A. Glycerol from biodiesel production: Technological paths for sustainability. *Renew. Sustain. Energy Rev.* **2018**, *88*, 109–122. [CrossRef]
36. Pinheiro, R.S. *Processos de Inovação Tecnológica Para a Glicerina Produzida no Processo de Obtenção de Biodiesel no Brasil*; Universidade Federal de Sao Carlos: Sao Carlos, Brazil, 2011.
37. Aguilar-Aguilar, F.A.; Nelson, D.L.; Pantoja, L.d.A.; Santos, A. Study of anaerobic co-digestion of crude glycerol and swine manure for the production of biogas. *Rev. Virtual Quim* **2017**, *9*, 2383–2403. [CrossRef]
38. Fountoulakis, M.; Petousi, I.; Manios, T. Co-digestion of sewage sludge with glycerol to boost biogas production. *Waste Manag.* **2010**, *30*, 1849–1853. [CrossRef] [PubMed]
39. Rivero, M.; Solera, R.; Perez, M. Anaerobic mesophilic co-digestion of sewage sludge with glycerol: Enhanced biohydrogen production. *Int. J. Hydrog. Energy* **2014**, *39*, 2481–2488. [CrossRef]
40. Dos Santos Ferreira, J.; Volschan, I.; Cammarota, M.C. Co-digestion of sewage sludge with crude or pretreated glycerol to increase biogas production. *Environ. Sci. Pollut. Res.* **2018**, *25*, 21811–21821. [CrossRef] [PubMed]
41. Razaviarani, V.; Buchanan, I.D.; Malik, S.; Katalambula, H. Pilot scale anaerobic co-digestion of municipal wastewater sludge with biodiesel waste glycerin. *Bioresour. Technol.* **2013**, *133*, 206–212. [CrossRef] [PubMed]
42. Nartker, S.; Ammerman, M.; Aurandt, J.; Stogsdil, M.; Hayden, O.; Antle, C. Increasing biogas production from sewage sludge anaerobic co-digestion process by adding crude glycerol from biodiesel industry. *Waste Manag.* **2014**, *34*, 2567–2571. [CrossRef] [PubMed]

43. Jensen, P.; Astals, S.; Lu, Y.; Devadas, M.; Batstone, D. Anaerobic codigestion of sewage sludge and glycerol, focusing on process kinetics, microbial dynamics and sludge dewaterability. *Water Res.* **2014**, *67*, 355–366. [CrossRef]
44. Obi, F.; Ugwuishiwu, B.; Nwakaire, J. Agricultural waste concept, generation, utilization and management. *Niger. J. Technol.* **2016**, *35*, 957–964. [CrossRef]
45. MacLellan, J.; Chen, R.; Kraemer, R.; Zhong, Y.; Liu, Y.; Liao, W. Anaerobic treatment of lignocellulosic material to co-produce methane and digested fiber for ethanol biorefining. *Bioresour. Technol.* **2013**, *130*, 418–423. [CrossRef] [PubMed]
46. Yue, Z.; Teater, C.; MacLellan, J.; Liu, Y.; Liao, W. Development of a new bioethanol feedstock–anaerobically digested fiber from confined dairy operations using different digestion configurations. *Biomass Bioenergy* **2011**, *35*, 1946–1953. [CrossRef]
47. Nayal, F.S.; Mammadov, A.; Ciliz, N. Environmental assessment of energy generation from agricultural and farm waste through anaerobic digestion. *J. Environ. Manag.* **2016**, *184*, 389–399. [CrossRef] [PubMed]
48. Xie, S.; Wickham, R.; Nghiem, L.D. Synergistic effect from anaerobic co-digestion of sewage sludge and organic wastes. *Int. Biodeterior. Biodegrad.* **2017**, *116*, 191–197. [CrossRef]
49. Maragkaki, A.; Fountoulakis, M.; Gypakis, A.; Kyriakou, A.; Lasaridi, K.; Manios, T. Pilot-scale anaerobic co-digestion of sewage sludge with agro-industrial by-products for increased biogas production of existing digesters at wastewater treatment plants. *Waste Manag.* **2017**, *59*, 362–370. [CrossRef] [PubMed]
50. Maragkaki, A.; Fountoulakis, M.; Kyriakou, A.; Lasaridi, K.; Manios, T. Boosting biogas production from sewage sludge by adding small amount of agro-industrial by-products and food waste residues. *Waste Manag.* **2018**, *71*, 605–611. [CrossRef] [PubMed]
51. Alagöz, B.A.; Yenigün, O.; Erdinçler, A. Ultrasound assisted biogas production from co-digestion of wastewater sludges and agricultural wastes: Comparison with microwave pre-treatment. *Ultrason. Sonochem.* **2018**, *40*, 193–200. [CrossRef]
52. Szaja, A.; Montusiewicz, A. Enhancing the co-digestion efficiency of sewage sludge and cheese whey using brewery spent grain as an additional substrate. *Bioresour. Technol.* **2019**, *291*, 121863. [CrossRef]
53. Montañés, R.; Pérez, M.; Solera, R. Mesophilic anaerobic co-digestion of sewage sludge and a lixiviation of sugar beet pulp: Optimisation of the semi-continuous process. *Bioresour. Technol.* **2013**, *142*, 655–662. [CrossRef]
54. Alagöz, B.A.; Yenigün, O.; Erdinçler, A. Enhancement of anaerobic digestion efficiency of wastewater sludge and olive waste: Synergistic effect of co-digestion and ultrasonic/microwave sludge pre-treatment. *Waste Manag.* **2015**, *46*, 182–188. [CrossRef]
55. Salama, E.-S.; Saha, S.; Kurade, M.B.; Dev, S.; Chang, S.W.; Jeon, B.-H. Recent trends in anaerobic co-digestion: Fat, oil, and grease (FOG) for enhanced biomethanation. *Prog. Energy Combust. Sci.* **2019**, *70*, 22–42. [CrossRef]
56. Fonda, K.D.; Hetherington, M.; Kawamoto, M.H. Dealing with FOG: A problem or an opportunity. *Proc. Water Environ. Fed.* **2003**, *2003*, 654–678. [CrossRef]
57. Kabouris, J.C.; Tezel, U.; Pavlostathis, S.G.; Engelmann, M.; Todd, A.C.; Gillette, R.A. The anaerobic biodegradability of municipal sludge and fat, oil, and grease at mesophilic conditions. *Water Environ. Res.* **2008**, *80*, 212–221. [CrossRef] [PubMed]
58. Masse, L.; Masse, D.; Kennedy, K.; Chou, S. Neutral fat hydrolysis and long-chain fatty acid oxidation during anaerobic digestion of slaughterhouse wastewater. *Biotechnol. Bioeng.* **2002**, *79*, 43–52. [CrossRef] [PubMed]
59. Cavaleiro, A.J.; Salvador, A.F.; Alves, J.I.; Alves, M. Continuous high rate anaerobic treatment of oleic acid based wastewater is possible after a step feeding start-up. *Environ. Sci. Technol.* **2009**, *43*, 2931–2936. [CrossRef] [PubMed]
60. Madigan, M.T.; Martinko, J.M.; Parker, J.; Brock, T.D. *Brock Biology of Microorganisms*, 11th ed.; Prentice Hall: Upper Saddle River, NJ, USA, 2006.
61. Pavlostathis, S.; Giraldo-Gomez, E. Kinetics of anaerobic treatment: A critical review. *Crit. Rev. Environ. Sci. Technol.* **1991**, *21*, 411–490. [CrossRef]
62. Canakci, M.; Van Gerpen, J. Biodiesel production from oils and fats with high free fatty acids. *Trans. ASAE* **2001**, *44*, 1429. [CrossRef]
63. Jeganathan, J.; Nakhla, G.; Bassi, A. Long-term performance of high-rate anaerobic reactors for the treatment of oily wastewater. *Environ. Sci. Technol.* **2006**, *40*, 6466–6472. [CrossRef]

64. Luostarinen, S.; Luste, S.; Sillanpää, M. Increased biogas production at wastewater treatment plants through co-digestion of sewage sludge with grease trap sludge from a meat processing plant. *Bioresour. Technol.* **2009**, *100*, 79–85. [CrossRef]
65. Girault, R.; Bridoux, G.; Nauleau, F.; Poullain, C.; Buffet, J.; Peu, P.; Sadowski, A.; Béline, F. Anaerobic co-digestion of waste activated sludge and greasy sludge from flotation process: Batch versus CSTR experiments to investigate optimal design. *Bioresour. Technol.* **2012**, *105*, 1–8. [CrossRef]
66. Pereira, M.; Pires, O.; Mota, M.; Alves, M. Anaerobic biodegradation of oleic and palmitic acids: Evidence of mass transfer limitations caused by long chain fatty acid accumulation onto the anaerobic sludge. *Biotechnol. Bioeng.* **2005**, *92*, 15–23. [CrossRef]
67. Rinzema, A. *Anaerobic Treatment of Wastewater with High Concentrations of Lipids or Sulfate*; Staff: Wageningen, The Netherlands, 1988.
68. Kuang, Y. *Enhancing Anaerobic Degradation of Lipids in Wastewater by Addition of Co-Substrate*; Murdoch University: Perth, Australia, 2002.
69. Davidsson, Å.; Lövstedt, C.; la Cour Jansen, J.; Gruvberger, C.; Aspegren, H. Co-digestion of grease trap sludge and sewage sludge. *Waste Manag.* **2008**, *28*, 986–992. [CrossRef]
70. Kabouris, J.C.; Tezel, U.; Pavlostathis, S.G.; Engelmann, M.; Dulaney, J.A.; Todd, A.C.; Gillette, R.A. Mesophilic and thermophilic anaerobic digestion of municipal sludge and fat, oil, and grease. *Water Environ. Res.* **2009**, *81*, 476–485. [CrossRef]
71. Silvestre, G.; Rodríguez-Abalde, A.; Fernández, B.; Flotats, X.; Bonmatí, A. Biomass adaptation over anaerobic co-digestion of sewage sludge and trapped grease waste. *Bioresour. Technol.* **2011**, *102*, 6830–6836. [CrossRef]
72. Silvestre, G.; Illa, J.; Fernández, B.; Bonmatí, A. Thermophilic anaerobic co-digestion of sewage sludge with grease waste: Effect of long chain fatty acids in the methane yield and its dewatering properties. *Appl. Energy* **2014**, *117*, 87–94. [CrossRef]
73. Alqaralleh, R.M.; Kennedy, K.; Delatolla, R.; Sartaj, M. Thermophilic and hyper-thermophilic co-digestion of waste activated sludge and fat, oil and grease: Evaluating and modeling methane production. *J. Environ. Manag.* **2016**, *183*, 551–561. [CrossRef]
74. Grady, C.L., Jr.; Daigger, G.T.; Love, N.G.; Filipe, C.D. *Biological Wastewater Treatment*, 3rd ed.; CRC Press: Boca Raton, FL, USA, 2011.
75. Stafford, D. The effects of mixing and volatile fatty acid concentrations on anaerobic digester performance. *Biomass* **1982**, *2*, 43–55. [CrossRef]
76. Karim, K.; Hoffmann, R.; Klasson, K.T.; Al-Dahhan, M. Anaerobic digestion of animal waste: Effect of mode of mixing. *Water Res.* **2005**, *39*, 3597–3606. [CrossRef]
77. Rizk, M.C.; Bergamasco, R.; Tavares, C.R.G. Anaerobic co-digestion of fruit and vegetable waste and sewage sludge. *Int. J. Chem. React. Eng.* **2007**, *5*. [CrossRef]
78. Stroot, P.G.; McMahon, K.D.; Mackie, R.I.; Raskin, L. Anaerobic codigestion of municipal solid waste and biosolids under various mixing conditions—I. Digester performance. *Water Res.* **2001**, *35*, 1804–1816. [CrossRef]
79. Latha, K.; Velraj, R.; Shanmugam, P.; Sivanesan, S. Mixing strategies of high solids anaerobic co-digestion using food waste with sewage sludge for enhanced biogas production. *J. Clean. Prod.* **2019**, *210*, 388–400. [CrossRef]
80. Wang, M.; Sun, X.; Li, P.; Yin, L.; Liu, D.; Zhang, Y.; Li, W.; Zheng, G. A novel alternate feeding mode for semi-continuous anaerobic co-digestion of food waste with chicken manure. *Bioresour. Technol.* **2014**, *164*, 309–314. [CrossRef]
81. Mahmoodi-Eshkaftaki, M.; Ebrahimi, R. Assess a new strategy and develop a new mixer to improve anaerobic microbial activities and clean biogas production. *J. Clean. Prod.* **2019**, *206*, 797–807. [CrossRef]
82. Monnet, F. An introduction to anaerobic digestion of organic wastes. *Remade Scotl.* **2003**, *379*, 1–48.
83. Heo, N.H.; Park, S.C.; Kang, H. Effects of mixture ratio and hydraulic retention time on single-stage anaerobic co-digestion of food waste and waste activated sludge. *J. Environ. Sci. Health Part A* **2004**, *39*, 1739–1756. [CrossRef]
84. Hagos, K.; Zong, J.; Li, D.; Liu, C.; Lu, X. Anaerobic co-digestion process for biogas production: Progress, challenges and perspectives. *Renew. Sustain. Energy Rev.* **2017**, *76*, 1485–1496. [CrossRef]
85. Siddique, M.N.I.; Wahid, Z.A. Achievements and perspectives of anaerobic co-digestion: A review. *J. Clean. Prod.* **2018**, *194*, 359–371. [CrossRef]

86. Khalid, A.; Arshad, M.; Anjum, M.; Mahmood, T.; Dawson, L. The anaerobic digestion of solid organic waste. *Waste Manag.* **2011**, *31*, 1737–1744. [CrossRef]
87. Li, Y.; Sasaki, H.; Yamashita, K.; Seki, K.; Kamigochi, I. High-rate methane fermentation of lipid-rich food wastes by a high-solids co-digestion process. *Water Sci. Technol.* **2002**, *45*, 143–150. [CrossRef]
88. Lee, M.; Hidaka, T.; Hagiwara, W.; Tsuno, H. Comparative performance and microbial diversity of hyperthermophilic and thermophilic co-digestion of kitchen garbage and excess sludge. *Bioresour. Technol.* **2009**, *100*, 578–585. [CrossRef]
89. Hansen, K.H.; Angelidaki, I.; Ahring, B.K. Anaerobic digestion of swine manure: Inhibition by ammonia. *Water Res.* **1998**, *32*, 5–12. [CrossRef]
90. Fountoulakis, M.; Manios, T. Enhanced methane and hydrogen production from municipal solid waste and agro-industrial by-products co-digested with crude glycerol. *Bioresour. Technol.* **2009**, *100*, 3043–3047. [CrossRef]
91. Vindis, P.; Mursec, B.; Janzekovic, M.; Cus, F. The impact of mesophilic and thermophilic anaerobic digestion on biogas production. *J. Achiev. Mater. Manuf. Eng.* **2009**, *36*, 192–198.
92. Ghatak, M.; Mahanata, P. Effect of Temperature on Biogas Production from Rice Straw and Rice Husk. In *IOP Conference Series: Materials Science and Engineering*; IOP Publishing: Bristol, UK, 2018.
93. Zeeman, G.; Wiegant, W.; Koster-Treffers, M.; Lettinga, G. The influence of the total-ammonia concentration on the thermophilic digestion of cow manure. *Agric. Wastes* **1985**, *14*, 19–35. [CrossRef]
94. Hanaki, K.; Matsuo, T.; Nagase, M. Mechanism of inhibition caused by long-chain fatty acids in anaerobic digestion process. *Biotechnol. Bioeng.* **1981**, *23*, 1591–1610. [CrossRef]
95. Hwu, C.-S.; Lettinga, G. Acute toxicity of oleate to acetate-utilizing methanogens in mesophilic and thermophilic anaerobic sludges. *Enzym. Microb. Technol.* **1997**, *21*, 297–301. [CrossRef]
96. Kim, S.-H.; Han, S.-K.; Shin, H.-S. Feasibility of biohydrogen production by anaerobic co-digestion of food waste and sewage sludge. *Int. J. Hydrog. Energy* **2004**, *29*, 1607–1616. [CrossRef]
97. Babaee, A.; Shayegan, J. Effect of organic loading rates (OLR) on production of methane from anaerobic digestion of vegetables waste. In Proceedings of the World Renewable Energy Congress-Sweden, Linkoping, Sweden, 8–13 May 2011; pp. 411–417.
98. Mata-Alvarez, J.; Llabrés, P.; Cecchi, F.; Pavan, P. Anaerobic digestion of the Barcelona central food market organic wastes: Experimental study. *Bioresour. Technol.* **1992**, *39*, 39–48. [CrossRef]
99. Ford, M.; Fleming, R. Mechanical solid-liquid separation of livestock manure. Literature review. In *Report to Ontario Pork, Case Study*; University of Guelph: Ridgetown, ON, Canada, 2002.
100. Massé, D.; Croteau, F.; Masse, L. The fate of crop nutrients during digestion of swine manure in psychrophilic anaerobic sequencing batch reactors. *Bioresour. Technol.* **2007**, *98*, 2819–2823. [CrossRef]
101. Tu, Q. *Fats, Oils and Greases to Biodiesel: Technology Development and Sustainability Assessment*; University of Cincinnati: Cincinnati, OH, USA, 2015.
102. Tu, Q.; McDonnell, B.E. Monte Carlo analysis of life cycle energy consumption and greenhouse gas (GHG) emission for biodiesel production from trap grease. *J. Clean. Prod.* **2016**, *112*, 2674–2683. [CrossRef]
103. Wallace, T.; Gibbons, D.; O'Dwyer, M.; Curran, T.P. International evolution of fat, oil and grease (FOG) waste management–A review. *J. Environ. Manag.* **2017**, *187*, 424–435. [CrossRef]

Review

Sustainable Waste-to-Energy Development in Malaysia: Appraisal of Environmental, Financial, and Public Issues Related with Energy Recovery from Municipal Solid Waste

Zi Jun Yong [1,2], Mohammed J.K. Bashir [1,*], Choon Aun Ng [1], Sumathi Sethupathi [1], Jun Wei Lim [3] and Pau Loke Show [4]

1 Department of Environmental Engineering, Faculty of Engineering and Green Technology (FEGT), Universiti Tunku Abdul Rahman, Kampar 31900, Perak, Malaysia; 930805tree@gmail.com (Z.J.Y.); ngca@utar.edu.my (C.A.N.); sumathi@utar.edu.my (S.S.)

2 Centre for Global Sustainability Studies (CGSS), Institute of Postgraduate Studies, Universiti Sains Malaysia, Penang 11800, Malaysia

3 Department of Fundamental and Applied Sciences, Centre for Biofuel and Biochemical Research, Institute of Self-Sustainable Building, Universiti Teknologi PETRONAS, Seri Iskandar 32610, Perak Darul Ridzuan, Malaysia; junwei.lim@utp.edu.my

4 Department of Chemical and Environmental Engineering, Faculty of Science and Engineering, University of Nottingham Malaysia, Jalan Broga, Semenyih 43500, Selangor, Malaysia; PauLoke.Show@nottingham.edu.my

* Correspondence: jkbashir@utar.edu.my; Tel.: +605-468-8888 (ext. 4559); Fax: +605-466-7449

Received: 16 July 2019; Accepted: 22 August 2019; Published: 1 October 2019

Abstract: As Malaysia is a fast-developing country, its prospects of sustainable energy generation are at the center of debate. Malaysian municipal solid waste (MSW) is projected to have a 3.3% increase in annual generation rate at the same time an increase of 3.3% for electricity demand. In Malaysia, most of the landfills are open dumpsite and 89% of the collected MSW end up in landfills. Furthermore, huge attention is being focused on converting MSW into energy due to the enormous amount of daily MSW being generated. Sanitary landfill to capture methane from waste landfill gas (LFG) and incineration in a combined heat and power plant (CHP) are common MSW-to-energy technologies in Malaysia. MSW in Malaysia contains 45% organic fraction thus landfill contributes as a potential LFG source. Waste-to-energy (WTE) technologies in treating MSW potentially provide an attractive economic investment since its feedstock (MSW) is collected almost for free. At present, there are considerable issues in WTE technologies although the technology employing MSW as feedstock are well established, for instance the fluctuation of MSW composition and the complexity in treatment facilities with its pollutant emissions. Thus, this study discusses various WTE technologies in Malaysia by considering the energy potentials from all existing incineration plants and landfill sites as an effective MSW management in Malaysia. Furthermore, to promote local innovation and technology development and to ensure successful long-term sustainable economic viability, social inclusiveness, and environmental sustainability in Malaysia, the four faculties of sustainable development namely technical, economic, environmental, and social issues affiliated with MSW-to-Energy technologies were compared and evaluated.

Keywords: Malaysia; Municipal Solid Waste (MSW); Waste-to-Energy (WTE); sustainability; technical; economic; environmental; social

---

1. Introduction

The aim of this paper is to provide readers with a better picture on the current environmental and energy issues arising from the continuous municipal solid waste (MSW) generation and economic development in Malaysia. The paper addresses how MSW-to-energy can play an important role in achieving the United Nation's sustainable development goals (UN-SDGs) and showcases the available MSW-to-energy plants in Malaysia. Moreover, a holistic analysis involving the four faculties in sustainable development, namely technical, economic, environmental, and social, together with the SWOT (strength, weakness, opportunity, and threat) analysis is discussed to select the most appropriate waste-to-energy (WTE) technology for ensuring long-term sustainability of WTE implementation and waste management practices in Malaysia can be achieved.

As the population increases, the need for more material and energy consumption increases. This contributes significantly to the amount of anthropogenic municipal solid waste (MSW) generation and greenhouse gases (GHGs) emissions resulting from electricity production to support the ever-increasing services needed by the growing population. Malaysia, as an emerging developing country, is currently facing the dilemma of MSW management with increasing GHGs especially methane gas resulting from the degradation of organic waste within landfills and the need for more energy (electricity) and material consumption to support the growing Malaysian population. Currently, the Malaysia's average solid waste generation per capita is 1.17 kg/day [1]. Major cities contribute toward higher waste generation per capita due to mass consumption and higher purchasing power. For instance, the capital of Malaysia, Kuala Lumpur, has an average generation rate of 1.35 kg/person/day and Penang has waste generation rate ranging from 1.5 to 1.8 kg/person/day [1,2]. The carbon dioxide (CO_2) emission per capita in Malaysia has increased from 5.42 metric tons in 2000 to 8.53 metric tons (MT) in 2016 which had an increase of 57.38% as shown in Figure 1; while the final electricity consumption combining agriculture, commercial, transport, industry, and residential in 2016 was 12,392 kilo-ton oil equivalent (ktoe)/person/year, which had increased 70.43% from the level in 2006 at 7271 ktoe/person/year as indicated in Figure 2. At present, the total solid waste generation in Malaysia is 33,130 tonnes per day, and is expected to increase further to 49,670 tonnes per day by 2030 [1]. The energy demand is expected to increase by 4.7% and there is an average annual growth of 8.1% in the final electricity consumption which will eventually contribute to more GHG emissions if the energy sector in Malaysia is still largely dependent on fossil fuel resources, namely diesel, coal, and natural gas for electricity generation [3]. Hence there is an urgent need to resolve multiple issues and dilemmas towards a more sustainable waste management, lowering per capita CO_2 emissions and clean energy production in Malaysia.

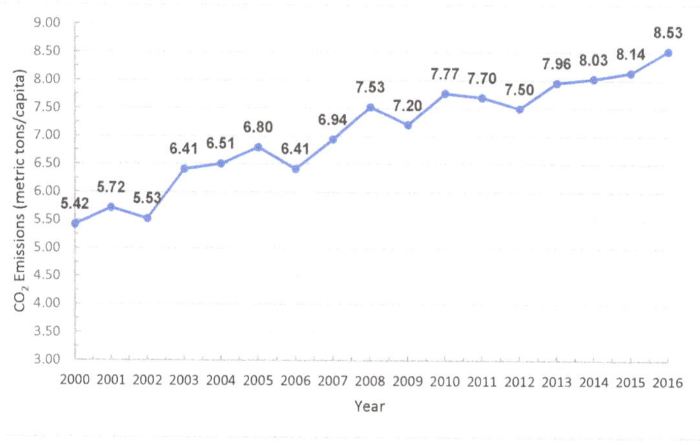

Figure 1. The increasing trend of CO_2 emissions per capita in Malaysia from 2000 to 2016 [4].

A total of 89% of the Malaysian MSW generated directly enter into landfills with minimal treatment, where only 1% of the total incoming MSW receive proper treatment [1]. In Malaysia, 50% of landfills are open dumping sites; 30% use-controlled tipping; 12% are controlled landfills with daily cover; 5% are sanitary landfills without leachate treatment facility; and the other 5% are sanitary landfills with leachate treatment. Within the coming 10 years (by 2030), over 80% of the Malaysian open dumping landfill sites are to be shut down due to reaching full capacity. The major MSW fractions generated in Malaysia are 45% organic material, 13% plastics, 12% diapers, 9% paper, 3% glass, 3% metal and others [1,5]. Figure 3 shows the composition of solid waste in Malaysia. It can be concluded that organic waste represents the largest portion of the total solid waste produced by Malaysians therefore making landfills a potential source of landfill gas (LFG) [6]. Nevertheless, decommissioning a landfill involves the process of obtaining other lands and is an environmentally challenging process. This will make land scarcer in the future. Employing MSW to energy can solve two problems at once, namely the demand for more energy and the continuous increase in MSW generation. Hence, waste is no longer an undesired product from the society but a new resource by treating non-recyclables and non-reusables from MSW to generate a substantial amount of energy for urban use while preserving scarce lands [7].

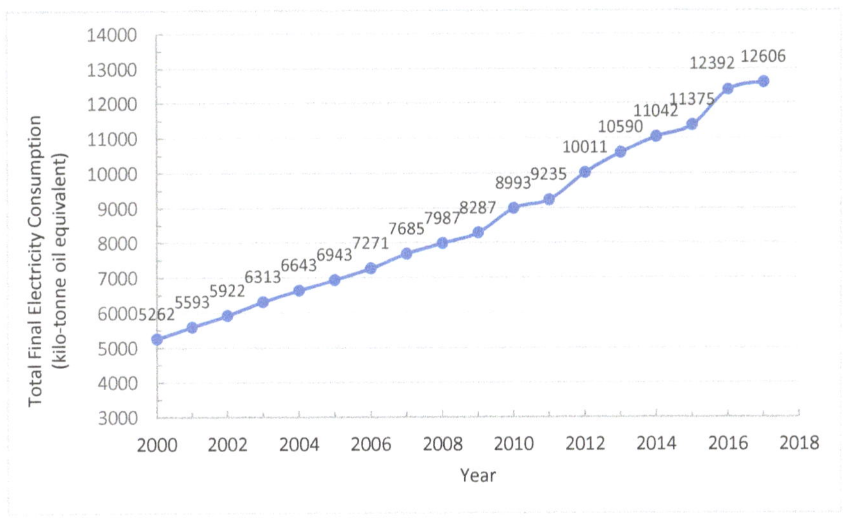

Figure 2. Final electricity consumption, kilo-tonne oil equivalent (ktoe) in Malaysia [8].

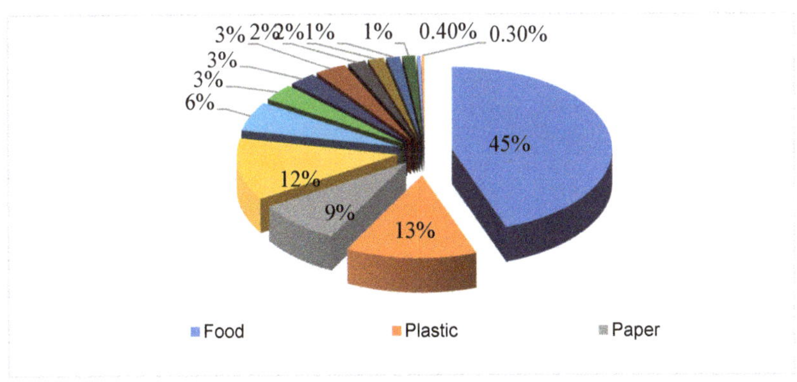

Figure 3. Malaysian waste fraction [8].

Malaysian electricity generation is still largely dependent on fossil fuel-based resources for which accounts for 77.27% or 23,518.10 MW installed capacity source from natural gas, coal, and diesel/Marine Fuel Oil (MFO) while large-hydro power contributes 18.78% or 5716.10 MW. Only a small fraction of 3.96% or 1205.20 MW are from renewable energy (RE) resources by excluding large-hydro namely biomass, biogas, solar, and geothermal as indicated in Figure 4.

Furthermore, the renewable energy installed capacity via the Feed-in Tariff (FiT) mechanism scheme, which only applies in Peninsular and Sabah set by the Sustainable Energy Development Authority (SEDA) Malaysia, is relatively low compared to other neighboring countries in terms of installed capacity as shown in Figure 5. According to Joshi [9], in 2015 the renewable energy installed capacity (excluding Sarawak) in Malaysia totaled 446 MW which has a short-fall of 45% of the aim to achieve 975 MW of RE targets in 2015 laid out in the Malaysian National Renewable Energy Policy and Action Plan (NREPAP). Within the same year, the electricity produced from renewable energy resources which is being fed into the electricity grid is well under 3%. At present, the total commissioned renewable energy installations as of the end of March 2019 in Malaysia is 604.44 MW which is still lower than the target fixed in 2015 [10].

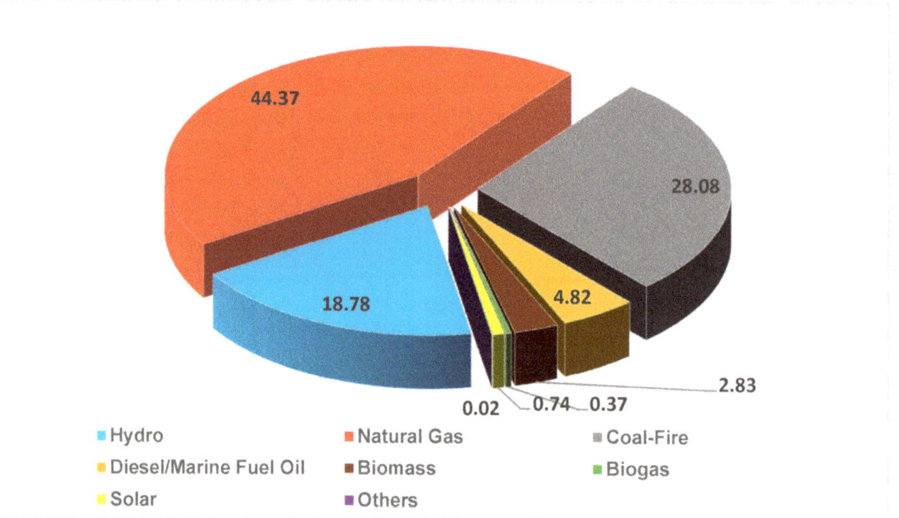

Figure 4. The percentage, %, of share in installed capacity of energy resources in the production of electricity [11].

For instance, the Feed-in Tariff renewables installed capacity as of 2017 in Malaysia totaled 528.06 MW [10] while in Vietnam is 2569 MW; Indonesia is 3833 MW, and Thailand is 6766 MW [12].

Lastly, this study investigates the key technical, economic, environmental, and social issues within Malaysia while realizing various WTE technologies. The opportunities and challenges in the implementation of MSW-to-energy are considered.

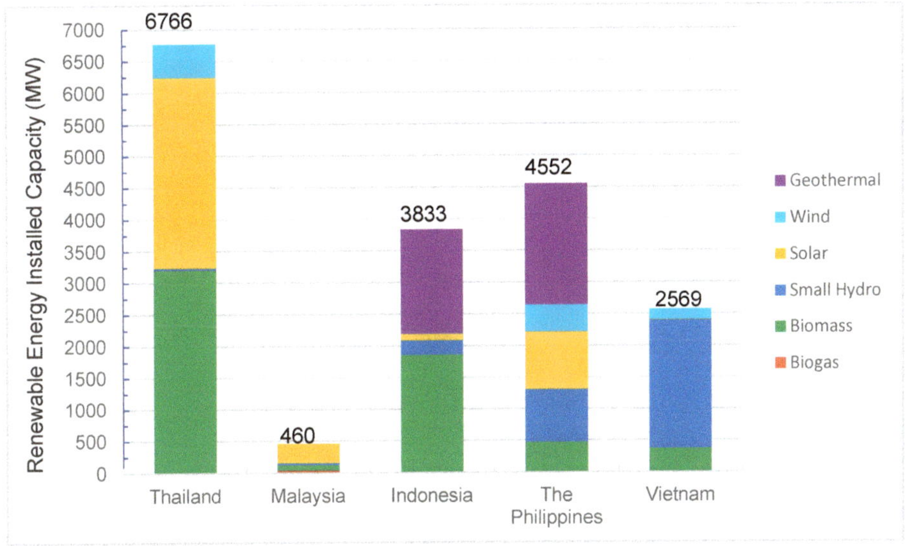

Figure 5. The renewable energy installed capacity in Mega-Watts (MW) of five Southeast Asia countries as of 31 March 2017 [12].

2. Municipal Solid Waste (MSW)-to-Energy in Malaysia

The government of Malaysia has included "waste minimization", "promotion of reuse", "developing a recycling-oriented", and "implementation of pilot projects for recycling" as some of its main policy goals, starting from the 8th Malaysian Plan. The continuation of waste reduction, recycling, reuse, and recovery with the focus of greater use of environmentally friendly products were emphasized in the 9th Malaysia Plan (2006–2010). Consequently, the National Solid Waste Management Department was established under the Ministry of Housing and Local Government [13]. Meanwhile, a new bill has been developed to implement the new Solid Waste and Public Cleansing Management Corporation Act 2007 (SWPCMC Act 2007) [14] which later in 2011 was followed by the enforcement and restructuring of the SWM Act [15]. This act also provides executive power to the Federal Government to implement municipal solid waste management and public cleansing activities throughout Peninsula Malaysia, Federal Territories of Putrajaya, and Labuan. Various small-scale technologies to recover material and energy from waste including composting, recycling, and burning have been identified and utilized for municipal solid waste management. Malaysia, like any other developing countries, is concerned about the increasing need for more energy to sustain the continuous economic and industry development. In search for a more holistic waste management at the same time towards the need for renewable energy, the government of Malaysia proposes the utilization of waste-to-energy technologies to recover energy from municipal solid waste [16]. This makes municipal solid waste as a new form of renewable source for energy production. WTE technology is an energy recovery process that converts chemicals from waste residues into practical forms of usable energy like electricity, heat, or steam. In general, Figure 6 shows the three main routes of MSW to energy recovery, namely bio-chemical, thermo-chemical, and mechanical-chemical.

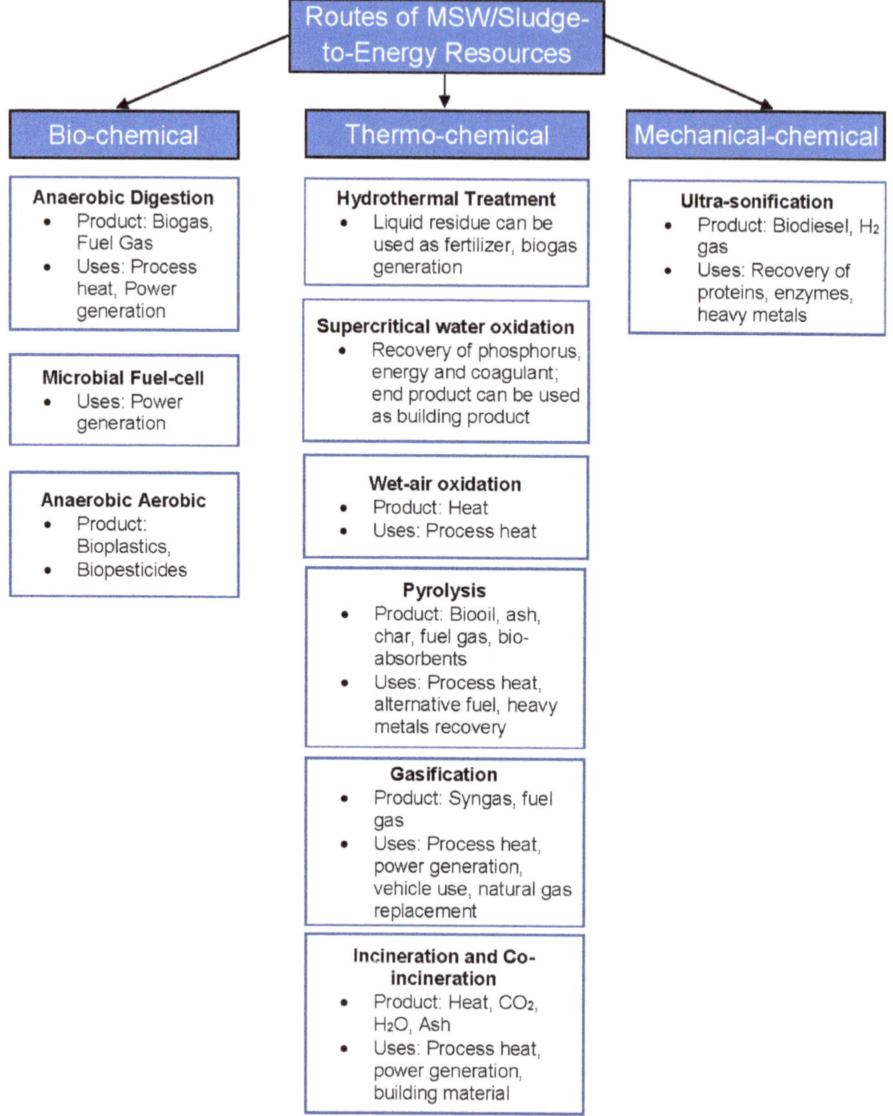

Figure 6. The various routes of waste-to-energy (WTE) treatment available [17].

The aim of WTE technology is to minimize the solid waste volume to be disposed into existing landfills despite the many differences of WTE technologies that are available in today's market. The WTE process can reduce the solid waste biodegradable fraction to zero, while at the same time generates valuable electricity and heat from non-recyclable waste. The major classifications of WTE technology are biological and thermal. Pyrolysis, plasma gasification, or gasification and combustion are a few examples of thermal treatment of waste where the waste is subjected to high temperatures with varying oxygen concentrations. Anaerobic digestion (AD) and microbial fuel-cell are examples of biological treatment of waste. Anaerobic digestion is normally used to recover energy to both forms of electricity and heat. Furthermore, anaerobic digestion produces nutrient rich fertilizer in the form of digestate from the wet and biodegradable waste streams (food and kitchen waste, sludge from wastewater treatment plants, manure from live-stock farming, high-energy yield crop from agricultural waste, and farm slurry). In addition, WTE involves the conversion of landfill gas (LFG) into energy through landfill technologies [18]. WTE technology is also one of many potential alternatives which can produce electricity from RE resources from the utilization of re-cycling potential of the degradable organic waste portion from various sources like household, commercial, and industrial and municipal wastewater treatment plants. Furthermore, energy extraction from waste can simultaneously increase the life-span of existing landfills and reduce the emissions of GHGs into the atmosphere as the energy produced from waste is a potential replacement to the equivalent sum of energy produced from fossil fuel resources. WTE technologies were established in consideration of the push towards a more sustainable and clean solid waste management together with the need for sustainable energy development. In Malaysia, the degradation of organic waste within landfill to produce biogas via LFG, incineration, and co-incineration of mixed MSW are present in converting waste into energy. In Malaysia, the idea of WTE resulted in the construction of an incineration plant at Langkawi and the establishment of biogas power plants within Peninsular Malaysia to convert MSW to electricity [16]. At the moment, Malaysia has four mini incinerators under various stages of implementation in Langkawi, Tioman, and Pangkor Island plus one in Cameron Highlands and one WTE plant located at the central region. Recently in 2018 within the 10th Malaysia Plan timeframe, another WTE plant is to be built and completed in Negeri Sembilan [19]. Energy can be recovered from MSW in Malaysia through one of the three ways in WTE treatment namely thermal, biological, and landfilling via LFG.

2.1. Complementing the United Nations Sustainable Development Goals (SDGs)

The implementation of WTE technologies and strategies contribute significantly in reducing total waste volume entering landfills, increasing the lifespan of existing landfills, and offsetting large repercussions associated with the burning of fossil fuels to generate electricity. This enables Malaysia to play a positive role internationally in offsetting the world's carbon emission by generating electricity from waste and to push towards a greater share of RE in the national electricity generation mix. WTE falls under SDG 7 for affordable and clean energy where technology advancement in WTE can provide the human race with sustainable, safe, low-cost, renewable, and secure energy sources in the future [20]. It is of utmost important to understand the targets involved in SDG 7 and to be able to link towards other appropriate SDGs to see how achieving SDG 7 can concurrently complement towards other SDGs. Table 1 indicates the five targets in SDG 7. Moreover, there are six identified SDGs which have a more direct relationship with SDG 7 in regards to the implementation of WTE in Malaysia, as represented in Figure 7 [21].

Table 1. All the five energy targets consisting of three substantive targets and two additional targets which are identified as a means of implementation in contributing towards achieving Sustainable Development Goal (SDG) No. 7 [22].

SDG 7: Ensure Access to Affordable, Reliable, Sustainable, and Modern Energy for All			
7.1	Achieve universal access to affordable, reliable, and modern energy services by 2030.	7.a	Enhance international cooperation to facilitate access to clean energy research and technology, including renewable energy, energy efficiency, and advanced and cleaner fossil-fuel technology, and promote investment in energy infrastructure and clean energy technology by 2030.
7.2	Increase substantially the share of renewable energy in the global energy mix by 2030.	7.b	Expand infrastructure and upgrade technology for supplying modern and sustainable energy services for all in developing countries, in particular least developed countries, small island developing States, and land-locked developing countries, in accordance with their respective programmes of support by 2030.
7.3	Double the global rate of improving in energy efficient practice by 2030.		

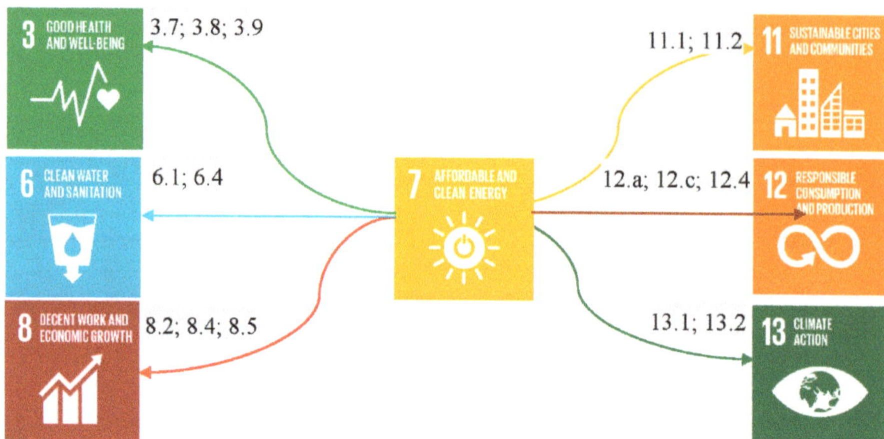

Figure 7. The relationship and interlinked of targets between SDG 7 with SDG 3, 6, 8, 11, 12, and 13 (Yong Zi Jun, 2019).

- **SDG 3** for good health and well-being for all can be achieved, as employing WTE can significantly reduce MSW entering landfills. This helps to prevent and reduce the generation of hazardous leachate containing persistent, carcinogenic compounds, and other pathogens from complex mixed waste within the landfill. Hence, this reduce the chances of water contamination from leachate infiltrating into nearby groundwater and surface water sources which maybe the source of water supply for downstream populations. This significantly reduces the impact on the number of cases of deaths and illness related to water pollution and water borne diseases (target 3.9). Moreover, WTE plants are usually located off city centers. The government can plan to integrate these WTE plants to provide a stable baseload electricity towards many remote clinics and hospitals located in sub-urban and rural areas. This can provide both rural and sub-urban residents access to universal health care (targets 3.7 and 3.8).
- **SDG 6** for ensuring access to water and sanitation for all can be secured via WTE process as WTE plants can supply clean electricity to both raw water and wastewater treatment plants for water related treatment processes such as blowers, aerators, pumping, chemical dosing equipment, meters, sensors etc., as well as to supply electricity to the treatment plant's centralized control and monitoring room. Hence, WTE plants power treatment plants to provide people adequate access to clean drinking water and sanitation, especially those who are living in remote areas (target 6.1). Moreover, preservation of water resources can be achieved via these advanced WTE plants. This is because WTE plants use the latest efficient boiler technology which has a higher efficiency in terms of using less water from tube-wells and rivers to produce the same amount of kWh of electricity (target 6.4).
- **SDG 8** for promoting sustained, inclusive, and sustainable economic growth, full and productive employment, and decent work for all. Embarking on WTE to produce electricity helps to diversify the energy industry and giving back value to MSW as a new resource to generate energy to drive the nation's economy (target 8.2). WTE plants help to decouple the economy in the energy and production sector from environmental degradation (target 8.4) as part of the fossil fuel-based resources in generating electricity is being substituted by WTE plants. Moreover, WTE can stimulate and generate more job opportunities in various stages of planning, design, constructing, commissioning, operating, and maintaining such WTE power plants (target 8.5).
- **SDG 11** for making cities and human settlements inclusive, safe, resilient, and sustainable. WTE has the potential to gradually substitute some of the nation's coal-fired and diesel power plants

to provide adequate base-load power supply to residential housing, hence providing access to adequate and safe housing to the people (target 11.1). Since WTE is decarbonizing the energy sector by replacing the conventional high air polluting coal-fired and diesel power plants, WTE emits lower GHGs allowing better urban air quality in cities. The carbon dioxide, sulfur dioxide, and nitrogen oxides emissions by WTE using MSW as fuel are 837, 0.8, and 5.4 pounds/MWh respectively which is much lower than coal-fired power plants which emits 2249, 13, and 6 pounds/MWh of electricity produced respectively [23]. Moreover, the implementation of WTE can divert away mixed MSW from entering landfills. This reduces the collective release of hazardous bad odor gases from landfills in Malaysia which result from the complex chemical reaction and degradation of waste within landfills. Hence, this reduces the adverse per capita environmental impact of cities by paying special attention to air quality and MSW management via WTE (target 11.6).

- **SDG 12** for ensuring sustainable consumption and production patterns. This is closely related to circular economy where the main intention of SDG 12 is to maximize the use of natural resources and to minimize the harmful waste generation and disposal towards the environment. WTE support this by reusing the waste stream to generate electricity hence (target 12.5) at the same time reducing the contamination of hazardous chemicals emitted by landfills into air, water, and land (target 12.4).
- **SDG 13** for climate action by taking urgent action to combat climate change and its impacts can be achieved via the use of more WTE as RE resources (target 7.2). At the same time, the use of more WTE plants to generate electricity for urban use crucially strengthen the adaptive capacity and resilience of cities towards climate-related hazards (target 13.1) as the burning of fossil fuel for energy is replaced with the conversion of municipal solid waste to energy. WTE technologies like LFG and AD capture and burn methane gas (CH_4) where methane is 21 times more potent in global warming potential (GWP) than CO_2 to generate heat and electricity. Hence, this contributes towards reducing GHGs emissions (targets 13.2). According to AlQattan [17], it was reported that 1/4 of a ton of high-quality coal or a barrel of oil can be offset for every metric ton of waste used in WTE to produce the same amount of electricity.

2.2. *Complementing the Malaysian Low Carbon Cities Framework (LCCF) Initiative: From Linear to Circular Cities Metabolism*

Currently, cities worldwide are responsible for consuming 70% of earth's energy and resources hence are the main source of generating solid waste and the generation of 70% of global carbon emission [24]. Moreover, half of the world's current population are already living in cities and the number is expected to rise to 60% by 2030 and up to 66% by 2050 [24]. Malaysia had voluntarily pledged to reduce 40% of the country's GHG emissions by 2020 compared to levels in 2005 during the 15th Conference of Parties (COP 15) at Copenhagen in Denmark, and pledged to reduce carbon emissions by 45% compared to levels in 2005 levels during the Paris UN Climate Change Conference in 2015 [25]. Turning MSW into energy is a key in creating a sustainable city by recycling all organic and inorganic waste streams, and at the same time serves as a key complement towards the Malaysian pledge during the COP 15 and moving towards a circular economy enabling value of products, materials, and resources to be maintained. WTE supports the initiative of moving from a linear to circular urban metabolism to minimize waste and pollution output to the environment as illustrated in Figure 8. Unfortunately, according to Abushammala et al. [26], Malaysia has a grand total of 212 disposal sites and only 14 sites are classified as sanitary landfills, as indicated in Table 2. This indicates that most Malaysian cities are currently adopting the conventional linear metabolism rather than a more environmentally sound approach of a closed-loop circular city metabolism where the reuse of waste into energy and new materials dominates the waste management practices. Non-sanitary landfills and open dumping are implemented as a main disposal method of MSW due to its low operating cost and convenience to dispose MSW, although this practice brings tremendous negative implications to the economy, environment, and towards the health of the people. With that in mind, according to the Sustainable Energy Development Authority (SEDA) Malaysia, there are a few MSW-to-energy plants or LFG-to-energy sites registered with SEDA for the biogas (landfill) Feed-in Tariff (FiT) scheme to produce and sell green electricity back to Tenaga National Berhad (TNB) which is later on fed into the national grid for urban electricity consumption.

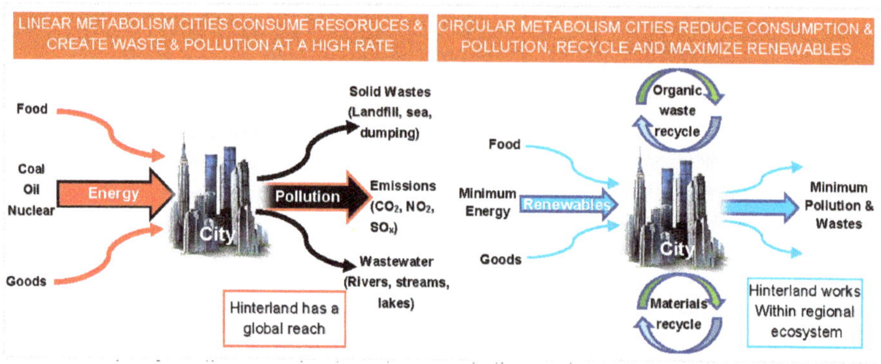

Figure 8. Moving from a linear to circular urban metabolism as a key component towards achieving sustainable cities, which assures the most efficient possible resource use and least generation of waste back into the biosphere [27].

Table 2. The types and number of disposal sites in Malaysia in respect to the amount of waste received [2,26].

Region	Number of Sites		Total Waste, Millions Metric Tons (mmt)	
	Sanitary Landfill	Disposal Site	Sanitary Landfill	Disposal Site
Northern	3	43	22.8	33.7
Central	4	23	6.5	23.0
Southern	2	33	6.6	21.1
East Coast	1	62	2.9	15.5
Borneo	4	51	1.6	9.3
Total	14	212	40.4	102

Moreover, WTE plants compliment the Malaysian Low Carbon Cities Framework (LCCF) and assessment system which was launched and adopted in 2009 by the Ministry of Energy, Green Technology, and Water (KeTTHA) and later had a LCCF version 2. This is part of Malaysia's initiative in playing a positive role in the world's climate agenda during the (COP) 15 in Copenhagen, Denmark in 2009. In general, there are four main performance aspects under the LCCF which can be quantified under the LCCF in the effort to lower and to keep track of the city's carbon emissions. The four LCCF aspects are:

- Urban Environment;
- Urban Transport;
- Urban Infrastructure;
- Urban Building.

Waste management has been classified as part of a component under the urban infrastructure, besides water management, energy, and other infrastructure provisions such as urban storm water management and flood mitigation, earth work management etc. In Malaysia, households are the main source of total solid waste generation which accounts for 65% compared to commercial and institutional at 28% and industry at 7% [1]. Therefore, WTE through the thermo-chemical waste treatment pathways like incineration, gasification, etc., plays a crucial role in household solid waste management in reducing nearly up to 95% of MSW by volume entering landfills.

3. WTE in Malaysia

3.1. Incineration Plant

In Malaysia, the use of incineration as WTE is only present in a very limited quantity at a small scale. A total of 80%–95% of MSW volume can be reduced by incineration. Back in 2011, several incineration projects with the expenditure of RM 187.74 million had been commenced by the Ministry of Local Government and Housing (MHLG) Malaysia to manage MSW. These included five units of small-scale incinerators of rotary kiln type that were erected in five tourism spots: Pulau Langkawi (100 ton/day), Pulau Labuan (60 ton/day), Cameron Highlands (40 ton/day), Pulau Pangkor (20 ton/day), and Pulau Tioman (10 ton/day) [28]. Table 3 summarizes the incineration plants in Malaysia, current status, and energy production rate.

Table 3. The summary of the five incineration plants which are funded by the Ministry of Local Government and Housing (MHLG) in Malaysia [28].

Location	Capacity (MSW tones/day)	Electricity Generation	Current Status	Year of Completion
Pulau-Pangkor	20	Nil	Active	2009
Pulau-Tioman	15	Nil	Closed	2010
Pulau-Langkawi	100	1 MW	Active	2010
Cameron Highlands	15	Nil	Active	2010
Labuan	60	Nil	Active	2010

In June 2011, XCN Technology Sdn Bhd constructed an incineration plant located on a 0.8 ha land area at Teluk Cempedak, Pangkor Island. The plant can reduce incoming waste volume up to 94% using the Autogenous Combustion Technology (ACT). In January 2012, the plant commenced operation and was handed over to MHLG. The plant is capable of combusting up to 20 tonnes of MSW per day. Two more mini incinerators located at Tioman Island's facility is capable of incinerating 3221 tons of MSW per year. Two other mini incinerators each were constructed in Cameron Highland and Labuan, where each incinerator has a capacity of combusting 15 and 60 tonnes of MSW per day respectively. None of the four-mentioned incinerator could utilize the MSW to produce energy and generate electricity rather than only to minimize the volume of the incoming MSW. The Langkawi incineration plant is the most comprehensive plant and was the first waste incineration plant in Malaysia to utilize the concept of WTE technology. The Langkawi plant was constructed at a cost of RM68 million and the plant is estimated to process 100 tons per-day of waste to generate one megawatt of electricity. There were two units of mini incinerator installed where the model of the incinerator is Hoval GG42. The operation of the two mini incinerators was monitored by the local authority of Langkawi Town Council. So far, the incinerator plant in Langkawi is the only incinerator plant in Malaysia to practice solid waste segregation at the site before the waste stream was fed into the incinerator. Waste which is not suitable for the incineration process due to high moisture content such as food waste and other wet waste was later dumped into a nearby landfill. The efficiency of the of Langkawi's plant is estimated to be around 80%–90% [29]. The incinerator disposes approximately 100 tons of waste per day and runs on power generated from the WTE technology. The waste heat generated from the incineration process is converted to 1 MW of electricity. The main features of the plant included waste separation systems, rotary kilns, waste heat recovery boilers, air pollution control systems, and a fully condensing steam turbine generator.

The Kajang WTE plant, which consists of a refuse-derived fuel (RDF) facility, is the most comprehensive incinerator as the system incorporates the refused-derived-facility (RFD) operation in Malaysia. WTE or refuse-derived-fuel technologies are both energy recovery initiatives currently emerging at Malaysia and are given priority in solid waste management. In 2008, the Kajang WTE facility was commissioned and has the capacity to process approximately 1100 tons of MSW daily into RDF which is used to produce approximately 8 MW of electricity daily. A total of 3 MW of electricity produced from the RDF plant is used to power the plant and the remainder 5 MW is fed and sold into the national electricity grid via the Feed-in-Tariff (FiT) mechanism [30]. At present, there is a plan to increase the export capacity of electricity generation from 5 to 6 MW. In addition, the current RFD technology only allows 77% of the energy stored within the MSW to be recovered in the form of fuel. It is anticipated that the energy recovered from MSW is targeted to increase to 83% by mixing biogas produce from organic waste via AD into the process, hence making the overall operation more efficient [30]. The increase in MSW generation from 6.37 to 13.38 Mt from year 2010 to 2030 greatly increases the potential of MSW application of WTE. The calorific value of the MSW at Malaysia is estimated between 1500 to 2600 kcal/kg, with energy potential from the incineration plant operating based on the provided calorific value incinerator plant at Malaysia which will have the potential to generate 640 kW/per day. Their study stated that there will be an increase of 60% in electrical generation from 2010 to 2030 by incineration plants as it is estimated that the increasing MSW as renewable source of energy in Malaysia has the potential to produce 5000 Gwh/year by 2020 and further increase to 6200 Gwh/y by year 2030 [16].

3.2. Landfill Gas (LFG)

MSW is one of the well-known biomass resources. By year 2020, the MSW generation is estimated to exceed 30,000 tones/day with 45% of the MSW to be processed at sanitary landfills. Decomposition of biodegradable components of MSW roughly produces 55%–65% of Methane (CH_4), 35%–45% of Carbon Dioxide (CO_2), 0%–3% of Nitrogen (N_2), 0%–2% of Oxygen (O_2), 0%–1% of hydrogen sulphide (H_2S), Hydrogen (H_2), and Ammonia (NH_3) [31]. In Malaysia, landfills are the major source of methane emission (53%); palm-mill effluent generates 38%; swine manure 6%; and other methane gas from industrial effluent 3% [32]. Figure 9 summarizes the generation of methane gas from landfill in Malaysia (ton/year). With that amount of potential methane gas estimated to be generated, there is great potential to harness energy from methane gas for renewable energy in Malaysia.

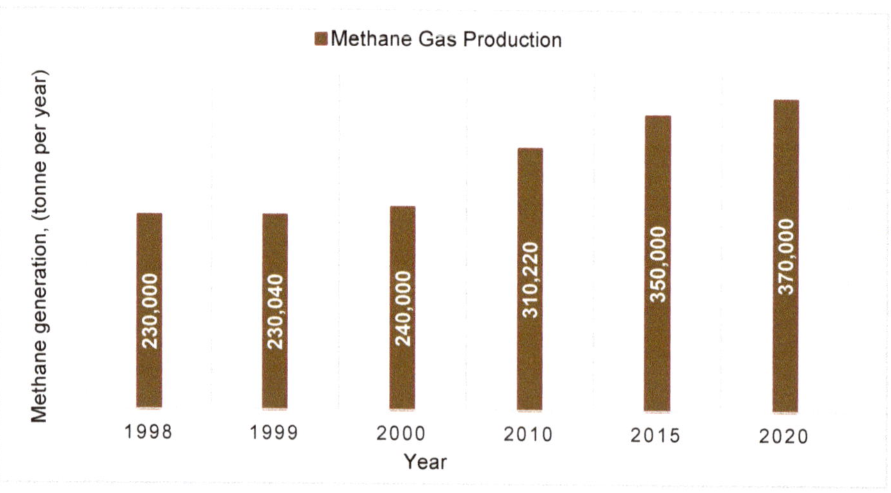

Figure 9. The generation of methane gas from 1998 to 2015 and the predicted value for 2020 [33].

The Bukit Tagar Sanitary Landfill (BTSL) is a practical example, which consist of 18 separate waste cells, which was developed under a parcel of s700-ha area at Hulu Selangor with a 120 million metric ton of waste capacity. At present, BTSL collects 2500 tons of MSW daily (with capacity to handle 5000 tons MSW daily) from Kuala Lumpur and Selayang and the landfill is projected to have a life span of another 130 years. At present, BSTL captures and harvests methane gas generated through horizontal wells from the first and second cell which contain 1.45 million tons and 2 million tons of MSW respectively. Both cells generate approximately 3600 cubic meters of LFG per hour with 60% methane concentration to generate 6 MW of electrical power. A 1.2 MW gas power engine to generate electricity was being installed in BTSL facility where the energy is produced to collected the methane gas. Currently, only part of the methane gas is captured and converted into electricity while the remaining gas is being flared. Currently, BSTL is one of the largest WTE projects which consist of a total of 10.5 MW gas engine capacity combined to generate electricity from LFG. Electricity generated from the site is sold and supplied back to Tenaga National Berhad (TNB), a utility company in Malaysia via FiT mechanism. Another example is the Air Hitam sanitary landfill which is located at Puchong. The landfill currently known as Worldwide Landfill Park (WLP) generates LFG containing a high concentration of methane gas from the degradation of organic waste which had accumulated over the past 10 years. The collected LFG is used to generate 2 MW of electricity monthly whereby 2 MW of capacity is able to power around 2000 houses. In 2006, the landfill was closed and rehabilitated and renamed as Worldwide Landfills Park, with the renewable energy project undertaken by Worldwide Landfills Sdn Bhd. The landfill is projected to continuously produce 2 MW of electrical capacity for at least another 16 years from the saturated 6.2 million metric tons of MSW from its past decade-old operations. According to Worldwide Landfills Sdn Bhd, the generated electric supply was sold to TNB at cost of RM 0.40 per kilowatt-hour (kWh). In addition, Jeram landfill with 160 acres of land area and MSW capacity of 2500 tons per day was established with a capacity to hold 6 million tons of MSW. Currently, the landfill collects approximately 2000 tons of waste daily and to date 4.1 million tons of waste has been added to the landfill. The landfill caters MSW from city councils of Shah Alam and Petaling Jaya, the municipal councils of Subang Jaya and Klang, the district council of Kuala Selangor, and private waste collectors. To date, the gas power plants at Air Hitam and Jeram sanitary landfills produce combined generated power of 6 MW which can power up to 6000 homes. The energy production from LFG is expected to increase from 3000 GWh/year in 2010 to 3300 GWh/year in 2020 and further increase to 5000 GWh/year by year 2030 [16]. An anaerobic digestion system can be combined with an LFG system to increase the total methane gas output hence increasing the efficiency of waste to energy production. Food waste has the higher percentage of MSW content, and a study on the food waste in Malaysia indicate the average food waste generated in year 2010 was a staggering 7600 Metric Ton/day, about 45% of the total MSW generated in 2016.

There are upcoming new LFG-to-Energy projects according to Table 4. At present, part of the biogas installed capacity in the BTSL will increase its capacity to 6.06 MW by installing new units of MWM (A Caterpillar Company) biogas engines on the same site, with each engine having a total electricity output of 2000 kW and electrical efficiency of 43%. Jeram landfill on the other hand will install the new biogas systems with an install capacity of 3.6 MW.

Table 4. List of operating and upcoming landfill biogas plants registered under the Feed-in-Tariff (FiT) scheme associated with municipal solid waste (MSW) landfill to energy (Source: SEDA, Sustainable Energy Development Authority Malaysia).

Year of FiTCD *	Landfill Name	Applicant/Company Name	Location, State	Total Installed Capacity (MW)
2012	Air Hitam Sanitary Landfill (Worldwide Landfill Park)	Jana Landfill Sdn Bhd	Puchong, Selangor Darul Ehsan	1.9572
2012	Bukit Tagar Sanitary Landfill	Kub-berjaya Energy Sdn Bhd	Hulu Selangor, Selangor Darul Ehsan	1.2
2013	Bukit Tagar Sanitary Landfill	Kub-berjaya Energy Sdn Bhd	Hulu Selangor, Selangor Darul Ehsan	3.2
2015	Jeram Sanitary Landfill	Jana Landfill Sdn Bhd	Kuala Selangor, Selangor Darul Ehsan	1.0
2015	Bukit Tagar Sanitary Landfill	Kub-berjaya Energy Sdn Bhd	Hulu Selangor, Selangor Darul Ehsan	2.0
2016	Seelong Sanitary Landfill	Swm Enviro Sdn Bhd	Kulai Jaya, Johor Darul Takzim	2.0
2016	Magenko IYO Alam Sekitar Bercham Landfill	Magenko Renewables (Ipoh) Sdn Bhd	Ipoh, Perak Darul Ridzuan	1.2
2018	Jeram Sanitary Landfill	Jana Landfill Sdn Bhd	Kuala Selangor, Selangor Darul Ehsan	1.202
New Project	Jeram Sanitary Landfill	Jana Landfill Sdn Bhd	Kuala Selangor, Selangor Darul Ehsan	3.6
New Project –Increase Capacity	Bukit Tagar Sanitary Landfill	Kub-berjaya Energy Sdn Bhd	Hulu Selangor, Selangor Darul Ehsan	6.06
New Project	Papan Sanitary Landfill	Selekta Selekta Spektra Sdn Bhd	Daerah Kinta, Perak Darul Ridzuan	2.05

FiTCD * = Feed-in-Tariff Commencement Date.

4. Assessment the Factors Affecting the Sustainability of MSW-To-Energy in Malaysia

4.1. Technical, Economic, Environmental, and Social Issues

Table 5 shows the factors which need to be considered holistically in terms of the four main categories, namely technology or technical, economic and finance, environment, as well as social and political factors. These factors and issues affect the final decision of selecting the appropriate WTE methods to deploy best suited to the Malaysian context.

Table 5. Summary of the technical, economic, environmental, and social issues relating to the MSW-to-Energy.

Technology	Environment
• Quality of Waste • Quantity of Waste • Consistent Supply of Waste	• Residual Management • Emissions Management • Location of Facility
Economic and Finance	**Social and Political**
• Waste and Electricity Management • Initial and Operating Cost • Revenue, Profit and Carbon Credits • Financial Assistance	• Public Health and Safety • Government Initiative and Political Will

4.1.1. Technical Issues

Quality of Waste

- There are different types of wastes that can be used as a potential source for renewable energy to achieve sustainable sound MSW management and towards the use of WTE [34]. A study indicates that MSW in Malaysia has a high calorific value of approximately 23,000 kJ/kg which is suitable and ready to be used as feedstock for energy production. However, non-organic waste which has lower moisture content is suitable for thermal treatment while MSW containing high organic waste which contributes towards higher moisture content in the overall MSW is not suitable for thermal treatment.
- The Malaysian MSW contains a high level of moisture as the largest 45% constituents is organic waste (food and kitchen waste), which reduces the calorific value for thermal incineration treatment between the range of only 1500 and 2600 kcal/kg, hence making the incineration process ineffective [35]. For example, the use of small-scale incinerators operating in the tourist islands in Malaysia were discontinued due to the unsorted MSW as feedstock which contains high moisture content as the presence of wet organic waste like food waste was reported in 2008 [36]. This results in a higher operational and maintenance cost in operating the incinerator in the long run as the incinerator is not functioning within the optimal condition which needs frequent maintenance.

Quantity of Waste

- Periathamby et al. [37] stated that in Malaysia, most landfills are small in capacity. These landfills hence were insufficient to generate adequate LFG volume viable for gas extraction. In addition, most of the Malaysian landfills are non-sanitary landfills and often open dumping sites which rely mainly on only natural clay lining. Furthermore, there is no proper infrastructure for the collection of LFG and leachate. Hence, the establishment of these disposal sites were mainly based on the most traditional and economical at the same time non-environmentally sound and responsible way of disposing of MSW. Therefore, although Malaysia has sufficient MSW for power generation, however, the previous MSW management style of these landfills are not designed with the intention to generate resources such as methane to profit the landfill owners and operators. Thus, most of the LFG generated escapes naturally from cracks and crevices within waste-cells in the landfill into the atmosphere.

Consistency Supply of Waste

- The consistency in supply of waste resources is of utmost importance to keep MSW-to-energy plants, particularly those associated with thermo-chemical treatment, operating optimally to supply a steady state of electricity into the grid [16]. Any fluctuation in the supply of waste would greatly reduce the plant's efficiency hence leading to technical and economic issues which will result in the reduction and affect the stability, hence the quality, of the electricity produced by these WTE plants. Incineration, pyrolysis, and plasma gasification WTE plants need huge volumes of waste, for instance a minimal of 100 tons per day (TPD) and best at 500 TPD of waste, to sustain the continuous combustion in the furnace of boilers to produce consistent heat to supply to boilers for steam production. For the bio-chemical pathway of WTE plants, particularly the AD biogas-to-energy plant, it is encouraged that other types of organic waste from other sources like the industry, agriculture, or live-stock farming can be incorporated with MSW as feedstock to reduce its susceptibility to fluctuating waste resources. The type of organic waste from various sources that can be mixed with MSW are palm oil residue, agriculture crop residue, tropical forest and garden waste, pulp and paper mill waste, animal manure etc. Moreover, the one advantage of AD WTE plants compared to any other WTE treatment methods is that AD process can accept a wide range of MSW volumes, ranging from 25 TPD to above 500 TDP, yet maintain a similar amount of efficiency in converting waste into energy. However, for LFG-to-energy applications at landfill sites, this will need at least a consistent volume of MSW of above 500 TPD to produce a steady stream of electricity supply for the grid [38].

4.1.2. Economic Issues

Investment in waste management systems and electricity generation systems

- In Malaysia, per-capita rate of MSW generation is ever increasing, whereas the energy demand by 2050 increases by a factor of four. In addition, there is an imbalance of existing energy production to the energy demand due to the increasing Malaysian population. According to Oh et al. [32] for each 1% growth on Malaysian GDP (Gross Domestic Product), there will be a 1.2%–1.5% increase in national energy demand. Thus, there is an urgent need to develop WTE technology to minimize MSW disposing to landfills, at the same time produce sufficient energy from MSW as an alternative resource to support the economic development in the coming decade in Malaysia.

Initial and operational cost

- An incineration project requires huge capital costs as the cost is mainly attributed to the need for advanced air pollution control technologies to filter the complex and hazardous flue gas from incinerator plants. In European nations, about two-thirds of the capital cost is spent on air pollution control systems and devices. Furthermore, higher fuel consumption and the need for professional and skilled employees required to operate and to maintain the incineration facility are the main contributors towards the high operational costs of incinerators. Kathirvale and Ahnathakrisnan. [39] reported that in Malaysia, the capital expenditure (CAPEX) and operational expenditure (OPEX) for an incinerator facility with capacity of 800–1000 tons of MSW per day is estimated within the range of RM 500–800 million and USD 40–100/tons of MSW respectively. Similarly, Yahaya. [40] reported that the CAPEX and OPEX of a 1000 ton/day capacity mass burn incinerator WTE facility is RM 360–550 million and RM 102–110 million/ton of MSW respectively. While, the CAPEX and OPEX for a plasma gasification WTE facility is RM650 million and 120 million/ton of MSW respectively. Moreover, the cost required to construct an incinerator compared to a sanitary landfill is estimated to be around 10 times and three times in terms of capital expenditure and operational expenditure, respectively.
- According to AlQattan et al. [17] more environmental efficient WTE systems have a co-generation or combined heat and power (CHP) system, although plants overall improve efficiency but have a lower profit margin due to high equipment and operation cost. Hence, this leaves a dilemma conflict between economic benefit and environmental sustainability, as the most environmentally efficient WTE systems may not always be the most financially prudent. For instance, the net profit

generated by a 150 kilo-tonne (kt) plant with electricity configuration is 25.4 Euro/kt of treated waste, which is significantly higher than a 300 kt plant with co-generative configuration where the profit generated was just 4 Euro/kt of treated waste.

- The cost for installing and running an LFG site is considerably high. In the case of LFG plants, the total cost for a plant are categorized into two categories with the listed details:

  (i) Total Initial Capital Cost

  - Digester equipment;
  - Gas engine;
  - Generator;
  - Balance of System (BOS);
  - Installation Cost;
  - Design and Consultancy;
  - Interconnection Cost (from plant to nearest TNB substation);
  - Preliminary Cost;
  - Other Capital Cost.

  (ii) Annual Operation and Maintenance Cost

  - Insurance Premium;
  - Operation and Maintenance Cost;
  - Fuel Cost at Site;
  - Other Operating Cost.

Table 6 indicates the typical total initial capital and operation and maintenance cost for an LFG WTE plant in Malaysia per MW of installation capacity. Comparing to the cost of other RE resource RM/MW installation, the total initial capital of LFG is lower than solar. However, the cost of LFG is becoming lower annually as more and more projects are being developed in Malaysia.

Table 6. The total initial capital and operation and maintenance cost per MW of LFG installation (Source: SEDA, Sustainable Energy Development Authority Malaysia).

FiTCD * (Year)	Total Initial Capital (RM/MW)	Operation and Maintenance Cost (RM/MW/Year)
2013	12.5 M *	0.63–0.78 M
2015	8 M	1.5–2.0 M
2016	6.67–7.50 M	1.0–1.2 M

FiTCD * = Feed-in-Tariff Commencement Date; M = Million.

Revenue, profit, and carbon credits

- The economic benefit from the operation of incineration plants which use MSW to produce energy were estimated through the sales of electricity and carbon credits. The revenue generated from the sale of electricity for these MSW incineration plants was around RM 5440 M/y (million/year) in year 2006 and RM 6280, and the amount is estimated to increase to RM 1800 M/y by 2020 and further increase to RM 2500 M/y by year 2030. Table 7 summarizes the potential revenue obtained from the operation of incinerator plants in Malaysia. The cost analysis of incineration in Malaysia is shown in Figure 10, where the figure indicates the net profit from the operation of incineration plant to generate energy from MSW. Based on Figure 10, MSW-to-energy via incineration has the potential to generate a net profit of RM 80 M/ton MSW while the combined process of using incineration together with LFG recovery system has the potential to generate RM 120–160 M/ton MSW [16].

Table 7. The potential revenue from the operation of incineration plants in Malaysia.

Year	Energy Production (G Wh/y)	Carbon Credits (M RM/y)	Electricity Sales (M RM/y)	Total Revenue (Carbon Credit + Electric) (M RM/y)
2006	3500	120	5320	5440
2010	3800	200	6080	6280
2020 *	4600	320	6840	7160
2030 *	6200	400	8740	9140

* Estimate.

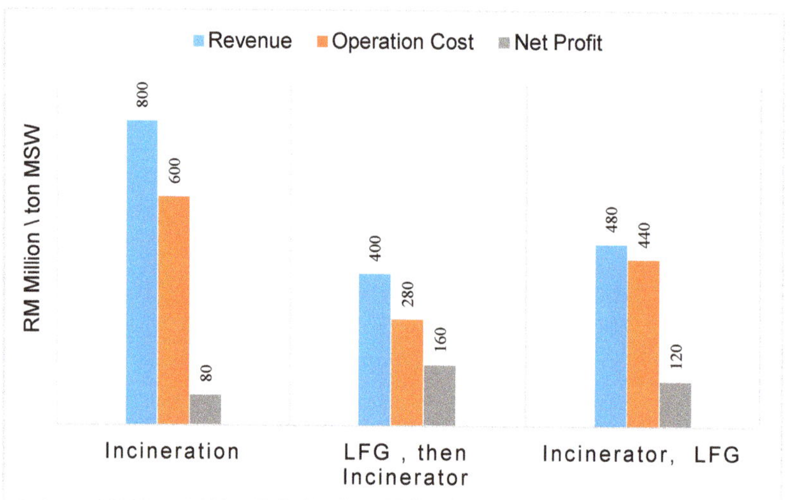

Figure 10. Summary of cost analysis of incineration operation [33].

- Johari et al. [33] reported that in Malaysia, 44.4% and 55.6% of methane emissions are from sanitary landfills and dumpsites, respectively. The usage of LFG technology promises low-operational cost and significantly lower emissions compared to incineration in generating the same amount electricity. According to Johari et al. [33], the revenue generated from converting LFG into energy is RM 998 million from selling both electricity and carbon credits by year 2020. The total revenue from LFG through the sales of carbon credit and electricity are shown in Figure 11.
- The economic analysis based on the net profit and revenue generated from the sales of electricity and carbon credits. From the year 2010 to 2030, LFG applications show an increase in electricity and carbon credit sales. In 2010, electricity generated from LFG has a revenue of RM 1600 M/year and is estimated to increase to RM 2000 M/year and RM 2140 M/year by 2020, respectively. Similarly, the sale of carbon credits are estimated to increase the revenue generated from RM 160 M/year to RM 200 M/year by year 2010 and 2030, respectively. The study performs cost analysis of the LFG system where it shows the operation of LFG power plants may reach a net profit of RM 120 M/ton of MSW from the sale of electricity within the next 10 years (by year 2030).

Lack of financial support to construct sanitary landfills

- The use of LFG technology in biogas sectors potentially mitigate various environmental issues and resolve energy resource shortages, however, this technology requires huge initial capital investment. The lack of financial support from the Government and from bank loans for waste management in Malaysia results in the lack of sustainable sanitary landfill practices. Moreover, due to this factor, only a few large operators and companies are more financially independent to construct and operate sanitary landfills with adequate landfill gas recovery systems and WTE technology from gas collection. This leaves the small-scale waste operators to not have the adequate initial capital to construct and improve their existing landfill facilities for LFG applications. Furthermore, the increase in land price poses a significant impact to the government and private companies in opening new sites for landfilling [41,42].

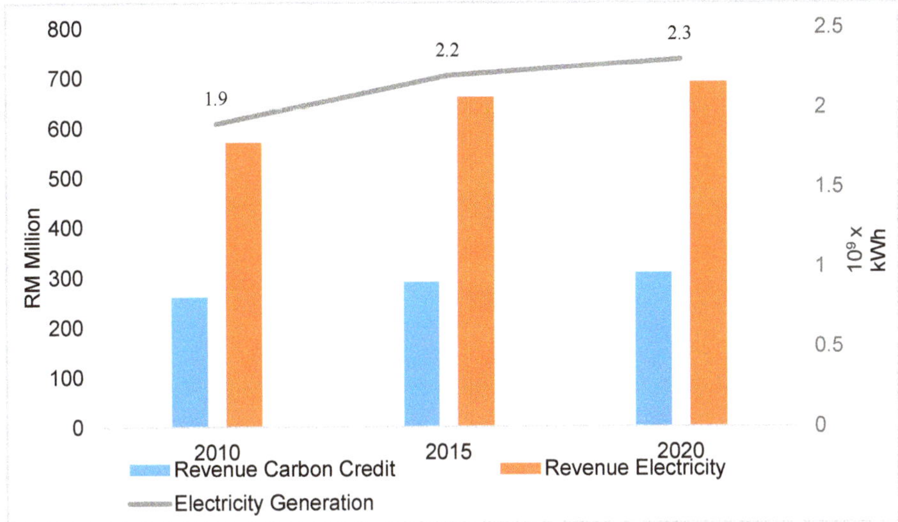

Figure 11. Electricity generation and revenue via landfill gas for the years 2010, 2015, and 2020 [33].

4.1.3. Environmental Issues

Residue management

- The incineration process plays a major role in effectively reducing the volume of MSW entering landfills and efficiently recovering energy from waste. The mass and volume of MSW could be reduced up to 75% and 90% respectively [38]. However, the retrieved ash needs to be handled after the incineration process whereby the ash consists of 10% by weight relative to the weight of the initial MSW fed into the incinerator before the combustion. These ashes are later disposed into the landfill. The disposal of ash into landfills remains a serious environmental issue as these ashes sourced from the incinerator's bottom ash and fly ash (from flue gas cleaning systems) contains 8%–12% ferrous metals and 0.5%–1.5% non-ferrous metals. This potentially brings great harm to the environment if there is no implementation towards the recycling and recovery of these precious yet hazardous heavy metals namely Zinc (Zn), Nickel (Ni), Lead (Pb), Mercury (Hg), Cadmium (Cd) and, among others, trace heavy metals which might be present [17].
- Compared to recovering LFG from landfill to generate energy, incineration (with energy recovery) has several environmental advantages and benefits, namely smaller plant footprint which requires a smaller area of land to operate, lower carbon emissions, minimal land contamination, by-product from incineration ash is inert (chemically stable without odor), and higher density of energy recovery per ton of MSW. Yet, the operation of incinerators in Malaysia encounters difficulties due to several issues. These issues comprise the opposition from non-governmental organizations and public against the health issues relating to the operation of incineration plants; unsuitability of available technology for incineration of local MSW; high capital expenditure; and stringent

emission standards compared to other combustion equipment. Additional difficulties encountered by present incinerator operators were failure to comply to environmental impact assessment standards; insufficient local expertise; incompetent local operators; and higher cost for incinerator maintenance and parts.

Emissions management

- There was no available emission standard for incinerators in Malaysia until the release of Environmental Quality (Dioxin and Furan) Regulation in 2004 by Department of Environment (DOE) Malaysia. However, the need for such regulation to be established in Malaysia must be grounded on the United States Environmental Protection Agency (USEPA) Method 23 and USEPA Method 8290 which are used for analysis in accordance with strict dioxin and furan discharge criteria (EQA,1974). Furthermore, stricter requirement needs for other pollutant release from the incinerator include heavy metals. As the laws can change and updates may be required in the air pollution controls, this could lead to much higher costs in the future [43].
- Dioxin emissions have stringent regulations in other developing and developed countries for instance China, Taiwan, Germany, and Japan, where dioxin emissions are limited to 1 ng/m^3 toxic equivalent for incinerators [44]. Furthermore, during the thermo-chemical treatment of MSW the complex composition of unsorted waste material may result in the formation of carcinogenic, complex, and environmentally persistent acidic gas pollutants, namely Sulphur Dioxide (SO_2), Polychlorinated biphenyls (PCBs), Volatile Organic Compounds (VOCs), Carbon Monoxide (CO), Hydrofluoric acid (HF), Hydrochloric Acid (HCl), and Hydrobromic acid (HBr), among others. If these gaseous pollutants were untreated or accidentally leaked into the environment, this will result in a long-term unforeseen negative impact to the people and the entire ecosystem [17]. These unforeseen negative impacts include the bioaccumulation of these carcinogenic compounds in the food chain and the development of diseases like lymphoma and sarcoma [45]. The emission standard of dioxin and furans is at 80 pg TEQ/m^3 and for mercury is at 20 μg/m^3 according to the department of Environmental Government of Nunavut [46].

Location of facility

- Currently, there is no standard listed in the Malaysian Environmental Quality Act, 1974 Act (127) and the Clean Air Regulation 2014 by the Department of Environment Malaysia (DOE) specifically stating the distance of incineration or WTE plants from and city centers or housing areas or having how many kilometers, in radius, of buffer zone to prevent the breach of toxic flue gas containing dioxin and furans. In the case of thermo-chemical treatment of MSW via incineration, gasification, and pyrolysis, according to Elliott et al. [38], the facility distance of up to 10 km threshold for exposed population for all MSW incinerators still gives a negative health impacts in the long-term. Table 8 below shows the adverse health outcomes associated with the distance of facility and dioxin emission level from various case studies in regards with WTE incinerators. The World Health Organization (WHO) recommended a minimum of 500–750 m buffer zone surrounding the facility to have a sufficient flue gas dilution ratio above 1000 which is based on ideal condition of relatively flat and unobstructed terrain. In reality, a more serious and holistic environmental impact assessment has to be conducted to determine the safe buffer zone as well as the distance of facility from residential housing area.

Table 8. The relation of peak dioxin levels from various incineration plants studies associated with health effects [44–50].

Country/Region	Peak Dioxin Emissions (pg/m^3 TEQ)	Distance of Affected Population from Facility (km)	Health Effects (Significant Association)
Besancon and four regions, France	16,000	N/A	Non-Hodgkin's lymphoma and soft tissue sarcoma
Rhone-Alpes Region, France	16,000	N/A	Clefts, renal dysplasia, 24 other anomalies
Southeast Region, France	16,000	= 10	Congenital urinary tract defects
Multiple sites, Japan	80,000	<2	Infant and neonatal death (due to congenital malformations, fetal death) and very low birth weight
Osaka prefecture, Japan	80,000	<4	Headache, stomach ache, wheeze, fatigue, atopic dermatitis and allergic rhinitis in school children

4.1.4. Social Issues

Public health and safety management

- Incineration can contribute positively to public health by eliminating harmful pathogens, bacteria, and other hazardous micro-organisms which accumulate in MSW. However, the byproducts from the incineration process, namely fly ash, bottom ash, and other residues from the incinerator's air-pollution control (APC) system in the form of sludge and slags from wet scrubber treatment process, pose an existential threat to public health via the carriage of toxic pollutants containing heavy metals such as lead, mercury, cadmium, along with other pollutants.
- In the case of MSW-to-energy via the route of thermo-chemical pathway, there were several studies which confirmed significant health effects. Giusti [50] reported that the flue gases from incinerators containing acidic gases (SO_2, NOx, N_2O, HCl, HF), certain metals (arsenic, beryllium, chromium, cadmium, lead and mercury), carbon monoxide (CO), carbon dioxide (CO_2), dioxins, furans, polyaromatic hydrocarbons (PAHs), polychlorinated biphenyls (PCBs), volatile organic carbons (VOCs), and odor, increases the chance of acquiring non-Hodgkin's lymphomas and soft tissue sarcoma. According to National Research Council [51], other potential health effects from incineration flue gas exposure are chest pain, dizziness, irritative symptoms, and poor coordination.
- From Table 8, it can be seen that infant and neonatal death occurs when the peak dioxin emission level from incineration plant is at 80,000 pg/m^3 in Japan [46]. Incineration plants will have to be situated far and must have least a 10 km buffer zone from cities centers and away from any residential and inhabitants' areas to prevent any risk of negative health implications caused by the release and exposure of toxic flue gases which contain furan, dioxin, and other heavy metals which is being blended in the emission stacks of these WTE plants.

Government Initiative and Political Will

- At present, the 11th Malaysia Plan (2016–2020) embraces the concept of responsible consumption and production which is the United Nations SDG 12 where the focus area is treating solid waste as a new resource which can be recovered materially via bio-chemical path (AD) or thermo-chemical path (Incineration, Pyrolysis, Plasma Gasification) for electrical generation and secondary fuel resources for energy production. Furthermore, the 11th MP aims to transform the Malaysian society towards a more energy and resource efficient society by managing waste holistically, increasing the percentage of RE in energy sector, as well as in the national electrical generation mix, creating a green market etc.

- The Renewable Energy Act in 2011 was designed to support the FiT mechanism to promote the use of RE. In the act, a 1.6% surcharge is imposed to the electricity bill of all consumers except for domestic consumers who consume electricity less than 300 kWh/month. The 1.6% surcharge is collected to fund the RE quota and development in Malaysia.
- The national renewable energy policy and action plan (2010) under thrust 1 where the FiT mechanism is implemented together with the establishment of the RE Fund and SEDA Malaysia is responsible as an authority to implement FiT projects for RE developers. The FiT mechanisms allow RE producers to generate income via the selling of electricity generated from RE resources (solar, biomass, biogas, and small hydro) to the national electricity grid based on different FiT premium rates over a certain period of tenure [46]. Table 9 shows the most updated FiT rates for biogas (landfill/waste) and biomass (solid waste) RE resources.

Table 9. The FiT rates for biogas and biomass as of 1 January 2019 [10].

<table>
<tr><th>Biogas (Landfill)</th><th>FiT Rates (RM/kWh)</th><th>Biomass (Solid Waste)</th><th>FiT Rates (RM/kWh)</th></tr>
<tr><td colspan="4">(a) Basic FiT rates of installed capacity of</td></tr>
<tr><td rowspan="3">(i) Up to and including 5 MW</td><td rowspan="3">0.2210–0.2814</td><td>(i) Up to and including 10 MW</td><td>0.3085</td></tr>
<tr><td>(ii) 10–20 MW</td><td>0.2886</td></tr>
<tr><td>(iii) 20–30 MW</td><td>0.2687</td></tr>
<tr><td colspan="4">(b) Bonus FiT rates having the following criteria (one or more)</td></tr>
<tr><td>(i) Use of gas engine technology with electrical efficiency of >40%</td><td>+0.0199</td><td>(i) Use of gasification technology</td><td>+0.0199</td></tr>
<tr><td>(ii) Use of locally manufactured or assembled gas engine technology</td><td>+0.0500</td><td>(ii) Use of steam-based electricity generation systems with overall efficiency of >20%</td><td>+0.0100</td></tr>
<tr><td rowspan="2">(iii) Use of landfill, sewage gas or agricultural waste including animal waste as fuel source</td><td rowspan="2">+0.0786</td><td>(iii) Use of locally manufactured or assembled boiler or gasifier</td><td>+0.0500</td></tr>
<tr><td>(iv) Use of solid waste as fuel source</td><td>+0.0982</td></tr>
</table>

- In 2009, the Malaysia's National Green Technology Policy (GT) was established to promote and integrate sustainable development into the national economic development. GT is defined as technology that promotes the use of RE which has zero or lower GHG emissions, improves environmental outcome by minimizing the degradation of the environment, conserves natural resources, and maximizes energy use by increasing efficiency.

4.2. *Assessing the Suitability and Sustainability Aspect of WTE in the Malaysian Context*

According to Aich and Ghosh. [38] any WTE project has long-term implications towards the sustainability aspect of the society and the development of a nation. It may contribute positively, and some may have an adverse effect towards the local economy, society, and the environment if selecting the wrong or not suitable technicality of the WTE systems to treat MSW. Hence, a SWOT (strength, weakness, opportunity, and treat) analysis of various WTE systems is required to be done for the selection of the most suitable technology to be adopted by the local context in Malaysia. By conducting SWOT analysis, this can help to reduce uncertainties and risk of WTE project by identifying weaknesses and threats of a WTE technology. Furthermore, the identification of new or alternative feed mix and the establishment of raw material supply chain can be done by identifying the strengths and opportunities of the WTE technology. Table 10 shows the general elements and guideline used to conduct the SWOT analysis for WTE technologies. In terms of technical, Aich and Ghosh [38] provide the technical parameter indicative chart which will help local authorities and decision makers identify the most suitable WTE technology. The main consideration in choosing the right technology must be based on both the waste characteristics as well as the local climatic condition in Malaysia.

Table 10. The general outline of the SWOT (strength, weakness, opportunity, threat) analysis to be used for WTE (Waste-to-Energy) projects [38].

Internal Factors	**Strength (S)** Characteristics that give advantages	**Weakness (W)** Characteristics that give disadvantages
	Availability of land area, location of facility, proximity to other urban centers, quantity and characteristics of Municipal Solid Waste (MSW), capital cost, operation and maintenance cost, risk factor of acquiring the right kind of MSW, scope of public and private participation, availability of indigenous technology, existing environmental issues, scope for natural resources conservation, and scope for backward and forward of technological integration.	
External Factors	**Opportunity (O)** Elements that can be exploited to provide additional benefit	**Threat (T)** Elements of the environment that bring barriers and trouble
	Local constraints, political situation, climatic condition, land restriction criteria, existing socio-economic and market condition, availability of raw material for auxiliary fuel/additive, national and international regulations, and initial capital cost.	

To better understand the Malaysia climatic condition, presented below is the Malaysian climate information [46]:

- Average Daily Temperature (°C) = 27–28;
- Average Annual Rainfall (mm/year) = 3318;
- Average Daily Wind Speed (km/h) = 5.5–8.6;
- Average Daily Humidity (%) = 79–80;

Table 11 shows the technical parameters indicative for various municipal solid waste (MSW)-to-energy technologies [38]. Hence, based on Malaysia's weather and climate conditions, the most suitable thermo-chemical treatment will be pyrolysis or plasma gasification. However, this needs a huge amount of funding assistance or initial capital either from the government or foreign investors as the initial capital cost in building up such plant is very high, as indicated. As indicated in Table 11, it must be noted that both pyrolysis and plasma gasification WTE plants require a consistent supply of waste to be fed into the system of a minimum 100 tonnes per day to maintain the facility operation in generating electricity. Regarding the most flexible system based on the plant size, anaerobic digestion is the best option in harnessing biogas for electricity production from organic waste in MSW. Malaysia can consider anaerobic digestion as the future WTE bio-chemical route while at the same time converting all possible open dump sites into LFG for methane capturing and energy generation.

Table 11. The technical parameter indicative chart for various MSW-to-Energy projects [38].

Indicators	Most Suitable (√) Moderate (M) Not Suitable (×)				
	WTE Technologies				
Technical Parameters	**Anaerobic Digestion (AD)**	**Incineration**	**Pyrolysis**	**Plasma Gasification**	**Landfill Gas Extraction (LFG)**
	Waste Characteristics				
High Calorific Value, >1200 Kcal/kg	×	√	√	√	×
High bio-degradable matter, >50%	√	M	M	√	√
Fixed carbon, <25%	√	N/A	N/A	√	√
Total Inert, >25%	×	×	×	√	M
C:N ratio, 20–30:1	√	N/A	M	√	N/A
Mixed with all types of waste	×	M	M	√	M
	Climate				
Hot Climate, >35 °C	√	√	√	√	√
Moderate Climate, 15–25 °C	M	√	√	√	√
High moisture content, >55%	√	×	M	√	×
High rainfall area	√	×	M	√	×
	Plant Size				
Up to 25 TPD	√	×	×	×	×
25–50 TPD	√	×	×	×	×
50–100 TPD	√	×	×	×	×
100–500 TPD	√	√	M	√	×
>500 TPD	√	√	√	√	√
	Economic Condition				
Capital Cost	Low to Moderate	High	High	Very High	Very High
Resource Conservation	√	√	√	√	√
Carbon Credit Advantages	√	M	M	√	M

5. Challenges and Future Prospects

In the effort of making WTE a profitable and effective project to support the decarbonization of the energy sector and the development of RE production in Malaysia, the following are some of the related improvements which can be considered.

- The government is responsible for playing crucial role in translating policies in papers into actual implementation by initiating behavior changes via a top to bottom approach. There are numerous policies related to solid waste management that have been established throughout the Malaysia Plan, however, there is a lack in the implementation and enforcement in both WTE policy and MSW in Malaysia. The stakeholders, mainly industrial players and academic institutions, must deliver adequate awareness and knowledge on the plans and initiatives developed in WTE policies.
- Effective and good communication, understanding, and cooperation is needed between the local and federal authorities regarding the planning and establishment of SWM policies. Furthermore,

there is also the need to establish an efficient governance in SWM policy and legislation. Creating a strong and reliable task force to which the authority can enforce the policy and legislative measure is greatly needed.

- To effectively treat and to maximize the energy potential from MSW and at the same time to save cost of constructing a facility to segregate MSW, both the government and local authorities need to educate and implement effective and practical household and commercial source segregation to separate organic waste and inorganic waste for WTE treatment processes. The organic waste can be treated via the AD biological process to recover both electricity and organic waste while inorganic waste can be treated thermally via pyrolysis and gasification to recover energy and secondary fuel and other products.
- The use of alternative approaches is encouraged, such as gasification and pyrolysis, rather than mass burn or incineration to treat either mixed waste or inorganic waste. Pyrolysis and gasification processes generate several useful secondary fuels, and at the same time heat and energy, which can be turned into electricity by reacting organic material fractions from the MSW at high temperature. Both processes treat MSW without direct combustion with precisely controlling the amount of oxygen and steam entering into the system. Syngas is produced at first in both processes where pollutants are removed before it is combusted, hence resulting in a cleaner emission level. Energy recovery from MSW via gasification and pyrolysis produces cleaner emissions compared to incineration and does not possess any threat to public health. Though the initial construction cost for gasification is higher compare to incineration, at 1.5 times the initial capital cost, however, gasification in terms of converting waste to energy is 30% more efficient than incineration [48]. Figure 12 compares different WTE technologies in terms of energy potential, cost, and GHG mitigation ability.

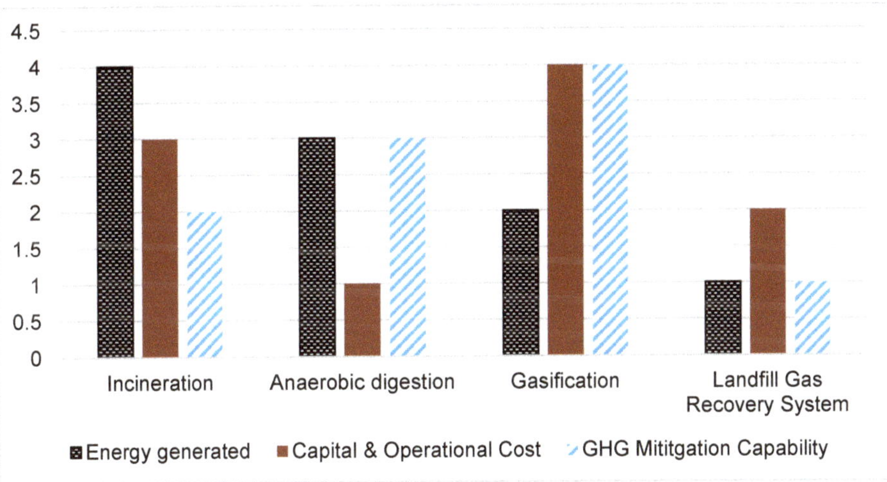

Figure 12. Comparison of different WTE technologies in terms of energy potential, cost, and greenhouse gas (GHG) mitigation ability [16].

6. Conclusions

In brief, due to the escalating increase in the generation of municipal solid waste (MSW), carbon dioxide (CO_2) emissions, and electricity consumption, there is an urgent need to develop and implement MSW-to-energy plants in Malaysia. MSW-to-energy plants can solve both environmental and energy issues by effectively minimizing waste volume, at the same time converting waste into energy. Hence,

MSW as a renewable source of clean energy in generating electricity into the national electricity grid via the national Feed-in-Tariff (FiT) program has great potential to decarbonize the energy and waste sector in Malaysia. The implementation of waste-to-energy (WTE) as part of the solution towards Malaysia's waste and energy problems contributes in achieving several of the 17 United Nations sustainable development goals (UN-SDGs). These goals include goal No. 3 (good health and wellbeing), 6 (clean water and sanitation), 7 (affordable and clean energy), 8 (decent work and economic growth, 11 (sustainable cities and communities), 12 (responsible consumption and production), and 13 (climate action). Moreover, WTE projects support the Malaysian Low Carbon Cities Framework (LCCF) which was launched and adopted in 2009 together with the voluntary pledge from the Malaysian government during the recent 15th conference of parties (COP) for the world's climate agenda.

The aim of all waste-to-energy (WTE) technologies is to reduce waste volume entering landfills at the same time generating energy from waste. In general, although there are many types of WTE technologies in the market, the main WTE routes can be classified into bio-chemical and thermo-chemical treatment methods. In Malaysia, landfill gas (LFG) as a bio-chemical pathway and incineration as a thermo-chemical pathway are the two most widely used WTE technologies to treat MSW and to recover energy in the form of electricity and heat. There are a number of sanitary landfills with LFG application, namely Air Hitam, Bukit Tagar, Jeram, and Seelong sanitary landfills which are engineered to collect and harness methane gas generated from biodegradable contents within the waste cells for electricity production under the FiT scheme. There are four small-scale incinerators mainly at tourist islands in Malaysia namely Cameron Highlands, Pulau Tioman, Pulau Pangkor, and Labuan, however, these four incinerators do not produce electricity and are only intended to be used for minimizing MSW volume. Only one larger scale incinerator with a capacity of treating 100 tonne/day of MSW at Pulau Langkawi was used to generate electricity from burning MSW.

The application of MSW-to-energy has great potential to be implemented in Malaysia as 89% of MSW is being disposed to landfills without treatment or any WTE applications. In order to ensure a long-term sustainable development of WTE in Malaysia, the four faculties of technology (quality or characteristic of waste; quantity of waste; consistency supply of waste), economy (waste and electricity management; initial capital together with operation and maintenance costs; revenue and carbon credits; financial support), environment (residual and emissions management; location of facility), and social (public health and safety; government initiative and political will) are being considered holistically within the Malaysia context to ensure the suitability at the same time to minimize any business risk of selected WTE technology yet to be implemented in Malaysia. Moreover, a SWOT (strength, weakness, opportunity, and treat) analysis of various WTE systems is being discussed to further complement in terms of technicality towards the sustainability of selecting the best possible WTE in the context of Malaysia.

Therefore, the best WTE technology in dealing with wet MSW, which contains higher amount of organic waste (food waste), is anaerobic digestion (AD) as AD can fully utilize the feedstock by effectively converting organic waste into both form of biogas which later can be converted into electricity and bio-fertilizer. Hence, AD leaves almost no secondary by-product which is hazardous and persistent to the environment if compared to incineration which produces subsequent hazardous ashes. AD is considered as one of the most flexible systems compared to other systems. AD can adsorb a wide range of tonnage between 25 tons per day (TPD) to more than 500 TPD while maintaining its optimal operation. The initial capital expenditure (CAPEX) and operational expenditure (OPEX) of AD systems is the cheapest among all the mentioned WTE technologies in this paper, is most suitable to be utilized in Asia as the process is suited to the warm climate of most of the Asia region, and the initial capital cost as well as the operational cost is the lowest among all WTE technologies where AD suits the financial strength and climatic condition of most Asia countries. However, the main challenge to a successful long-term AD operation is that the MSW must be first sorted or segregated either at the source or at a sorting facility at the plant. This is to ensure only organic waste entering the digester, as other foreign non-organic substances like trace heavy metals and other carcinogenic and hazardous compounds which might bring toxicity possibly affect the biological process in the digestor.

The most sustainable WTE technology to treat mixed MSW containing higher amount of inorganic waste is plasma gasification rather than conventional incineration process. This is because plasma gasification generates high amount of energy, and heat at the same time can produce multiple secondary fuels which later can be turned into energy compared to incineration in which the process only generates heat. Moreover, plasma gasification can accept all kinds of mixed waste and the process generates flue gas which is much cleaner than in conventional incineration. Although the main challenge of plasma gasification is the initial capital costing, which is 1.5 times of that to an incinerator, however, in the long-run plasma gasification has 30% more efficiency compared to the incineration process, therefore generating more output out for the same amount of tonnage of MSW.

Lastly, Malaysia, as a developing country with a vision towards transitioning into a first world country and had pledged to contribute towards climate action and achieving the UN-SDGs, needs to embark towards a circular city metabolism where MSW-to-energy is widely used. National MSW-to-energy practices gives back value to waste as a renewable source for electricity production at the same time minimizing anthropogenic waste and pollutants entering the natural environment. Through nationwide MSW-to-energy implementation and realization, the socio-economic development in Malaysia can be more environmentally sound and efficient, hence enabling Malaysia to achieve sustainable development.

Author Contributions: Conceptualization, Z.J.Y. and M.J.K.B.; formal analysis, C.A.N.; resources, S.S.; writing—original draft preparation, Z.J.Y. and J.W.L.; writing—review and editing, P.L.S.; supervision, M.J.K.B.; funding acquisition, M.J.K.B. and J.W.L.

Funding: This research was funded by Universiti Tunku Abdul Rahman (UTAR), under the UTARRF research grant number PSR/RMC/UTARRF/2018 C1/M04. One of us, Jun Wei Lim also wishes to thank the funds received from Universiti Teknologi PETRONAS via YUTP-FRG (0153AA-E48) and Ministry of Education Malaysia under HICoE.

Acknowledgments: The authors would like to thank the colleagues Faculty of Engineering and Green Technology (FEGT), Universiti Tunku Abdul Rahman for their help.

Conflicts of Interest: The authors declare no conflict of interest.

References

1. MHLG. *Ministry of Housing and Local Government Malaysia, 2015*; Final Lab Report: Solid Waste Management Lab; MHLG: Pemandu, Malaysia, 2015. Available online: http://www.kpkt.gov.my/resources/index/user_1/Attachments/hebahan_slider/slaid_dapatan_makmal.pdf (accessed on 25 April 2019).
2. Mahamud, M.A.; Hidefumi, I.; Akio, O.; Chan, N.W.; Samat, N. A Strategic Partnership in Implementing Waste-To-Energy (Wte) Technology in Penang, Malaysia. *Acad. Strateg. Manag. J.* **2018**, *17*, 1–8.
3. Bong, C.P.C.; Ho, W.S.; Hashim, H.; Lim, J.S.; Ho, C.S.; Tan, W.S.P.; Lee, C.T. Review on the renewable energy and solid waste management policies towards biogas development in Malaysia. *Renew. Sustain. Energy Rev.* **2017**, *70*, 988–998. [CrossRef]
4. The World Bank Group. *CO2 Emissions (Metric Tons Per Capita): Malaysia*; The World Bank Group: Washington, DC, USA, 2019; Available online: https://data.worldbank.org/indicator/EN.ATM.CO2E.PC?locations=MY (accessed on 21 February 2019).
5. Budhiarta, I.; Siwar CBasri, H. Current status of municipal solid waste generation in Malaysia. *Int. J. Adv. Sci. Eng. Inf. Technol.* **2012**, *2*, 129–134. [CrossRef]
6. Noor, Z.Z.; Yusuf, R.O.; Abba, A.H.; Hassan, M.A.A.; Din, M.F.M. An overview for energy recovery from municipal solid wastes (MSW) in Malaysia scenario. *Renew. Sustain. Energy Rev.* **2013**, *20*, 378–384. [CrossRef]
7. World Energy Council. *World Energy Resources Waste to Energy*; Report; World Energy Council: London, UK, 2016.
8. Suruhanjaya Tenaga (Energy Commission Malaysia). *Malaysia Energy Information Hub*; Suruhanjaya Tenaga (Energy Commission Malaysia): Putrajaya, Malaysia, 2019. Available online: https://meih.st.gov.my/statistics (accessed on 3 February 2019).

9. Joshi, D. *Evaluating the Performance of the Sustainable Energy Development Authority (SEDA) and Renewable Energy Policy in Malaysia*; Penang Institute: Pulau Pinang, Malaysia, 2018; Available online: https://penanginstitute.org/wp-content/uploads/2018/06/Evaluating-the-Performance-of-SEDA-and-RE-Policy-in-Malaysia_PI_Darshan_5-June-2018.pdf (accessed on 3 February 2019).
10. SEDA. *Sustainable Energy Development Authority Malaysia*; SEDA: Putrajaya, Malaysia, 2019. Available online: http://www.seda.gov.my/ (accessed on 1 March 2019).
11. Suruhanjaya Tenaga (Energy Commission Malaysia). *Malaysia Energy Statistics Handbook*; Suruhanjaya Tenaga (Energy Commission Malaysia): Putrajaya, Malaysia, 2017.
12. Rohatgi, A. *2H 2017 Southeast Asia Renewables Market Outlook*; Bloomberg New Energy Finance: UK, 2017.
13. Nadzri, Y. The Way Forward: Solid Waste Managment in Malaysia. In Proceedings of the 10th Annual Waste Management Conference Exhibition, Putrajaya, Malaysia, 18–19 July 2012; pp. 10–19.
14. Manaf, L.A.; Samah, M.A.A.; Zukki, N.I.M. Municipal solid waste management in Malaysia: Practices and challenges. *Waste Manag.* **2009**, *29*, 2902–2906. [CrossRef] [PubMed]
15. Moh, Y.C.; Manaf, L.A. Overview of household solid waste recycling policy status and challenges in Malaysia. *Resour. Conserv. Recycl.* **2014**, *82*, 50–61. [CrossRef]
16. Tan, S.T.; Ho, W.S.; Hashim, H.; Lee, C.T.; Taib, M.R.; Ho, C.S. Energy, economic and environmental (3E) analysis of waste-to-energy (WTE) strategies for municipal solid waste (MSW) management in Malaysia. *Energy Convers. Manag.* **2015**, *102*, 111–120. [CrossRef]
17. AlQattan, N.; Acheampong, M.; Jaward, F.M.; Ertem, F.C.; Vijayakumar, N.; Bello, T. Reviewing the potential of Waste-to-Energy (WTE) technologies for Sustainable Development Goal (SDG) numbers seven and eleven. *Renew. Energy Focus* **2018**, *27*, 97–110. [CrossRef]
18. Gumisiriza, R.; Hawumba, J.F.; Okure, M.; Hensel, O. Biomass waste-to-energy valorisation technologies: A review case for banana processing in Uganda. *Biotechnol. Biofuels* **2017**, *10*, 11. [CrossRef]
19. Zainu, Z.A.; Songip, A.R. Policies, challenges and strategies for Municipal waste management in Malaysia. *J. Sci. Technol. Innov. Policy* **2017**, *3*, 18–22.
20. Dada, O.; Mbohwa, C. Energy from waste: A possible way of meeting goal 7 of the sustainable development goals. *Mater. Today Proc.* **2018**, *5*, 10577–10584. [CrossRef]
21. United Nations Economic and Social Commissions (UNESCAP). *Visualization of Interlinkages for SDG 7*; UNESCAP: New York, NY, USA, 2016; Available online: https://www.unescap.org/sites/default/files/Visualisation%20of%20interlinkages%20for%20SDG%207.pdf (accessed on 27 February 2019).
22. UN Environment. *GOAL 7: Affordable ad Clean Energy*; UN Environment: Nairobi, Kenyan, 2019; Available online: https://www.unenvironment.org/explore-topics/sustainable-development-goals/why-do-sustainable-development-goals-matter/goal-7 (accessed on 4 February 2019).
23. O'Brien, J.K. Comparison of Air Emissions from Waste-to-Energy Facilities to Fossil Fuel Power Plants. In Proceedings of the 14th Annual North American Waste-to-Energy Conference, Tampa, FL, USA, 1–3 May 2006; pp. 69–78.
24. United Nations. *Tracking Progress Towards Inclusive, Safe, Resilient and Sustainable Cities and Human Settlements. SDG 11 Synthesis Report: High Level Political Forum 2018*; The United Nations: New York, NY, USA, 2018.
25. KeTTHA. *Ministry of Energy, Green Technology and Water Malaysia., Low Carbon Cities Framework*; Version 2, KeTTHA: Putrajaya, Malaysia, 2017; Available online: https://www.greentechmalaysia.my/media/LCCF_Book-Version-2-2017.pdf (accessed on 23 February 2019).
26. Abushammala, M.F.; Basri, N.E.A.; Basri, H.; El-Shafie, A.H.; Kadhum, A.A.H. Regional landfills methane emission inventory in Malaysia. *Waste Manag. Res.* **2011**, *29*, 863–873. [CrossRef] [PubMed]
27. Girardet, H. Regenerative cities. In *Green Economy Reader*; Springer: Cham, Switzerland, 2017; pp. 183–204.
28. Bashir MJ, K.; Ng, C.A.; Sethupathi, S.; Lim, J.W. Assessment of the Environmental, Technical and Economic Issues Associated with Energy Recovery from Municipal Solid Waste in Malaysia. *IOP Conf. Ser. Earth Environ. Sci.* **2019**, *268*, 012044. [CrossRef]
29. Sulaiman, M.R.; Kadir, S.A.; Aishah, S.; Ibrahim, R.; Husin, M. A study on the problems of the usage of incinerators in Malaysia. *Sci. Res. J.* **2007**, *4*, 1–12. [CrossRef]
30. Neville, A. Kajang Waste-to-Energy Plant, Semenyih, Malaysia. *Power* **2010**, *154*, 35–36. Available online: http://www.powermag.com/renewables/waste_to_energy/Top-Plant-Kajang-Waste-to-Energy-Plant-Semenyih-Malaysia_3225 (accessed on 23 February 2019).

31. Balat, M.; Balat, H. Biogas as a renewable energy source—A review. *Energy Sources Part A* **2009**, *31*, 1280–1293. [CrossRef]
32. Oh, T.H.; Pang, S.Y.; Chua, S.C. Energy policy and alternative energy in Malaysia: Issues and challenges for sustainable growth. *Renew. Sustain. Energy Rev.* **2010**, *14*, 1241–1252. [CrossRef]
33. Johari, A.; Ahmed, S.I.; Hashim, H.; Alkali, H.; Ramli, M. Economic and environmental benefits of landfill gas from municipal solid waste in Malaysia. *Renew. Sustain. Energy Rev.* **2012**, *16*, 2907–2912. [CrossRef]
34. Chua, K.H.; Sahid, E.J.M.; Leong, Y.P. Sustainable municipal solid waste management and GHG abatement in Malaysia. *ST-4 Green Energy Manag.* **2011**, *4*, 1–8.
35. Kalantarifard, A.; Yang, G.S. Energy potential from municipal solid waste in Tanjung Langsat landfill, Johor, Malaysia. *Int. J. Eng. Sci. Technol.* **2011**, *3*, 8560–8568.
36. Aja, O.C.; Al-Kayiem, H.H. Review of municipal solid waste management options in Malaysia, with an emphasis on sustainable waste-to-energy options. *J. Mater. Cycles Waste Manag.* **2014**, *16*, 693–710. [CrossRef]
37. Periathamby, A.; Hamid, F.S.; Khidzir, K. Evolution of solid waste management in Malaysia: Impacts and implications of the solid waste bill, 2007. *J. Mater. Cycles Waste Manag.* **2009**, *11*, 96–103. [CrossRef]
38. Aich, A.; Ghosh, S.K. Application of SWOT Analysis for the Selection of Technology for Processing and Disposal of MSW. *Procedia Environ. Sci.* **2016**, *35*, 209–228. [CrossRef]
39. Kathiravale, S.; Ahnantakrishnan, J.C. Municipal Solid Waste-Management Cost and Opportunities, 1–13. 2007. Available online: http://dspace.unimap.edu.my/xmlui/bitstream/handle/123456789/13777/08-10-12-13-msw%20management.pdf?sequence=1 (accessed on 3 February 2019).
40. Yahaya, N.B. *Solid Waste Management in Malaysia: The Way Forward*; Ministry of Housing and Local Government (MHLG): Putrajaya, Malaysia, 2012; Available online: http://ensearch.org/wp-content/uploads/2012/07/Paper-13.pdf (accessed on 6 August 2019).
41. Agamuthu, P.; Fauziah, S.H. Challenges and issues in moving towards sustainable landfilling in a transitory country-Malaysia. *Waste Manag. Res.* **2011**, *29*, 13–19. [CrossRef] [PubMed]
42. Omran, A.; Mahmood, A.; Aziz, H.A. Current practice of solid waste management in malaysia and its disposal. *Environ. Eng. Manag. J.* **2007**, *6*, 295–300. [CrossRef]
43. Yunus, M.N.M. *Waste Incineration–Challenges and Possible Application in Malaysia*; Malaysia Institute for Nuclear Technology Research (MINT): Bangi, Malaysia, 2000. Available online: http://web.kpkt.gov.my/thermal/incenerator2.cfm (accessed on 27 February 2019).
44. Elliott, C.; Campbell, A.; Balen, E.V. *Health Assessment for Thermal Treatment of Municipal Solid Waste in British Columbia: Evidence Review and Recommendations*; British Columbia Centre for Diesease Control (BCCDC): Vancouver, BC, Canada, 2012.
45. Department of Environment Government of Nunavut. *Environmental Guideline for the Burning and Incineration of Solid Waste*; Department of Environment Government of Nunavut: Iqaluit, NU, Canada, 2012. Available online: https://www.gov.nu.ca/sites/default/files/guideline_burning_and_incineration_of_solid_waste_2012.pdf (accessed on 28 February 2019).
46. Wong, S.L.; Ngadi, N.; Abdullah, T.A.T.; Inuwa, I.M. Recent advances of feed-in tariff in Malaysia. *Renew. Sustain. Energy Rev.* **2015**, *41*, 42–52. [CrossRef]
47. METMalaysia. *Malaysian Meteorological Department. 2016*; Annual Report; METMalaysia: Petaling Jaya, Malaysia, 2016.
48. Stantec. *Waste to Energy: A Technical Review of Municipal Solid Waste Thermal Treatment Practices*; British Columbia Ministry of Environment: Burnaby, BC, Canada, 2011. Available online: http://www.env.gov.bc.ca/epd/mun-waste/reports/pdf/BCMOE-WTE-Emissions-final.pdf (accessed on 3 February 2019).
49. Knoema. *World Data Atlas: Malaysia-CO2 Emissions per Capita*; Knoema: Washington, DC, USA, 2018; Available online: https://knoema.com/atlas/Malaysia/CO2-emissions-per-capita (accessed on 6 August 2019).
50. Giusti, L. A review of waste management practices and their impact on human health. *Waste Manag.* **2009**, *29*, 2227–2239. [CrossRef]
51. National Research Council. *Waste Incineration and Public Health*; National Academies Press: Washington, DC, USA, 2000.

Correction

Correction: Tham, P.E., et al. Recovery of Protein from Dairy Milk Waste Product Using Alcohol–Salt Liquid Biphasic Flotation. *Processes* 2019, 7, 875

Pei En Tham [1], Yan Jer Ng [1], Revathy Sankaran [2], Kuan Shiong Khoo [1], Kit Wayne Chew [3,*], Yee Jiun Yap [3], Masnindah Malahubban [4], Fitri Abdul Aziz Zakry [4] and Pau Loke Show [1,*]

1 Department of Chemical and Environmental Engineering, Faculty of Science and Engineering, University of Nottingham Malaysia, Jalan Broga, Semenyih 43500, Selangor Darul Ehsan, Malaysia; peien0405@gmail.com (P.E.T.); yanjer98@hotmail.com (Y.J.N.); kuanshiong.khoo@hotmail.com (K.S.K.)

2 Institute of Biological Sciences, Faculty of Science, University of Malaya, Jalan Universiti, Kuala Lumpur 50603, Malaysia; revathysankaran@ymail.com

3 School of Mathematical Sciences, Faculty of Science and Engineering, University of Nottingham Malaysia, Jalan Broga, Semenyih 43500, Selangor Darul Ehsan, Malaysia; YeeJiun.Yap@nottingham.edu.my

4 Faculty of Agriculture and Food Sciences, Universiti Putra Malaysia Sarawak Campus, Bintulu 97008, Sarawak, Malaysia; masnindah@upm.edu.my (M.M.); zakryfitri@upm.edu.my (F.A.A.Z.)

* Correspondence: KitWayne.Chew@nottingham.edu.my or kitwayne.chew@gmail.com (K.W.C.); PauLoke.Show@nottingham.edu.my or showpauloke@gmail.com (P.L.S.); Tel.: +60-3-8924-8605 (P.L.S.)

Received: 12 March 2020; Accepted: 16 March 2020; Published: 25 March 2020

We were not aware of some errors made in the proofreading phase; therefore, we wish to make the following corrections to the mathematical equations in the text.

(1) The bar at some units are placed on the bottom instead of the top, the correct representation for the following equations should be: Flux $\overline{J}$ is defined as the number of bubbles passing though an area A per unit time. Assuming no diffusion, i.e., only convection,

$$\overline{J} = \sigma\overline{s} \tag{2}$$

where σ and $\overline{s}$ are the density and velocity of the bubbles respectively. Putting Equation (2) into (1) gives

$$\frac{\partial\sigma}{\partial t} = -\nabla\cdot\overline{J}. \tag{3}$$

The bubbles are assumed to move into the region R, resulting in a negative $\nabla\cdot\overline{J}$, and the negative sign is to make it positive. To solve Equation (16), Laplace transform is applied to Equation (16), giving

$$\begin{aligned} f(z)\overline{\sigma}_z(z,s) + s\overline{\sigma}(z,s) - \sigma(z,0) + F(z)\overline{\sigma}(z,s) &= 0 \\ f(z)\overline{\sigma}_z(z,s) + [s + F(z)]\overline{\sigma}(z,s) - \sigma(z,0) &= 0 \end{aligned} \tag{17}$$

Since $\sigma(z,0) = 0$, Equation (17) becomes

$$f(z)\overline{\sigma}_z(z,s) + [s + F(z)]\overline{\sigma}(z,s) = 0 \tag{18}$$

To solve Equation (18), we divide Equation (18) by $f(z)$ to give

$$\overline{\sigma}_z(z,s) + \frac{s + F(z)}{f(z)}\overline{\sigma}(z,s) = 0 \tag{19}$$

Multiplying Equation (19) with integrating factor $e^{\int \frac{s+F(z)}{f(z)}dz}$ yields

$$\frac{d}{dz}\left[e^{q(z,s)}\overline{\sigma}(z,s)\right] = 0 \tag{20}$$

Integrating Equation (20) gives

$$\overline{\sigma}_z(z,s) = Ce^{Q(z,s)} \tag{22}$$

(2) The cylindrical coordinates $\varnothing$ should have the terms as followed: $\varnothing = \tan^{-1}\frac{y}{x}$.

(3) There is an additional 'ln' term in both the equations which should not be present. The correct equations are as shown below:

$$q(z,s) = \frac{9sv}{8\rho g\alpha^2}\left(\frac{\rho^2 g^2 z^3}{3} - \rho g z^2 P_{ex} + (P_{ex})^2 z\right) + ln\left|\frac{8\rho g\alpha^2}{9v(\rho gz - P_{ex})^2}\right| \tag{21}$$

Transforming Equation (22) from the s domain back to the t domain yields

$$L^{-1}\{\overline{\sigma}_z(z,s)\} = \sigma_z(z,t) = \delta\left[t - \frac{9v}{8\rho g\alpha^2}\left(\rho g z^2 P_{ex} - (P_{ex})^2 z - \frac{\rho^2 g^2 z^3}{3}\right) - ln\left|\frac{8\rho g\alpha^2}{9v(\rho gz - P_{ex})^2}\right|\right] \tag{23}$$

where L is the Laplace transform.

Reference

1. Tham, P.E.; Ng, Y.J.; Sankaran, R.; Khoo, K.S.; Chew, K.W.; Yap, Y.J.; Malahubban, M.; Aziz Zakry, F.A.; Show, P.L. Recovery of Protein from Dairy Milk Waste Product Using Alcohol-Salt Liquid Biphasic Flotation. *Processes* **2019**, *7*, 875. [CrossRef]

MDPI
St. Alban-Anlage 66
4052 Basel
Switzerland
Tel. +41 61 683 77 34
Fax +41 61 302 89 18
www.mdpi.com

Processes Editorial Office
E-mail: processes@mdpi.com
www.mdpi.com/journal/processes

www.ingramcontent.com/pod-product-compliance
Lightning Source LLC
LaVergne TN
LVHW071526170726
843515LV00008B/2073

9783039365197